Data Analysis in Vegetation Ecology

3rd Edition

Data Analysis in Vegetation Ecology

3rd Edition

Otto Wildi

WSL Swiss Federal Institute for Forest, Snow and Landscape Research
CH-8903 Birmensdorf
Switzerland

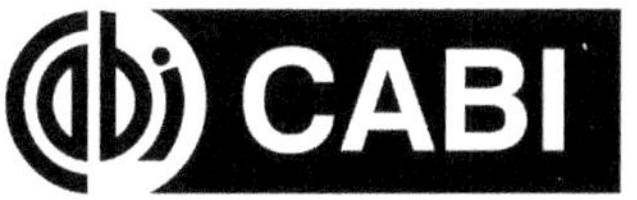

CABI is a trading name of CAB International

CABI
Nosworthy Way
Wallingford
Oxfordshire OX10 8DE
UK

Tel: +44 (0)1491 832111
Fax: +44 (0)1491 833508
E-mail: info@cabi.org
Website: www.cabi.org

CABI
745 Atlantic Avenue
8th Floor
Boston, MA 02111
USA

T: +1 (617)682 9015
E-mail: cabi-nao@cabi.org

A catalogue record for this book is available from the British Library, London, UK.

Library of Congress Cataloging-in-Publication Data

Names: Wildi, Otto, author.
Title: Data analysis in vegetation ecology / Otto Wildi, WSL Swiss Federal Institute for Forest, Snow and Landscape Research, CH-8903 Birmensdorf, Switzerland.
Description: 3rd edition. | Wallingford, Oxfordshire, UK ; Boston, MA : CABI, [2017] | Includes bibliographical references and index.
Identifiers: LCCN 2017022420 (print) | LCCN 2017023412 (ebook) | ISBN 9781786394231 (ePDF) | ISBN 9781786394248 (ePub) | ISBN 9781786394224 (pbk : alk. paper) | ISBN 9781786394231 (e-book) | ISBN 9781786394248 (e-pub)
Subjects: LCSH: Plant communities--Data processing. | Plant communities--Mathematical models. | Plant ecology--Data processing. | Plant ecology--Mathematical models.
Classification: LCC QK911 (ebook) | LCC QK911 .W523 2017 (print) | DDC 581.7/820285--dc23
LC record available at https://lccn.loc.gov/2017022420

ISBN-13: 978 1 78639 422 4 (pbk)
978 1 78639 423 1 (e-pdf)
978 1 78639 424 8 (e-pub)

Commissioning editor: Ward Cooper
Editorial assistant: Emma McCann
Production editor: Tim Kapp

Printed and bound in the UK from copy supplied by the author by CPI Group (UK) Ltd, Croydon, CR0 4YY

Plants are so unlike people that it's very difficult for us to appreciate fully their complexity and sophistication.

Michael Pollan, *The Botany of Desire*

Contents

List of Figures

List of Tables

Preface to the first edition

When starting to rearrange my lecture notes I had a 'short introduction to multivariate vegetation analysis' in mind. It ended up as a 'not so short introduction'. The book now summarizes some of the well-known methods used in vegetation ecology. The matter presented is but a small selection of what is available to date. By focussing on methodological issues I try to explain what plant ecologists do, and why they measure and analyse data. Rather than just generating numbers and pretty graphs, the models and methods I discuss are a contribution to the understanding of the state and functioning of the ecosystems analysed. But because researchers are usually driven by their curiosity about the functioning of the systems I successively began to integrate examples encountered in my work. These now occupy a considerable portion of this book. I am convinced that the fascination of research lies in the perception of the real world and its amalgamation in the form of high-quality data with hidden content processed by a variety of methods reflecting our model view of the world. Neither my results nor my conclusions are final. Hoping that the reader will like some of my ideas and perspectives, I encourage them to use and to improve on them. There is a considerable potential for innovation left.

The examples presented in this book all come from Central Europe. While this was not intended originally, I became convinced the topics they cover are of general relevance, as similar investigations exist almost everywhere in the world. An example is the pollen data set: pollen profiles offer the unique chance to study vegetation change over millennia. This is the time scale of processes such as climate change and the expansion of the human population. Another, much shorter time series than that of pollen data is found in permanent plot data originating from the Swiss National Park that I had the opportunity to look at. The unique feature of this is that it dates back to the year 1917, when Josias Braun-Banquet personally installed the first wooden poles, which are still in place. Records of the full set of species have been collected ever since in five-year steps. A totally different data set comes from the Swiss Forest Inventory, presented in the last chapter of this book. Whereas many vegetation surveys are merely preferential collections of plot data, this data set is an example of systematic sampling on a grid encompassing huge environmental gradients. It helps to assess which patterns really exist, and whether some of those described in papers or textbooks are real or merely reflect the imagination or preference of researchers scanning the landscape for nice locations. In this case the data set available for answering the question is still moderate in size, but handling of large data sets will eventually be needed in similar contexts. I used the Swiss wetland data set as an example for handling data of much larger size, in this case with $n = 17,608$ relevés. Although this is outnumbered by others, it resides on a statistical sampling design.

Some basic knowledge of vegetation ecology might be needed to understand the examples presented in this book. Readers wishing to acquire this are advised to refer, for example, to the comprehensive volumes *Vegetation Ecology* by Eddy van der Maarel (2005) and *Aims and Methods of Vegetation Ecology* by Mueller-Dombois and Ellenberg (1974), presently available as a reprint. The structure of my book is influenced by Orlóci's (1978) *Multivariate Analysis in Vegetation Research*, which I explored the first time when proofreading it in 1977. Various applications are found in the books of Gauch (1982), Pielou (1984) and Digby and Kempton (1987) and many multivariate methods used in vegetation ecology are introduced in Jongman *et al.* (1995). To study statistical methods used in this book in more detail, I strongly recommend the probably most comprehensive textbook existing today, the second edition of *Numerical Ecology* by Legendre and Legendre (1998). Several books provide an introduction to the use of statistical packages, which are referred to in the appendix. For many reasons I decided to omit the software issue in the main text; upon the request of several reviewers I added a section to the appendix where I reveal how I calculated my examples and mention programs, program packages and databases.

I would like to express my thanks to all individuals that have contributed to the success of this book. First of all Rachel Wade from Wiley-Blackwell, who strongly supported the efforts to print the manuscript in time and organized all the technical work. I thank Tim West for careful copy-editing, and Robert Hambrook for leading through the production process. My colleagues Anita C. Risch and Martin Schütz revised the entire text, providing corrections and suggestions. Meinrad Küchler helped in the computation of several examples. André F. Lotter provided the pollen data set. I cannot remember all the people who had an influence on the point of view presented here: many ideas came from László Orlóci through our long lasting collaboration, others from Madhur Anand, Enrico Féoli, Valério de Patta Pillar, Janos Podani and Helene Wagner. I particularly thank my family for encouraging me to tackle this work and for their tolerance when I was working at night and on weekends to get it completed.

Birmensdorf, 1 December 2009

Preface to the second edition

Successful attempts to include instructions in $\mathcal{R}$ motivated me to prepare a second edition of the book while keeping it basically unchanged in style and content. Hence, I hoped to circumvent yet another introduction into a software environment as done earlier for MULVA-5 (Wildi and Orlóci 1996), which I previously used in many of my examples. I found the syntax of $\mathcal{R}$ to be close to ordinary mathematical notation allowing technical instructions to be minimized. Finally, this book is not an introduction to $\mathcal{R}$. There are many others doing this, such as Crawley (2005), Venables and Ripley (2010), Adler (2010) or for advanced users Borcard *et al.* (2011), all highly recommended and referenced. The instructions I included in this second edition are aimed to serve the unexperienced in $\mathcal{R}$, getting technical help from colleagues or experts in installing and initializing $\mathcal{R}$ and loading some packages and functions, including the one I specifically provide for this book (package `dave`). Unintendedly, doing the examples explained in this second edition may even act as a beginners' course in $\mathcal{R}$, hopefully at minimum effort.

Writing this second edition was a delicate task too. First, various results had to be reproduced by an entirely different or newly developed software. Only after revising the very last chapter it was clear that all this could be done in $\mathcal{R}$. It is well known that many scientists using $\mathcal{R}$ love it, those who avoid it, fear it. My objective is to encourage newcomers to do the examples and I put every effort into most parsimonious solutions. The instructions and functions I prepared for the book look and hopefully feel simple, hiding the tremendous complexity of the $\mathcal{R}$ environment. In this context I thank my colleagues who gave me technical advice, Thomas Dalang, Dirk Schmatz, Meinrad Küchler and Alan Haynes. The attendees of a course held with an early version of the book helped me to identify bugs and traps, namely Angéline Bedolla, Elizabeth Feldmeyer, Ulrich Graf, Julia Haas, Alan Haynes, Caroline Heiri, Martina Hobi, Christine Keller, Meinrad Küchler, Helen Küchler, Mathieu Lévesque, Anna Pedretti, Kathrin Priewasser, Anita C. Risch, Marcus Schaub, Martin Schütz, Anna Schweiger, Andreas Schwyzer, Bastian Ullrich and Sonja Wipf. Again, Anita C. Risch and Martin Schütz were willing to read the whole text critically.

All examples in the book are derived in $\mathcal{R}$ version 2.15.2 (R Development Core Team 2012). Whenever a specific method was missing I wrote a new function to avoid overloading readers with cumbersome code. On the downside every new function represents yet another black box. In the present state the reader will find solutions for all methods presented in the book, although figures may appear a little different: for the book I adapted these to layout requirements using an extended set of plot parameters explained in $\mathcal{R}$ when typing `?plot.default` and further screening for `par`. In the end I devise an $\mathcal{R}$ package for this book: `dave`, the name composed of initials of the book title (Appendix 14). An integrated part of `dave` consists of the many data sets listed in Appendix 14. I would like to express my thanks to all authors cited there, for giving the right to access these, as far as yet unpublished. Many are real world examples, although, with respect for ongoing research, fairly aged.

While elaborating this second edition I got trapped by the temptation to extend the panoply of methods where functions of other packages were ready to use. This concerns, for example, resemblance measures, classification techniques and ordination methods. In the modelling part I replaced my old fashioned heuristic approach by the now widely used logistic regression techniques including instructions for scenario building, considered important in the time of global change. For newcomers in $\mathcal{R}$ I highly recommend following the instructions quite carefully: $\mathcal{R}$ is very much like a programming language and for the average human brain it is extremely difficult to exactly remember all the details to get the examples running. To support proper use I extended the index considerably to facilitate quick access to all major methods covered in this book. The latter will work only when all packages required are loaded, namely `dave`, `stats`, `vegan`, `labdsv`, `nnet`, `cluster` and all considered 'related' upon downloading from a CERAN repository found in the Internet.

I would again like to thank the publications team of Wiley-Blackwell for all their encouragement and support I experienced throughout this revision. We agreed that the new edition shall serve users not only in theory but now also in practice, a combination adding to the complexity of publication. Finally, I express my thanks to my host institution, the Swiss Federal Institute for Forest, Snow and Landscape Research WSL, for providing access to its computer network and literature databases needed to complete this work.

Birmensdorf, 1 October 2012

Preface to the third edition

Revising a book like this is a delicate task. The corresponding software, the 'dave' package in this case, had to be updated while avoiding conflicts with previous versions. This is why the basic structure of this third edition is unchanged although all chapters have been revised and often extended to implement further topics deemed attractive to readers.

While this book still intends to communicate skills required for data analysis, in this third edition I try to move a step further in the direction of finding a best practice, or at least, finding a good one. When it comes to choosing a method or a combination of methods the reader will eventually encounter case studies where results are visually or quantitatively compared. This concerns data transformation (of vegetation data only), the use of resemblance functions as well as clustering and ordination methods. Many of these evaluations are embedded in advanced chapters mainly in the context of ecological models. That is of course a tightrope walk ('Gratwanderung') because the best solution is frequently context dependent and sometimes a matter of personal preference. Nevertheless I no longer hesitate to present my own view in various sections concluding some main chapters.

Using mathematics to investigate natural systems is not always self-evident as much we do bases on paradigms now briefly addressed in the introduction. The most fundamental among these may be similarity theory, summarizing various assumptions once taken for granted but only recently explicitly expressed by Feoli and Orlóci (2011).

A notable extension of content is a new chapter on the handling, analysis and interpretation of plant traits and indicator values. This introduces a general framework for including supplementary information about plant taxa. Some experts in the field are convinced that considering plant functional traits is a big step ahead in vegetation science although I rather focus on the pitfalls when using statistical inference when comparing alternative sets of variables. The chapter on classification is extended by presenting an example of divisive clustering and a section introducing revisions of classifications. In a new section a panoply of clustering methods is compared graphically. An analogous comparison is implemented in the ordination chapter in the hope to facilitate the readers' choice.

Carefully revised are the instructions given for the $\mathcal{R}$ environment. Due to overall thematic extensions more of these are found now, that is to say, there was no way to avoid some proliferation. In other words, this now third edition fosters what has been avoided in the first one: giving support and advice in technical terms to the handling and analysis of vegetation and site data in the hope that the reader will appreciate this.

The Swiss Federal Institute for Forest, Snow and Landscape Research, WSL, provided infrastructure required to fulfil this complex task, most of all access to the literature databases. Klaus Ecker helped to supplement one of the examples with environmental information and Elizabeth Feldmeyer provided her succession data for a new time series example.

I am much obliged to CABI and most of all to Ward Cooper and his team for their willingness to publish this now pretty large book in colour, to Emma McCann for organizing the publication process and specifically to Tim Kapp and Angela Whittaker for their tremendous effort in copy editing the entire text.

Birmensdorf, 28 August 2017

1 Introduction

Irrespective of whether our interest in the vegetation of the earth evolves from naive curiosity or from the endeavour to conserve this natural resource, the study of vegetation ecology inevitably follows the rules of epistemology and it is shaped by paradigms.

1.1 Epistemology

This book is about understanding vegetation systems in a scientific context, a topic of vegetation ecology. It is written for researchers motivated by the curiosity and ambition to assess and understand vegetation dynamics. Vegetation, according to van der Maarel and Franklin (2013) 'can be loosely defined as a system of largely spontaneously growing plants'. What humans grow in gardens and fields is hence excluded. The fascination of investigating vegetation resides in the mystery of what plants 'have in mind' when populating the world. The goal of all efforts in plant ecology, as in other fields of science, is to learn more about the rules governing natural systems. These rules are causing patterns, and the assessment of patterns is the recurrent theme of this book. Albeit sometimes patterns are striking, their proper recognition can be elusive for various reasons.

We are aware that access of humans to the *real world* is rather restricted and – as we know from experience – differs among individuals. Our sins provide us with a limited set of signals and even this is far too voluminous to be comprehensively processed by our brain. We usually implicitly accept that the complexity of this world exceeds our imagination. Furthermore, the extents of our target systems are frequently excessive. For example, the data set addressed in Chapter 13 uses data collected within plots of 200 square metres, yet spaced by as much as 4 kilometres. And finally, humans are part of the natural system and not independent observers as would be needed for truly unbiased recognition.

To achieve progress in research an image of the real world is needed: the *data world*. In this we get a description – an image – of the real world in the form of numbers. (An image can be a spreadsheet filled with numbers, a digital photograph, a digital terrain model, or other.) Assuring that such an image reflects properties of the system is a challenge – it is the task of sampling design, as yet under-represented in the literature of vegetation ecology. The point in mind is to avoid personal bias, resulting in conclusions triggered by the investigator. All examples in this book result from attempts to avoid bias, except when mentioned explicitly. Data also deserve various kinds of treatment, like data management, quality control and of course analysis. But what constitutes the final issue of data analysis?

Upon analysis we develop our *model world* describing our understanding of the real world. This understanding is nothing else but a hypothesis about the state and functioning of reality. The model world too is often represented in the form of numbers, although typically less voluminous and complex than the numerical description of the real world. It can therefore also be the subject of further analysis. An advanced issue of data analysis is in relating the data world to the model world and vice versa. Finding globally valid models reflecting the real world is a difficult task due to the complexity of systems (Orlóci 1993; Anand 1997). Complexity has its origin in some fairly well-known phenomena, one being the scale effect. Any pattern in ecosystems will emerge at a specific spatial and temporal scale only: at short spatial distance competition and facilitation among plants can be detected (Connell and Slatyer 1977); these would remain undetected over a range of kilometres. In order to study the effect of global climate change (Orlóci 2001; Walther *et al.* 2002) the scale revealed by satellite imagery is probably more promising than the same of a local survey. Choosing the best scale for an investigation is a matter of decision, experience and often exploration. For this a multi-step approach is needed, in which intermediate results are used to evaluate the next decision in the analysis. Poore (1955, 1962) called this successive approximation, Wildi and Orlóci (1991) flexible analysis and Albert *et al.* (2010) model-based sampling. Hence, the diversity and flexibility of methods is nothing else but a response to the complex nature of systems.

Due to complexity there exist alternatives of valid models, for instance, amalgamating different spatial scales. And once a proper scale is found there is still a need to simultaneously consider an 'upper' and a 'lower' level, because knowing sensitivity to change in scale is essential. Parker and Pickett (1998) discuss this in the context of temporal scales and interpret the interaction as follows: 'The middle level represents the scale of investigation, and processes of slower rate act as the context and processes of faster rates reflect the mechanisms, initial conditions or variance.'

Another source of complexity is uncertainty in data. Data quality often suffers from practical constraints, for example from limitation in time, money and accessibility (Albert *et al.* 2010). A detailed vegetation survey is time-consuming, and while sampling, change may already be underway (Wildi *et al.* 2004). Such data will therefore exhibit an undesired temporal trend. A specific bias causes variable selection. For example, it is easier to measure components above ground than below ground (van der Maarel and Franklin 2013, p. 6), a distinction vital in vegetation ecology. Once the measurements are complete they may reflect random fluctuation or chaotic behaviour (Kienast *et al.* 2007) blurring deterministic components. It is a main objective in data analysis to distinguish random from deterministic, either linear or nonlinear effects. Nonlinearity would not be a problem if we knew the kind of relationship that was hidden in the data (e.g. Gaussian, exponential, logarithmic; Austin 1987), but finding a proper function is usually a non-trivial problem.

Spatial and temporal interactions add much to the complexity of vegetation systems. In space, the problem of direction arises, as the order of objects depends on the direction considered. In most ecosystems, the environmental conditions, for example elevation or humidity, change across the area. Biological variables responding to this will likely behave similarly and therefore become space-dependent (Legendre and Legendre 2012). Even if there is no general trend in space, a local phenomenon may exist: spatial autocorrelation. This means that sampling units in close neighbourhood are more similar than one could expect from ecological conditions. Similarly, correlation over time exists as well. In analogy to space, there occur temporal dependence and temporal autocorrelation. Many processes are temporally continuous and the systems will usually change gradually only, causing two subsequent states to be similar. Finally, time and space are not independent, but linked. Spatial patterns too tend to change continuously over time. Therefore, a time series observed at one point in space is expected be similar to another series observed nearby.

In summary, all knowledge we generate by analysing the data world contributes to our model world which in the end is aimed at serving society. When translating this into practice we meet yet another world, the man-made *world of values*. This is people's perception and valuation of the world, which we know from experience is continuously changing. The results we derive in the course of analysis carry the potential to deliver input into value systems, but we should keep in mind what Diamond (1999) mentioned when talking about accepting innovations: 'Society accepts the solution if it is compatible with the society's values and other technologies.' Assessing the existence of global warming, as an example, can be a matter of modelling. Convincing people of the practical relevance of the problem is a question of evaluation and communication, skills not addressed in this book.

1.2 Paradigms ruling analysis

To date a comprehensive and proprietary theory of the plant community is still missing (McGill *et al.* 2006; Feoli and Orlóci 2011). However, in the course of the history of data analysis in vegetation science (which may have had a starting point with the invention of a similarity coefficient by Jaccard in the year 1901) various theories have been inherited from other fields of science and new paradigms evolved and became widely accepted. As can be seen below these are usually related, sometimes depicting the same idea from a different point of view. They contribute considerably to our present understanding of the functioning of plant-environment systems.

Similarity theory. Feoli and Orlóci (2011) recently emphasized the universal role of 'similarity theory' in the analysis of multivariate biological data. This theory has equal importance in other fields of science like genetics or meteorology, and it encompasses various axioms, as for instance:

1. The similarity of any two objects (plots, for instance) described by many variables (species, chemical or physical measurements, etc.) can be measured or estimated.

2. Similarity applied to entire sets of objects (addressed as samples in Section 2.3) constitutes the similarity pattern of these sets (see Section 4.5).

3. Similarity (including similarity patterns) 'can express linkages to factors which control the existence of the objects' (Feoli and Orlóci 2011). This relies on the assumption that any two objects that are similar in terms of vegetation are expected to be similar in terms of growth conditions as well, and vice versa.

Similarity theory, according to Feoli and Orlóci (2011), plays a crucial role in the joint analysis of biological, environmental, spatial and temporal data. To this end it also yields the justification to research on similarity functions (Chapter 4). Needless to say that the majority of methods described in this book rely on similarity or its complement, distance. But similarity relates to various other paradigms as well.

Convergent evolution. This encapsulates the observations that 'plants of different species develop similar morphology to adapt to similar environment' (Feoli and Orlóci (2011), a fact discussed in many papers, such as Orlóci and Orlóci (1985), Feoli and Orlóci (1991), Díaz (1995), Pillar (1999), Tobias *et al.* 2014). A famous example is the succulent life form of the *Cactaceae* in the New World and its counterpart of *Euphorbiaceae* in the Old World (Wildi and Orlóci 2007). Convergent evolution justifies the use of characters other than taxonomic, that is, genetic, morphologic or functional (Chapter 8), for example. As a consequence of convergent evolution the principles of similarity theory equally apply to plant traits used for vegetation description.

Convergence and divergence. According to similarity theory similar plant communities correspond to similar physical-chemical environments (see above). Therefore, in the course of vegetation change, patterns of plant communities and environmental factors tend to converge (that is, tend to get more similar) and convergence may become an element of our model world (MacArthur and Levins 1967; Pillar *et al.* 2009). Convergence is sometimes used in a static context, simply expressing that patterns in plant communities and the environment are similar. But there are also opposing forces causing divergence. Species requiring similar environmental conditions for survival are competing for the same ecological niche. Hence, the least fit may be expelled by competition. Any final plant composition is therefore the result of convergence and divergence acting simultaneously.

Vegetation dynamics. Vegetation change too follows similarity rules and it is therefore potentially predictable. Notwithstanding environmental conditions, Feoli and Scoppola (1980) hypothesize that 'a given plant association has the highest probability of transition to the one with which it shares maximal similarity'. This principle also explains traditional succession which is therefore just a special case of vegetation dynamics. And, as mentioned by Feoli and Orlóci (2011), it is also applicable in the opposite direction, when disturbance counteracts convergence, for example.

Filtering. Filtering is a popular term explaining the constrained set of species or species traits in a stable plant community (Keddy 1992; Fridley 2003). The principle is sketched in Figure 1.1 with three different filters modulating four species pools. Such a filter, according to Keddy (1992), does nothing but natural selection on the community level. Filtering invokes some general principles in the formation of plant communities:

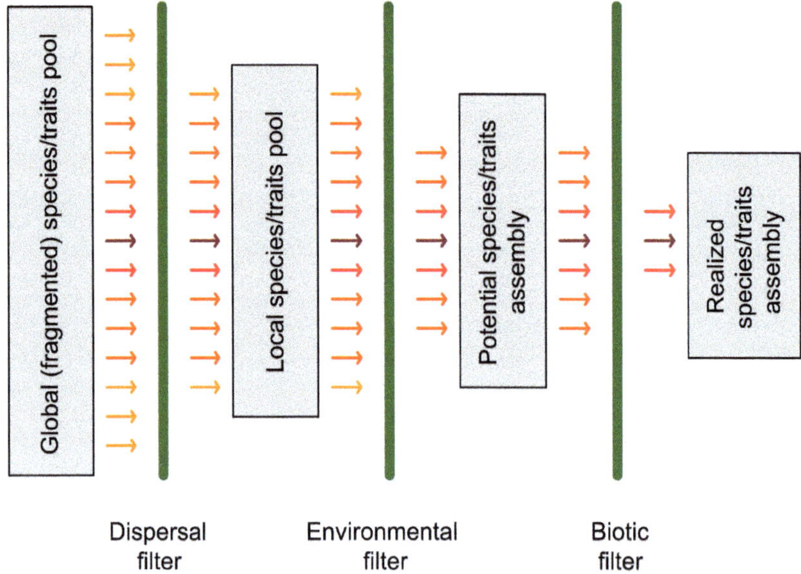

Figure 1.1: Principles of the filter concept. A species or traits pool is reduced stepwise by different kinds of processes called filtering.

1. The formation of any plant community is strictly limited by the pool of species or the pool of traits.
2. If a pool is changed, for instance by local evolution, this will directly translate into the final composition of plant communities. This aspect is not included in the scheme of Figure 1.1.
3. The idea of filtering perfectly describes what can happen when alien species ('neophytes') invade. First, invasion extends the species pool and second, it affects biological filters (competition, for example) to eventually generate entirely new combinations of species and traits.

Several of these principles are sometimes subsumed under the term 'assembly rules' despite the fact that filtering is a selecting rather than a combining process. But as in all paradigms mentioned so far, it is another nice outset for generating hypotheses. It is the task of analysis to accept or reject the latter using methods of data description and statistics.

2 Patterns in vegetation ecology

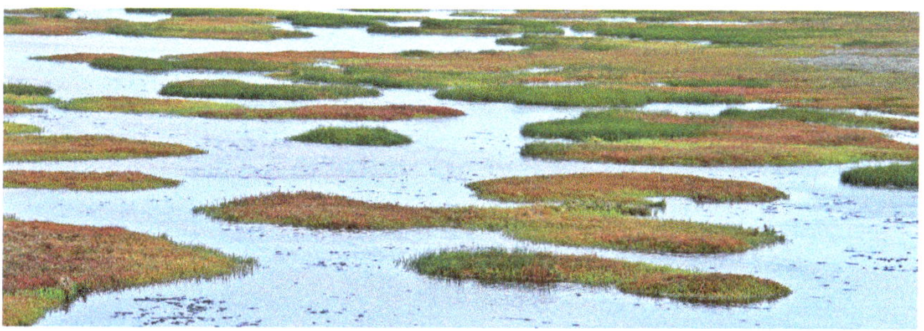

Patterns are recurring regularities, and pattern recognition is a target of data analysis in vegetation ecology. A proper sampling design is required to deliver data reflecting properties of the real system. Similarity patterns eventually occur in any sub-system, in resemblance space, in geographical space and in time.

2.1 Pattern recognition

Why search for patterns in vegetation ecology? Because the spatial and temporal distribution of species is nonrandom. The species behaviour is governed by rules causing detectable, recurring patterns that can be described mathematically, such as by a straight line (a regression line, for example) or a hyperbola-shaped point cloud. In a temporal context, we may observe a continuous increase, a decrease or an oscillation of population size, for example. Pattern recognition typically reveals not only one, but various alternative patterns emphasizing different properties of the system. In all these cases it can easily happen that one is dominating and thereby impeding the recognition of others. Figure 2.1 is an example for this. Panel (a) is the original picture of the flower of a water lily (*Nymphaea alba*). Panel (b) is the same reduced to 25 by 25 pixels forming a square grid pattern.

This is so dominant that our brain has trouble identifying it as a flower. Recognition is considerably improved in panel (c) where the same pixels are additionally blurred and our eyes are no longer misled by the grid pattern.

In ecosystems too, patterns are frequently superimposed, and this is even the rule. One of the aims of pattern recognition is therefore to identify and simultaneously distinguish superimposed patterns, for example, by partitioning the data in an appropriate way.

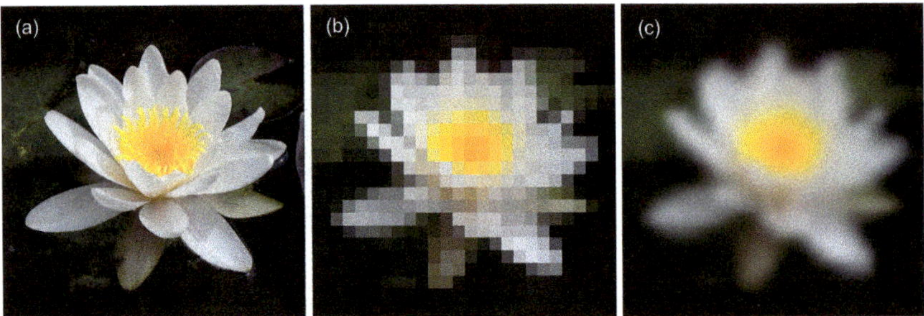

Figure 2.1: Picture of a flower of a water lily (*Nymphaea alba*). Panel (a) is the original, reduced to 25 by 25 pixels in panel (b). Panel (c) is the same as (b), but additionally blurred.

A typical application of pattern recognition is vegetation mapping. The usually inhomogeneous and complex vegetation cover of an area is reduced to a limited number of types. Figure 2.2 shows the vegetation pattern along the alpine timber line, predominantly caused by respective climate conditions (Körner 1999). Three vegetation types according to altitudinal level are distinguished. Before drawing such a map the types have to be assessed, a task discussed in Chapter 5. In the terminology of Chapter 1, Figure 2.2(a) is a picture of the real world and as such represents the data world. The vegetation map of Figure 2.2(b) represents the model world, that is, the way we understand the real world.

Patterns are often obscured not just by overlay, but by random variation (sometimes referred to as statistical noise) hiding the regularities. Methods are needed to partition the total variation into two components, one containing the regularity and one representing randomness. Probably the best known parameters describing a set of measurements are the mean ($\overline{x}$) and its variance (s^2):

$$\overline{x} = \sum_{i=1}^{n} x_i \qquad s^2 = \sum_{i=1}^{n} (x - \overline{x})^2$$

The mean could be interpreted as the deterministic component and the deviations from it as the random component of a measurement. Even in the simplest natural systems the existence of a deterministic pattern and a random component can be expected. A typical example in vegetation ecology

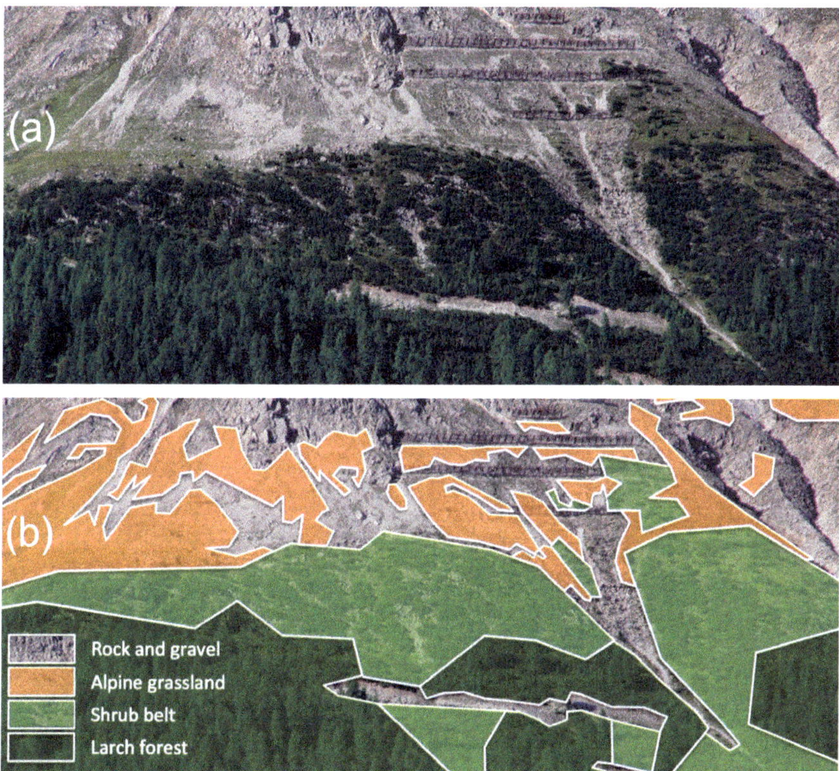

Figure 2.2: Vegetation mapping as a method for assessing a pattern (vegetation belts along the alpine timber line, Bergün, Switzerland). (a) represents raw data whereas (b) represents the model world, an abstraction of the real system.

is the representation of a vegetation gradient as an ordination, that is, a graph expressing the resemblance pattern of plots involved (Chapter 6). A continuous gradient of underlying conditions, time or environmental factors yields an observational sequence. When a vegetation gradient of this type is analysed, it will not manifest as a straight line but as a curve instead, also known as a horseshoe (see Section 6.5). What deviates from this can be considered statistical noise, but it can also come from yet another superimposed pattern. The issue is sketched in Figure 2.3 with data set `sveg` extensively used in Chapter 6, for example. It is easy to see that the data points (the plots) are nonrandomly arranged in the diagram. Two alternative patterns are highlighted. In panel (a) the points are assumed to form a horseshoe-shaped series suggesting the presence of a gradient structure in the data. It is shown in Chapter 7 that the existence of this kind of pattern is strongly supported by various analytical methods. Alternatively, a group pattern can be distinguished, graph (b). Confidence ellipses illustrate this kind of interpretation and there is analytical evidence for the existence of group structures (see Section 9.8). Which of these patterns is preferred typically depends on the purpose of the analysis. For creating a vegetation map (Figure 2.2, for example) assigning a group structure is indispensable. It is worth mentioning that Figure 2.3 addresses the famous community versus continuum controversy dominating vegetation ecology in the first half of the 20th century (Clements 1916; Gleason 1926, 1939).

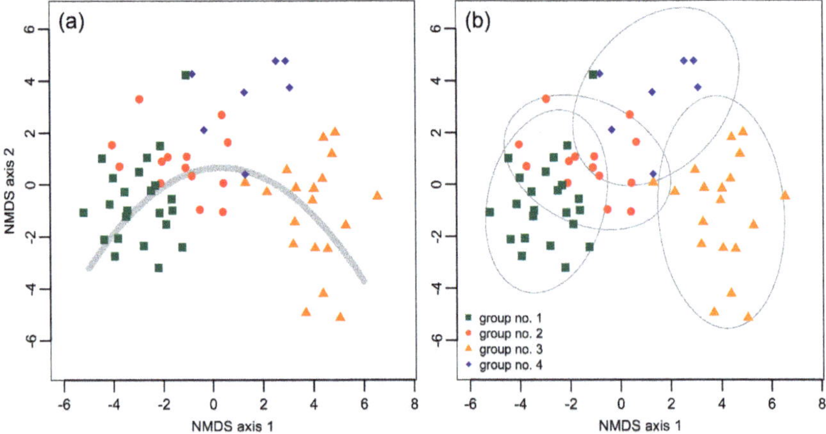

Figure 2.3: Ordination of a typical horseshoe-shaped vegetation gradient in a wetland (data set sveg, see Figure 2.7 and Chapter 6). Panel (a) accentuates the gradient, (b) the group pattern.

Vegetation patterns are not always hidden, but sometimes apparent even to the inexperienced eye as can be seen from the examples in Figure 2.4. From the scientific point of view patterns in panels (a) and (b) are of general relevance only if the same phenomena are recurrent, reflecting a rule rather than a special case. With regard to the sampling design (Section 2.3) this means that replications stemming from different locations are required.

Even if a pattern is visible, pattern recognition remains the main tool of analysis. First, the strength of the pattern has to be quantified. Second, the same has to be done with the environmental data and third, the two patterns have to be compared. A strong agreement suggests that plant species or plant traits respond to the environmental conditions. In Figure 2.4(a) avalanches are preventing part of the slope from being forested and young trees are unable to stop the process. In contrast, patterns in peat bogs, as shown in Figure 2.4(b), often evolve as a periodic pattern when the peat layer is shrinking. According to the theory of Watt (1947) this can be a transitional state in a cyclic process where the hummocks are decaying to

Figure 2.4: Spatial vegetation patterns. (a) Tree pattern caused by avalanches. (b) Peat bog pattern.

become wholes and the wholes are slowly overgrown by mosses until they form new hummocks.

Temporal patterns are sometimes apparent as well, even though mostly from time series data (Chapter 10). The analysis of temporal data also resides on similarity theory. Subsequent states in time are compared and the resulting patterns are temporal by nature. Temporal data differ from spatial in various ways. This is the reason why a separate chapter is devoted to the issue.

What is the cause of vegetation change and the resulting patterns in time and space? The answer is often found when considering environmental impacts, that is, exogenous driving forces. Analysing impacts is often the only way to assess causes. Ideally, these are completely under the control of the investigator who then is able to perform an experiment. Examples are shown in Figure 2.5. In panels (a) through (c) natural events are shown which cannot be controlled. A survey can be organized only after the impact has taken place (unless it is simulated by an artificial intervention). In contrast, panels (d) through (f) show human impacts. These allow the state of the system to be analysed prior to the event. And as mentioned in Chapter 10 reference plots can be implemented into the design to simultaneously investigate how the system evolves in the absence of an impact.

2.2 Multivariate pattern analysis

This book deals with the analysis of patterns in multivariate data. As mentioned in Chapter 1 all this makes sense only if the data sets are representations of the real world. In Section 2.3 I refer to the sampling problem, a big issue, as proper sampling only can yield data of the required quality. Any further step is based on similarity theory (Section 1.2). Mathematical analysis starts with Chapter 3 on transformation, allowing adjustment of the

Figure 2.5: Impacts causing patterns. (a) A natural event – forest gap caused by storm *Lothar*, 26 December 1999. (b) Animal interaction: beaver activity. (c) Snow slide causing soil erosion on steep slopes. (d) Man-made – a hay meadow just below the timber line. (e) Fertilizing an Alpine meadow in experimental research. (f) Shaping a riverbank to initiate a future vegetation gradient.

data to the objective of the investigation, while also overcoming restrictions imposed by measurement tools. First, transformations concern individual measurements (scalars), such as species cover, abundance or biomass, for which I frequently use the neutral term *species performance*. Second, entire vectors can be subjected to transformation. A relevé vector is an example, encompassing all measurements belonging to a plot, including species performance scores and site factors. A species vector considers performance scores in a relevé sample. In a traditional synoptic table (Section 7.6) a relevé vector is a column and a species vector a row.

In Chapter 4 multivariate comparison is presented. When comparing two relevés, one has to consider all species or all site factors simultaneously. This can be done in many different ways. The same applies to the species vectors, depicting species occurrence across relevé samples, and the site vectors addressing site factors. The resemblance pattern is then assessed by comparing all pairs of species or relevé vectors. If the number of vectors involved is equal to n then the resemblance matrix of all pairwise similarities has $m = n * (n - 1)/2$ elements. Because of the huge size of this matrix, further analysis can be time consuming.

Many of the subsequent analyses directly access similarity matrices, such as classification (Chapter 5), showing groups instead of single relevés, ordination (Chapter 6), showing similarity in reduced dimensional space, and ranking (e.g. Section 6.7), allowing the identification of relevés or species considered important in the given context. These three approaches unveil patterns. Chapter 7 is devoted to the comparison of patterns, being biological, environmental, spatial or temporal. In Chapter 8 this is extended to various alternative vegetation descriptors, such as plant traits and indicator values. The analysis of temporal patterns is shown in Chapter 10 and it is related to static (Chapter 9) and dynamic (Chapter 11) modelling, of which the very basic principles as well as examples are shown. Finally, two case studies illustrate practical issues through specific data sets: the revision of a classification of wetland vegetation in Switzerland in Chapter 12 as an example of using databases, and the analysis of forest vegetation data in Chapter 13, focusing on the interpretation of ecological patterns.

2.3 Sampling for pattern recognition

2.3.1 Getting a sample

The aim of data sampling is to generate a numerical description of the real system we wish to analyse. That is what a 'good' sampling design does. With a 'bad' design there exists the risk of uncovering a pattern not present in the real world. Generating a sampling design means that the sampling elements have to be chosen, which is explained below. In this section, sampling theory is not presented in detail. The elements are introduced because they determine the organization of the data sets. Their choice is a central issue in any investigation as it will determine the range of the validity of the results. There exist few guidelines helping to find a good sampling design and much is left to the intuition of the researcher. An outline of challenges and known pitfalls of popular practices is presented by Albert *et al.* (2010).

The terminology used throughout this book is shown in Table 2.1 and applied to an object in Figure 2.6. Many of the terms are prone to confusion, for example, a sample in some textbooks is the same as a sampling unit in others. Official terms given in Table 2.1 are openly available (International Statistical Institute 2009). The first step in sampling is the delineation of the *population* (not to be confounded with the population in the biological sense), which is the object to be investigated; for example, the full investigation area. As explained by Albert *et al.* (2010), delineation can take place in a geographical context, but alternatively also in an environmental context like precipitation or temperature, for example. The results will be valid for

Table 2.1: Terms used in sampling design (International Statistical Institute 2009).

Terms	Meaning	Example
Population (universe)	Target of investigation, all measurable items	All plants in a delineated investigation area
Sample	All measurements taken within the investigation area	A vegetation table
Sampling unit	One element of a sample	A relevé
Attribute	Descriptors of the sampling units	Plant species, site factors
Sampling plan	Location of units, size and shape	Sampling grid
Stratum	Subset of the sample	Relevés between 600 m and 800 m a.s.l.

this population in terms of time, space and content. In theory, all items belonging to the population could be measured, such as the diameter and height of all trees in a forest, for example. In practice, however, the costs of such a strategy (termed *full enumeration*) would be excessive and much of the energy and money be wasted. Instead, a selection of all measurable items is taken: the *sample* with carefully chosen sample size (Orlóci and Pillar 1989). In the terminology used here, the sample is the full set of measurements taken from the population. It provides an estimate of real means and variances of parameters of interest. The sample consists of *sampling units*. In vegetation science, a sampling unit is often a plot of predefined size and shape (Kent 2012), as illustrated in Figure 2.6. Each sampling unit is characterized by *attributes*, such as percentage cover of one or all species. One can measure just one attribute per sampling unit. But in practice, the number of attributes is often rather high. This is the case when relevés are taken where all the species occurring in the plot are recorded.

There are many more decisions needed to accomplish a full sampling design, one of them being the sampling plan [for which Albert *et al.* (2010) reserve the term sampling design]. Plots can be arranged systematically, as seen in Figure 2.6; for other applications, a random arrangement is the best choice. An example of systematic sampling, widely used throughout this book, is the 'Schlaenggli' data set `sveg` (Wildi 1977), encompassing 63 relevés and 119 species. Figure 2.7 shows two aerial views of the investigation area, using ordinary and infrared colour film respectively. As will be seen from many results the systematic sample in this case is an ideal choice to reveal various aspects of an underlying continuous pattern.

In more complex situations, a *stratification* of the entire population is suggested. When stratifying its surface, the investigation area is divided into subspaces, the *strata*, which are delineated based on prior information on the investigation area, such as a thematic map. To increase the efficiency of sampling, different sampling plans can be applied to the individual strata. If small strata are more intensely sampled than large strata, the number of sampling units eventually becomes equal for all strata. Not mentioned in Table 2.1 are plot size, plot shape and the time of sampling.

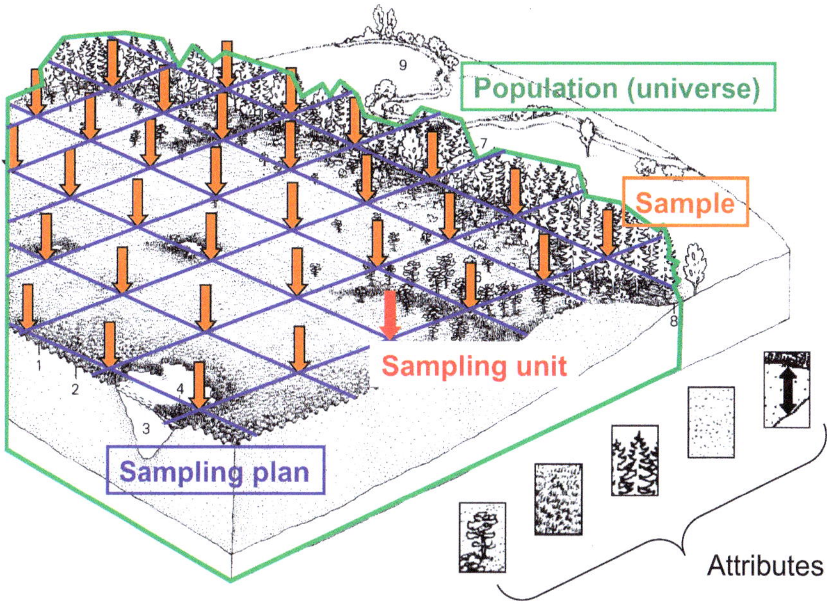

Figure 2.6: The elements of sampling design. In this example, a systematic sampling plan is applied to geographic space to assess the state of a peat bog.

2.3.2 Organizing the data

At first glance, organizing the data appears to be a technical matter only: in vegetation ecology the sample is usually presented in a rectangular matrix, where the rows are reserved for the sampling units and the columns are the attributes (or vice versa). Moreover, in space–time systems, the variables can be grouped by type. For this, the concept of space is used. A data table of the kind presented here forms the *data space*. As will be shown later (Chapter 4), there are other, even more abstract spaces such as the resemblance space.

At this point, some subtypes of data space are distinguished:

The biological space. This consists of the attribute vectors describing the biotic part of the system, such as plant species, plant cover, animal species, population sizes, life forms, functional traits, etc. In many models of data analysis these act as dependent variables.

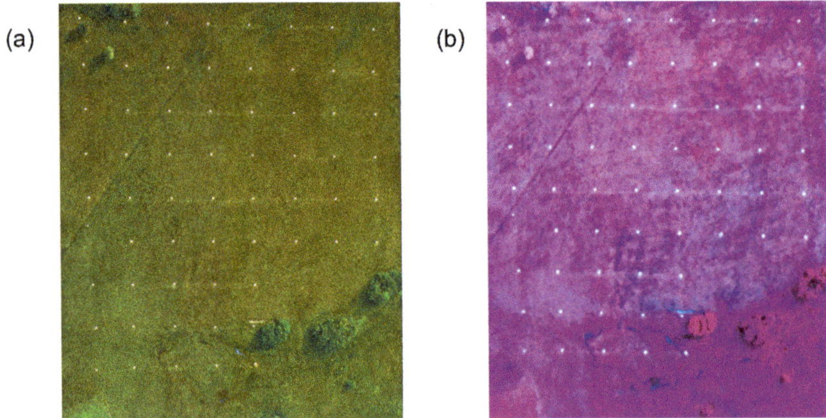

Figure 2.7: Sampling plan of the 'Schlaenggli' data set sveg. Photographs taken in 1974. Panel (a): normal colour, panel (b) infrared film used. Grid width is 10 by 10 m, plot size is 1 m^2.

The environmental space. The attributes involved measure the environmental conditions, like climate, nutrients, the substrate or disturbance such as fire or land use. They are often considered explanatory or independent variables.

The physical space in two or three dimensions. In the sample space, each sampling unit is described by its x-, y- and z-coordinate. By assigning this, the sampling plan also becomes part of the sample. Specific methods exist for the analysis of spatial data.

Temporal space. This has just one dimension, the time axis. As in physical space, there are special methods to analyse time series data.

In traditional phytosociology (Braun-Blanquet 1964; Dengler *et al.* 2008) there is a convention to put ecological, spatial and temporal attributes on top of data tables. The biological ones are then added in the form of species lists, a practice illustrated in Table 7.1. In the realm of $\mathcal{R}$ (and the world of statistics and informatics in general) matrices are usually used and displayed in transposed orientation: the sampling units are in the rows and the attributes are the columns. Furthermore many users store their biotic and environmental data in separate files as illustrated in Figure 2.2. It is vital that the rows in both files refer to the same relevés labelled r.1 through r.n. From such files import to $\mathcal{R}$ is done as explained in Section 2.4.

Manipulation of data sets is a big issue in vegetation ecology. This may come as a surprise, because samples are considered an unbiased numerical representation of the system to be investigated. But there are exceptions to

Table 2.2: Organization of vegetation (a) and site data (b) in $\mathcal{R}$. Rows of both data sets have the same names.

(a) **Biotic data**

	Spec.name.1	Spec.name.2	Spec.name.3	Plant.trait.1	Plant.trait.2
r.1	.	.	.	.	.
r.2	.	.	.	.	.
r.3	.	.	.	.	.
.	.	.	.	.	.
r.n	.	.	.	.	.

(b) **Site data**

	Site.factor.1	Time	x-coordinate	y-coordinate	Stratum
r.1	.	.	.	.	.
r.2	.	.	.	.	.
r.3	.	.	.	.	.
.	.	.	.	.	.
r.n	.	.	.	.	.

this. For example, there is sometimes a need to split relevé sets – if only a portion is to be analysed – or merge sets, if joint comparison is intended. Many of the methods presented in this book aim to reduce (or extend) data sets and the operations to do the task are found in various sections:

Removing low frequency species. Species occurring in one or in a few relevés of a sample usually do not contribute much to the overall pattern. How to remove species vectors with low frequency is explained in Section 7.6.3. However, this can be a dangerous strategy as it may hide outlier relevés (Section 12.3.1). Removing empty species vectors, on the other hand, is highly recommended as some methods fail to succeed otherwise.

Removing species with low resolving power. The term 'resolving power' is used in the context of relevé groups, that is, vegetation types: it may be of interest to retain those species helping to identify vegetation types while deleting others. There are two methods proposed, one in Section 7.2.2, and an alternative in Section 7.2.3.

Removing species based on redundancy. If any two or more species occur jointly, they carry redundant information. Redundancy is removed, for example, by retaining just one out of these. Orthogonal ranking does this as explained in Section 6.7.1 where eliminating redundant species is demonstrated.

Removing relevés based on redundancy. The method introduced in Section 6.7.1 is also applicable to relevés. It results in a reduced set of relevés accounting for a maximum of variation in the parent vegetation sample. In Section 6.7.2 it is shown how such a selection works and how it is used for optimized long-term vegetation surveys.

Selecting large groups only. This is helpful to revise classifications existing already. An example including technical instructions is given in Section 12.3.2.

Merging relevé samples. When merging vegetation data the resulting joint sample will include all relevés, but some of the species usually are common to both fusion candidates, whereas others are unique to either of these. A useful merge function will recognize this based on the common species names. An example handling this situation is found in Section 13.3.2.

In the context of ecological investigations files containing ecological information as in Figure 2.2(b) have to be reduced or extended parallel to vegetation data. This ensures all resulting matrices have the same number of rows.

2.4 Pattern recognition in $\mathcal{R}$

Various programs and program packages can be used to do analysis of the kind presented in this book. Choosing $\mathcal{R}$ to provide the computer instructions has some advantages. $\mathcal{R}$ is a free software environment permanently supported by a worldwide community of experts in their fields (Adler 2010, R Development Core Team 2017). But it is also a simple desktop calculator as well as a programming language. One can learn $\mathcal{R}$ with the aid of one out of many workshops found in the Internet. The instructions provided in the following chapters assume no previous knowledge of $\mathcal{R}$. They intend to serve users inclined to reproduce the examples in the book, and of course to do the same with their own data. Luckily the $\mathcal{R}$ syntax is partly self-explaining and often close to standard mathematical notation allowing technical explanations to be kept at a minimum.

Like any other software $\mathcal{R}$ must be installed on the computer, best done by an experienced person. The R code (for Mac, Linux and Windows) is found on sites accessed when searching for CRAN in the Internet. When finally running $\mathcal{R}$ a window opens with a command-line interface, an old-fashioned prompt where you enter the instructions followed by hitting the return key. This provides access to some pre-selected $\mathcal{R}$ libraries where widely used functions are available by default. It is good practice to create a specific folder for accessing and storing data in the course of analysis. In my computer this is `DataAnalysis`. After starting the $\mathcal{R}$ session the path pointing to this folder has to be set. In my computer it is:

```
setwd("/Users/ottowildiair/Documents/Buero_Owi/R/DataAnalysis")
```

This is the line I am running whenever I restart $\mathcal{R}$. Function `setwd()` sets what is called the working directory. The experienced expert will recognize the operating system I am using and DataAnalysis being the dedicated folder. To reproduce all the examples explained in this book one additional package has to be downloaded from CRAN: `dave`. Package `dave` is the one specifically written for this book and it loads some more required, like `vegan` and `labdsv`. It is activated by typing:

```
library(dave)
```

This will automatically load packages `dave`, `vegan` and `labdsv`, implementing all functions listed in Appendix 14 as well as those in packages `vegan` and `labdsv`. Equally important, all the data sets used in the book and listed in Appendix 14 are now present in the current $\mathcal{R}$ session.

But what do we do when running $\mathcal{R}$? Well, we process what is called the *data world* in Chapter 1. And the results contribute to the *model world*, the one constituting our understanding of the real world. When using data not included in `dave` the first step is reading data from external sources, for example by directly typing into the $\mathcal{R}$ console:

```
vec<- c(1,3,5,12)
vec
```

```
[1] 1 3 5 12
```

Function `c()` (combine) assigns numbers 1, 3, 5 and 12 to a vector, `vec`, of length four. The small font text below indicates output of $\mathcal{R}$ to the console. This kind of input is used in the case of very small examples, whereas for larger data sets direct typing is not practical. Especially in the case of real world data, files organized as shown in Figure 2.2(a) and (b) are read from disk. This is required in cases where readers want to import their own data files into the $\mathcal{R}$ session, preferably generated by spreadsheet programs and stored, for example, in csv format. For example, after loading the `dave` package, there is the fairly small `nveg` vegetation data set available in the current $\mathcal{R}$ session (Appendix 14). The following will generate the csv file `nzzveg.csv` in the working directory (the working folder) of the computer:

```
write.csv(nveg,file="nzzveg.csv")
```

The new file `nzzveg.csv` can be used for import into a spreadsheet program. In practice, the opposite operation, getting access to a csv file within the current $\mathcal{R}$ session, is even more important:

```
nveg2<- read.csv(file="nzzveg.csv")
```

The data are assigned to object `nveg2` to avoid overwriting the original (although the two versions are identical).

It is crucial for the newcomer to recognize `nveg` as an 'object', as $\mathcal{R}$ is object-oriented. Object `nveg` could be a scalar representing just one variable, a one-dimensional vector or, as in this case, a data frame with row and column names. We can check this by asking for the 'class' using function `class()` and the dimensions of `nveg` by function `dim()`:

```
class(nveg) ; dim(nveg)

[1] "data.frame"
[1] 11 21
```

In this two-dimensional array part of the object can be addressed by indices, written in square brackets. We can display, as an example, all 11 rows but 5 columns only by typing `nveg[,1:5]`. In $\mathcal{R}$, omitting the first index means taking all rows, whereas writing 1:5 selects columns 1 through 5. The power of $\mathcal{R}$ lies in the object-oriented operations. If we type, for example, `nveg*2` without specifying any indices, then all 11 by 21 elements are multiplied by 2. In the majority of the applications entire objects are processed this way.

The ecological data belonging to `nveg` [Figure 2.2(b)] is the data frame `nsit` with the following column names:

```
names(nsit)

[1] "PH"        "ALTITUDE"  "SLOPE.deg" "X.AXIS"    "Y.AXIS"    "EXPOSURE"
[7] "YEAR"      "GROUP_NO"
```

Like all other data frames used in the examples throughout this book `nsit` is listed in Appendix 14 where a short description of content is included and the origin of data referenced. Examples of vegetation data sets listed in Appendix 14 are:

`nveg` – a small artificial vegetation data set
`sveg` – a somewhat bigger, real world wetland vegetation data set
`sn6veg` – a vegetation data set from the Swiss National Park

And the corresponding site factors belonging to these examples are:

`nsit` – artificial site factors
`ssit` – site factors measured in the field
`sn6sit` – site information from the Swiss National Park

The simplest way to inspect the content of a data frame is by typing its respective name into the $\mathcal{R}$ console and hitting the return key. In most cases it is more convenient to inspect a first portion only, such as the first 8 lines when using function `head()`:

```
head(nsit)

    PH ALTITUDE SLOPE.deg X.AXIS Y.AXIS EXPOSURE YEAR GROUP_NO
2  4.4      450      10.0      1      1        S   97        1
4  6.2      500       0.5      3      4        E   90        2
6  4.8      420       4.5      2      1        E   98        1
9  5.6      580       6.0      3      3        E   94        3
10 6.5      560       4.0      2      4        N   89        2
18 6.0      400       2.5      3      2        E   93        3
```

What happens to these data in the course of analysis? The prevailing operation on data frames of vegetation and site factors is some kind of data reduction to reveal patterns in complex data sets addressing questions, such as:

- Are there variables (for example releveś, species) to omit because they contain similar properties? This seeks for redundancy in the data.
- Is the similarity pattern in the data efficiently represented by groups instead of single vectors? This may require classification.
- Can a resemblance pattern be displayed in reduced multidimensional space? Ordination in this case is an approach of choice.
- Is there a formal interaction between vegetation and environment, space or time? Various statistical models address this question.

Analysis in $\mathcal{R}$ always proceeds along the same scheme: applying a function to one or several objects and directing the result to an object using a backwards arrow:

```
o.pca<- pca(nveg)
```

In this example `pca()` is the function used and it performs this on data frame `nveg`. The name of the output list is `o.pca` (output from pca), a name chosen by the user. In this book this notation is used consistently: letter o (for 'output'), followed by a period and an abbreviation of the name of the function yielding the results. A list in $\mathcal{R}$ is an object containing various items that can be accessed by name. Item names are displayed by function `names()`:

```
names(o.pca)
```

```
[1] "scores"   "loadings" "sdev"     "totdev"
```

The meaning and content of this list is usually described in the help page of the respective function, displayed when typing a question mark followed by function name, `?pca`, as an example. Content is displayed by typing the list name followed by a dollar sign and the name of the item:

```
o.pca$totdev
```

```
[1] 32.30909
```

Provided we understand the basics of principal component analysis (PCA, Chapter 6) we reckon this being the total variance of `nveg` just by guessing.

An important and practical feature of many $\mathcal{R}$ functions is the availability of 'methods'. A method accesses output objects of functions and extracts specific results from these. In the case of function `pca()` there are methods such as:

```
plot(o.pca)
summary(o.pca)
varplot.pca(o.pca)
scores.pca(o.pca)
loadings.pca(o.pca)
```

The first two of these, `plot()` and `summary()`, inherit from generic functions included in $\mathcal{R}$. All methods are either provided by the author of the corresponding function or written by authors of different packages.

3 Transformation

Measurements in vegetation ecology take place at scales determined by measurement tools, conventions and traditions. Because subsequent analysis is considered evidence-based, data transformations must adapt the scale to the aim of the investigation and further statistical requirements.

3.1 Data types

As mentioned in Chapter 1, the aim of measurement is to generate a numerical description of the real world. This sounds like a merely technical issue; on closer inspection, however, data only capture part of the reality and often mirror the tool that has been used for measuring. We measure what we can measure and we omit what we cannot. Sometimes we also have a choice in the method we use to obtain some particular information, as for example when measuring the colour of light. We can either use a scale with discrete states (red, blue, yellow, etc.) or measure the wavelength of electromagnetic radiation. In the first case the measurement addresses a type of colour on a nominal scale, in the second we get a number on a ratio scale, a totally different data type, metric in this case. It is vital to consider different data types because their numerical analyses follow different rules. As will be shown below the transformation of one type into another is an important but controversially debated issue in vegetation ecology. There are textbooks distinguishing many data types (Gan *et al.* 2007); alternatively, a rather parsimonious version would be the one Jongman *et al.* (1995) propose:

Nominal data are recorded on a nominal scale, that is, a list of possible states. In Chapter 8 plants are described by discrete traits, for example. Table 3.1 illustrates an alternative to species performance as vegetation descriptors: life form (Cornelissen *et al.* 2003; Landolt *et al.* 2010). Data of this type restrict the application of mathematical operations. Growth forms are either the same or different and the operations making sense are $=$ or $\neq$.

Ordinal data are measurements on a rank scale. While this consists of discrete states just as nominal data it also implies an order. Indicator values introduced by Ellenberg (1974) and Landolt (1977) are an example illustrated in Table 3.1 with a five-step scale used by Landolt *et al.* (2010). At first glance this looks like a continuous variable with range one to five. But the authors unequivocally define the scale in a discrete manner. The R value, for example, distinguishes states like extremely acid (1), acid (2), weakly acid to weakly neutral (3), etc. In view of this the operations applicable to ordinal data are the same as those for nominal data. In addition, calculating a difference in ranks makes sense: a large difference in ranks usually means low similarity of any two elements. In Chapter 8 it is explained that there is a common (and successful) practice to disregard the ordinal nature of indicator values by routinely calculating means and standard deviations (Diekmann 2003; Wildi 2016).

Interval data are continuous measurements allowing the calculation and comparison of means, standard deviations and differences. But the position of the zero level is arbitrary. A typical example is temperature measured in degrees centigrade. There we can say that the difference between 25 °C and 30 °C is the same as between 10 °C and 15 °C, but 20 °C degrees is not twice as hot as 10 °C. Another example is the seasonal flowering period of plants shown in the right-hand column of Table 3.1.

Ratio data are measurements of distance, volume, weight, force and so on. Because there is a defined zero point it is possible to calculate ratios. Using a ratio scale gives maximum flexibility in the analysis. In vegetation ecology most environmental data are of this type, but also population size or cover scores. The latter are frequently expressed as percentages (Section 3.4).

One simple rule for the transformation of data types concerns the direction in which this is done. It is easy to transform from ratio to interval,

Species	Life form	Indicator values T	L	R	N	Months of flowering
Larix decidua	Phanerophyte	2	4	2	2	1 2 3
Picea abies	Phanerophyte	2.5	1	X	3	5 6
Dryas octopetala	Woody chamaephyte	1.5	5	5	2	5 6
Crocus albiflorus	Geophyte	2.5	4	4	4	5 6 7
Moinia arundinacea	Hemi-cryptophyte	3.5	3	4	3	5 6 7
Nymphaea alba	Pleustophyte	4	4	3	3	5 6 7

Table 3.1: Selected data types used for plant description. Nominal data: Life forms. Ordinal data: Indicator-values T, L, R and N. Interval data: Months where plants are flowering. Data taken from Landolt *et al.* (2010).

ordinal and further to nominal (with loss of information, however), but the opposite direction requires additional assumptions about the meaning of the measurements. This is a common practice when analysing plant cover-abundance data, as will be shown in Section 3.4. The transformations presented in the following two sections apply to ordinal, ratio and metric data only. In classical statistics (Sampford 1962) there are formal rules that have to be considered when using transformation, such as the need to correct for non-normal distributions of the data. In fact, transformation is often used to adapt data to statistical models. Yet, I present a slightly different view here: attributes are measured at a specific scale (given by the measuring device used). This scale does not necessarily serve the objective of the investigation. In this case transformations are a means to adapt measurements to the requirements of ecological models.

The $\mathcal{R}$ language distinguishes data types too, but it operates based on 'class' type. Class type is normally assessed automatically, but it may differ from what we have in mind. It is good practice to check for class and change it if needed. As an example, we type a vector of text variables into the command line of our $\mathcal{R}$ console window and name it `exposure`:

```
exposure<- c("N","N","E","N","S","E")
```

Typing the name of the vector displays the content:

```
exposure
```

```
[1] "N" "N" "E" "N" "S" "E"
```

What is the class of `exposure`? We can check this by function `class()`:

```
class(exposure)
```

```
[1] "character"
```

To consider the content of `exposure` as a nominal variable we better use data type `factor` in $\mathcal{R}$. Thereafter function `table()` displays useful statistics of content with variables ordered alphabetically and with corresponding frequency:

```
exposure<- as.factor(exposure)
table(exposure)
```

```
exposure
E N S
2 3 1
```

Nominal variables can be composed of letters, numbers or other signs. In a rank scale, however, one better uses numbers to get access to all possible arithmetical operations. As an example we assess a vector `slope` for inclination of a plot with states 'flat', 'inclined' and 'steep', but using numbers instead. We check for data type `numeric` using function `is.numeric()`:

```
slope<- c(1,1,2,1,3,2)
is.numeric(slope)
```

```
[1] TRUE
```

Again, function `table()` gives a summary of content:

```
table(slope)
```

```
slope
1 2 3
3 2 1
```

This tells us that three of our plots are flat, two are inclined and just one is steep. $\mathcal{R}$ considers vectors `exposure` and `slope` as objects to be accessed, for example, by vector operations as shown below.

3.2 Scalar transformation and the species enigma

When transformations are applied to individual measurements, I call them scalar transformations. Scalar transformation means that the scale used for measuring is adjusted according to our intention. Such transformations are widespread in environmental science. Often a relationship between two variables only emerges after proper transformation. Figure 3.1 illustrates this in a biological example. It is generally assumed that the survival of plant and animal populations depends on appropriate environmental conditions. When the conditions are favourable, populations may grow. Under less favourable conditions, they are likely to remain small. A small population may, for example, consist of five individuals. But 'large' is not, say, 20, but 100 or even more. When correlating population size with an environmental variable, temperature for example, a transformed number of individuals may be a better measure of population size. When taking $n' = n^{0.25}$, we adopt a more qualitative view of the size: 5 will become 1.49 (small), 20 will be 2.23 (average) and 100 is 3.16 (large). All the operations used in Figure 3.1 are readily available in $\mathcal{R}$. First, we read population size from the $\mathcal{R}$ console:

```
pop.size<- c(5,25,100)
pop.size
```

```
[1]   5  25 100
```

Then we apply the transformation mentioned above:

```
approx.rank<-  pop.size^0.25
approx.rank
```

```
[1] 1.495349 2.236068 3.162278
```

To get the rank scale shown in Figure 3.1 we truncate all decimal numbers using function `round()`:

```
exact.rank<- round(approx.rank,digits=0)
exact.rank
```

```
[1] 1 2 3
```

An alternative to this, with similar effect, is logarithmic transformation, for example by function `log10()` which uses base 10:

```
pop.log<- log10(pop.size+1)
pop.log
```

```
[1] 0.7781513 1.4149733 2.0043214
```

According to common practice adding 1 to the population size sets zeros to one and it avoids function `log10(0)` when species are absent.

Cover-abundance, but also population size, is sometimes transformed to *presence–absence* to consider occurrence of species only. This is done by function `sign()`:

```
pop.pa<-sign(pop.size)
pop.pa
```

```
[1] 1 1 1
```

The results of the transformations shown so far are stored in vectors `pop.size`, `approx.rank`, `exact.rank`, `pop.log` and `pop.pa` for further use. For example, correlating these values with temperature could easily yield a good linear relationship.

Another way of reasoning is that scalar transformation changes the perspective of objects. In Figure 3.2, graph (a), the coordinates of all 12 trees are untransformed and all have the same size. In graphs (b) and (c), x- and y-axes coordinates are root transformed with the effect that trees located near the upper right corner get less weight (they appear to be smaller). In graph (d) the opposite is achieved, that is, taking the square of all coordinates increases the weight of large values. Logarithmic transformation, graph (e), has a similar effect as root transformation. Finally, using the inverse of the original scores, graph (f), turns the trees upside down. In summary, the kind of reasoning illustrated in Figure 3.2 plays an important role when handling cover-abundance scores in vegetation data (see Section 3.4).

Biological population				
Population size, n	0	5	25	100
Nominal	Void	Small	Medium	Large
Rank	0	1	2	3
$n' = n^{0.25}$	0	1.49	2.23	3.16
$\log(n+1)$	0	0.78	1.41	2.00
Signum transformation	0	1	1	1

Figure 3.1: Scalar transformation of population size to improve correlation with environmental factors.

Transformation may sometimes contribute to the solution of problems inherent in ecosystems, such as poor correlation of species occurrence under similar site conditions. Despite the hope of many practitioners that species

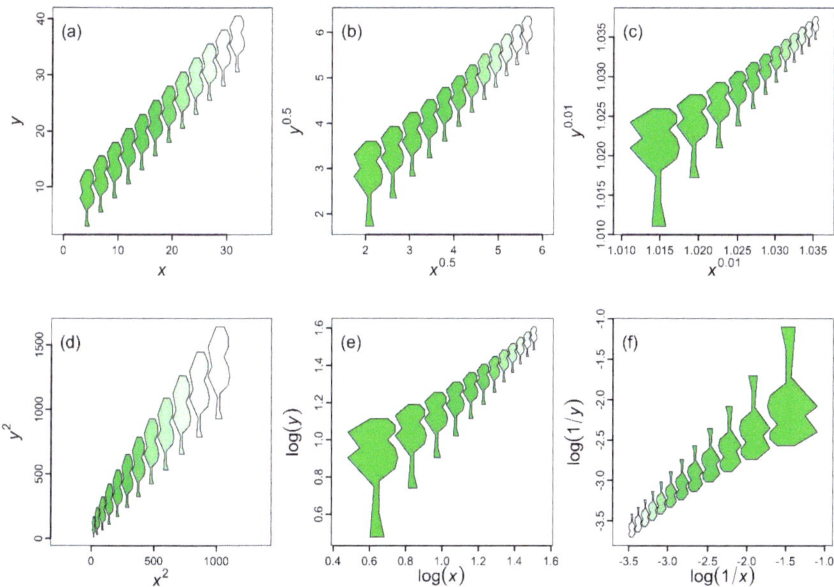

Figure 3.2: Scalar transformations of the coordinates of a graph. These affect the perspective adopted in the course of the analysis.

will form groups, thereby simplifying the identification of vegetation types, reality differs. When inspecting synoptic tables (Section 7.6) many species overlap nicely, but they hardly ever occupy the same niche. Even worse, species apparently tend to avoid common dispersion (Clarke 1993): a consequence of divergence (Section 1.2) As claimed by Gleason (1926, 1939) in his 'individualistic concept of the plant association', species behave like loners. And in fact, if the formation of an ecological niche is the result of a Darwinian struggle for life, then species are prone to ecological differentiation. I attempt to sketch a typical case of two overlapping species in Figure 3.3. The response of both species to the hypothetical gradient is Gaussian. Figure 3.3(a) illustrates the small overlap area when using cover scores. In Figure 3.3(b) the scores are transformed to presence–absence. This affects *relative* overlap, which is enlarged. Such a modification may be most welcome as co-occurrence measures of species are often unpleasantly low. Additional advice on how to profit from this effect is given in Section 3.4.

3.3 Vector transformation

As shown in Section 2.3.2 data are organized in two-dimensional data matrices. For processing in $\mathcal{R}$ the row vectors are the sampling units and the

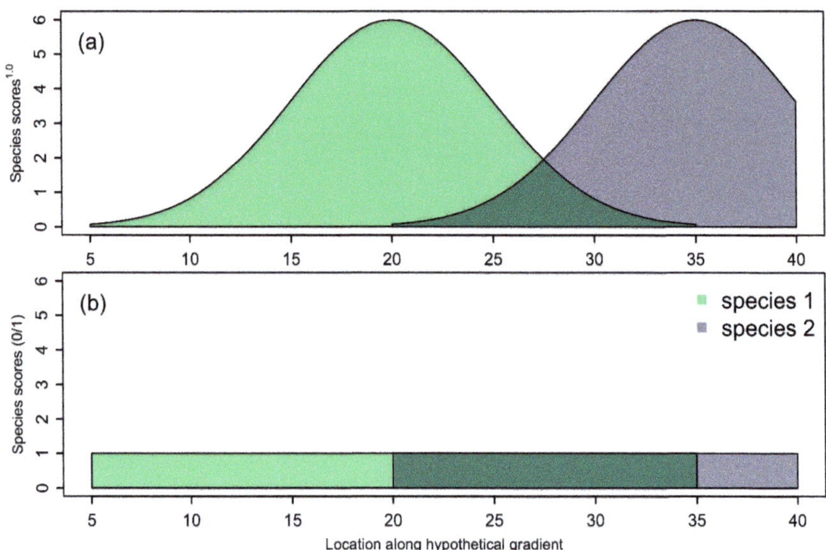

Figure 3.3: Overlap of two species with Gaussian response along a hypothetical gradient. (a) Species scores on a 0–6 performance scale. (b) The same scores, but transformed to presence–absence.

column vectors the attributes. Transformation of vectors may concern rows, columns or both simultaneously. The aim in either case resides in adjusting properties of vectors. When sampling units are transformed, a goal may be to give equal weight to all samples. Attribute transformation may result in balancing their weight in describing the sampling units. Some of the most commonly applied vector transformations are described in Table 3.2, with a numerical example given in Table 3.3.

Table 3.2: Effects of different vector transformations on the properties of data.

Term	Formula	Explanation
Centring	$x_i' = x_i - \overline{x}$	Adjusts mean to zero
Normalizing	$x_i' = \frac{x_i}{\sqrt{\sum_{x^2}}}$	Adjusts vector length to 1.0
Standardizing	$x_i' = \frac{x_i - \overline{x}}{\sqrt{\frac{1}{n-1}\sum_{(x_i - \overline{x})^2}}}$	Adjusts mean to zero and variance to 1.0
Range adjustment	$x_i' = \frac{x_i - x_{min}}{x_{max} - x_{min}}$	This is a fuzzy transformation (range 0.0–1.0)
Hellinger transformation	$x_{ij}' = \sqrt{\frac{x_{ij}}{x_i}}$	Adjusting relevé vectors to sum of squares 1.0

A first step, rarely used alone, is *centring*. The mean of the vector is deduced from each element. As a result, the new mean and the new sum both become zero. The sum of squares also changes, without becoming zero. The variance, however, remains unchanged. In $\mathcal{R}$ we first type the raw scores in Table 3.3 column by column using function `c()` and declare this a matrix with two rows. The backwards arrow names this `raw` and its content is displayed when typing the name:

```
raw<- matrix(c(2,0,0,0,5,1,4,2,6,2),nrow=2)
rownames(raw)<- c("r1","r2")
colnames(raw)<- c("s1","s2","s3","s4","s5")
raw
```

```
   s1 s2 s3 s4 s5
r1  2  0  5  4  6
r2  0  0  1  2  2
```

The row and column names assigned to this matrix will remain unchanged whenever a transformation is applied. Below we use function `scale()` for *centring* the rows of a matrix, but this by definition operates on columns only. To transform the rows, matrix `raw` has to be transposed [function `t()`] before, and to get it back into the proper orientation also after applying centring:

```
cent<- t(scale(t(raw),center=TRUE,scale=FALSE))
cent
```

```
     s1   s2  s3  s4  s5
```

Table 3.3: Two different vectors as example showing the effect of vector transformation.

x_1	x_2	x_3	x_4	x_5	$\sum$	$\overline{x}$	$\sum_{x^2}$	$\sqrt{\sum_{x^2}}$	S_x
Raw data									
2.00	0.00	5.00	4.00	6.00	17.00	3.40	81.00	9.00	2.15
0.00	0.00	1.00	2.00	2.00	5.00	1.00	9.00	3.00	0.89
Centred									
−1.40	−3.40	1.60	0.60	2.60	0.00	0.00	23.20	4.82	2.15
−1.00	−1.00	0.00	1.00	1.00	0.00	0.00	4.00	2.00	0.89
Normalized									
0.22	0.00	0.56	0.44	0.67	1.89	0.38	1.00	1.00	0.24
0.00	0.00	0.33	0.67	0.67	1.67	0.33	1.00	1.00	0.30
Standardized									
−0.58	−1.41	0.66	0.25	1.08	0.00	0.00	4.00	2.00	1.00
−1.00	−1.00	0.00	1.00	1.00	0.00	0.00	4.00	2.00	1.00
Fuzzyfied									
0.33	0.00	0.83	0.67	1.00	2.83	0.57	2.25	1.50	0.36
0.00	0.00	0.50	1.00	1.00	2.50	0.50	2.25	1.50	0.45
Hellinger transformation									
0.34	0.00	0.54	0.49	0.59	1.96	0.39	1.00	1.00	0.24
0.00	0.00	0.45	0.63	0.63	1.71	0.34	1.00	1.00	0.32

```
r1 -1.4 -3.4 1.6 0.6 2.6
r2 -1.0 -1.0 0.0 1.0 1.0
...
```

Setting center=TRUE but scale=FALSE accords with the goal we intend, that is, dividing each element by the mean of the corresponding vector.

Normalizing is a different method of transformation. Each element of the vector is divided by its (Euclidean) length. The vector sums and the vector means change and the vector lengths become 1.0. As shown in Table 3.3, the vectors become more similar in many ways while the variances still differ. Normalizing the rows of a matrix in $\mathcal{R}$ is easier than centring because package vegan offers the function decostand():

```
norm<- decostand(raw,method="norm")
norm
```

```
          s1 s2        s3        s4        s5
r1 0.2222222  0 0.5555556 0.4444444 0.6666667
r2 0.0000000  0 0.3333333 0.6666667 0.6666667
...
```

A rigorous transformation is *standardizing*. This is a combination of centring and normalizing. As a result, the vector mean is zero and the standard deviation (and the variance) becomes 1.0. The length of the vector is equal to the square root of the number of elements minus 1. Standardization is used to compare differently scaled measurements, such as temperature and the height of trees, for example. However, standardization has a downside: if the information is hidden in the variance then it will be lost. For computation in $\mathcal{R}$, function decostand() provides a method 'standardize'. This by default operates on columns and we have to change the default by setting MARGIN=1:

```
stand<- decostand(raw,MARGIN=1,method="standardize")
stand
```

```
           s1        s2        s3        s4       s5
r1 -0.5813184 -1.411773 0.6643638 0.2491364 1.079591
r2 -1.0000000 -1.000000 0.0000000 1.0000000 1.000000
...
```

Fuzzyfying is a simple transformation (Boyce and Ellison 2001). The elements are adjusted to range from zero (lowest score) to 1.0 (highest score). It should be used only if one really intends to adopt this view of the data. Aberrant values can set the extreme values in an undesirable way, deteriorating the observations completely. Fuzzy transformation is not an alternative to normalizing or standardizing, but it can be applied in combination with these. Fuzzy transformation is yet another option in function decostand(), named 'range'. This by default operates on columns and we get access to the rows by setting MARGIN=1:

```
fraw<- decostand(raw,method="range",MARGIN=1)
fraw

          s1 s2        s3        s4 s5
r1 0.3333333  0 0.8333333 0.6666667  1
r2 0.0000000  0 0.5000000 1.0000000  1
...
```

Hellinger transformation is a specific case devised for comparison of species vectors by Legendre and Gallagher (2001). We know from experience that cover of a species not only depends on ecological conditions, but also reflects a property of the species itself. Some, for instance *Phragmites australis*, are able to form huge, closed stands, whereas many tiny species will never reach high cover values. Hence, Hellinger transformation adjusts the species scores to range from zero to one adopting a view closer to presence–absence. Relevé vectors are adjusted to equal sum of squares of 1.0. The method is also available in function `decostand()`:

```
hell<- decostand(raw,method="hellinger")
hell

          s1 s2        s3        s4        s5
r1 0.3429972  0 0.5423261 0.4850713 0.5940885
r2 0.0000000  0 0.4472136 0.6324555 0.6324555
```

Legendre and Gallagher (2001) suggest using this transformation in conjunction with Euclidean distance thereby obtaining Hellinger distance (Chapter 4).

3.4 Example: Transformation of plant cover data

In phytosociology, Braun-Blanquet (1932) established a scale for measuring the quantity of plant species – that is, species performance – in vegetation relevés. He released his first comprehensive book on that topic in 1928 (English version in 1932). From the point of view of modern data analysis this scale (the so-called cover-abundance scale) is a mixture of form and content. At lower species densities, it expresses the abundance of individuals. At high densities, it directly translates to plant cover percentage. As shown in Table 3.4, it starts with a nominal notation in the form of the symbol '−', followed by '+'. Then it continues with a rank scale from 1 to 5. In the past 100 years, huge data sets have been collected all over the globe using this scale (Dengler *et al.* 2008). Handling it is therefore an issue in data analysis. Table 3.4 demonstrates how transformation could be done based on ideas published in a review paper by van der Maarel (1979).

In a first step the code is transformed into a proper rank scale with a range from 0 to 6 (column three in Table 3.4). Even though "the points on this scale are by no means 'equidistant' " (van der Maarel 1979) the ranks are

Table 3.4: Transformation of cover-abundance values in phytosociology. $x^{1.0}$ is the vector of rank scores replacing the Braun-Blanquet code. Transformation $x^{'} = x^{2.5}$ approximates cover percentage.

Code	Cover %	$x^{1.0}$	$x^{0.1}$	$x^{0.25}$	$x^{0.5}$	$x^{2.5}$
–	0	0	0	0	0	0
+	< 1	1	1	1	1	1
1	5	2	1.07	1.19	1.41	5.65
2	17.5	3	1.12	1.31	1.73	15.58
3	37.5	4	1.15	1.41	2.00	32.00
4	62.5	5	1.17	1.50	2.24	55.90
5	87.5	6	1.19	1.57	2.45	88.18

then treated as if they were metric. The justification for this is shown in the right-hand columns, where the rank scale is further transformed according to:

$$x^{'} = x^{y} \tag{3.1}$$

where x' is the transformed score. When $y < 1$ the data approach a binary state $\{0,1\}$. Near $y = 2.5$ it approximates the initial cover percentages. By choosing the appropriate value for y the scope of the analysis can hence be altered to emphasize either the qualitative or the quantitative aspect. For many applications, choosing $y = 0.25$ would be a good compromise as this expresses the qualitative view while considering the quantitative as well.

The rank scale in Table 3.4, but also the use of cover percentage, allows for flexibility in the course of analysis, because transformation can be applied to entire vectors and even two-dimensional matrices in just one step:

```
rank.scale<- c(0,1,2,3,4,5,6)
trans.scale<- rank.scale^0.25
trans.scale
```

```
[1] 0.000000 1.000000 1.189207 1.316074 1.414214 1.495349 1.565085
```

I use this method in many examples throughout this book. It also allows the approximation of a rank scale if the original scores are cover percentages:

```
cover.scale<- c(0,1,5.6,15.6,32,56,88)
rank.scale2<- cover.scale^0.4
rank.scale2
```

```
[1] 0.000000 1.000000 1.991935 3.000888 4.000000 5.003515 5.995054
```

As outlined so far choosing y properly allows the implementation of various alternative views of plant cover data. But which one is the 'best'? The answer is context dependent. An example of an optimization procedure is presented in Section 7.5, Table 7.8, where the best transformation is the one maximizing the agreement between vegetation pattern and ecological factors.

Alternative methods to alter step widths in cover scales exist. One of these, also introduced by van der Maarel (1979), is Clymo's function,

$$x' = \frac{1 - \mathrm{e}^{-ax}}{1 - \mathrm{e}^{-a}} \tag{3.2}$$

where x are original scores adjusted to range from 0 to 1, e is the base of the natural logarithm and a is the transformation exponent. For values of a approaching zero x' and x are almost linearly related. When a is large (100, for example), presence–absence transformation is approached:

```
x<- c(0,1,2,3,4,5,6) ; a<- 10
x<- x/max(x)
x.dot<- (1-exp(-a*x))/(1-exp(-a))
round(x.dot,digits=2)
```

```
[1] 0.00 0.81 0.96 0.99 1.00 1.00 1.00
```

The second line of the code adjusts the score to range from 0 to 1. Rounding serves readability only. To apply this transformation to a vegetation data set `x` is replaced by the corresponding object name, `nveg`, for example.

Lengyel and Podani (2015) used Clymo's function to evaluate, among other parameters, the effect of transforming cover-abundance scores on the results of classification (Chapter 5). They found that classifications result considerably when varying parameter a. Only the choice of the clustering method was even more influential.

The practice of metric handling of the Braun-Blanquet scale is quite common, yet violating statistical purism. In Podani (2006) comparisons are presented with truly ordinal methods. There, ordinal treatment turns out to be inferior to metric in most cases.

3.5 Which transformation?

There are statistical methods where specific transformations are prescribed, for instance, to change the distribution of a variable from skewed to symmetric. In most cases, however, the decision is left to the user. Although there are no 'right' nor 'wrong' transformations it is logical that in the end they will affect the result and eventually also interpretations. Unfortunately there is no stand-alone criterion by which a choice could be justified nor will the reader find any in this section. The performance of transformation can only be evaluated in view of a target, such as the correlation of two variables, for example. In this case the transformations yielding the highest correlations are candidates for the 'best choice'.

The first instance in this book illustrating the role of transformation in action is found in Chapter 5, Section 5.8, demonstrating that different classification methods yield different solutions for grouping sampling units. This, however, is done in combination with the choice of transformation. Figure 5.8 visually demonstrates that varying transformation of cover-abundance scores can lead to totally different solutions given the same combination of clustering method and distance measure.

The same procedure is possible with ordination methods, Section 6.8 in Chapter 6. This is a visual comparison too (Figure 6.15), but there I argue that ordination should be sensitive to data transformation such that the result reflects different ways of viewing data. Some combinations of ordination methods and resemblance measures do this impressively well, while others do not: resemblance measures that transform data implicitly may override any previous transformation.

A next example is a linear model expressing the interaction of vegetation and site factors. It is nonparametric multiple analysis of variance (NP-MANOVA) and the results are presented in Section 7.5.2. These suggest that when selecting pH as the explaining variable the best model is achieved when transforming cover-scale close to presence–absence (Table 7.8). As in previous cases the choice of the distance measure has to be considered as well.

Within the same chapter, in Section 7.5.3, the example from Section 5.8 is revisited. Here, the performance of clustering is measured quantitatively, displayed in Figure 7.9. The quality criterion is the proportion of variance that is explained by the classes. Nevertheless one should keep in mind that the underlying classifications may express different properties of pattern, quantitative versus qualitative, for instance.

4 Multivariate comparison

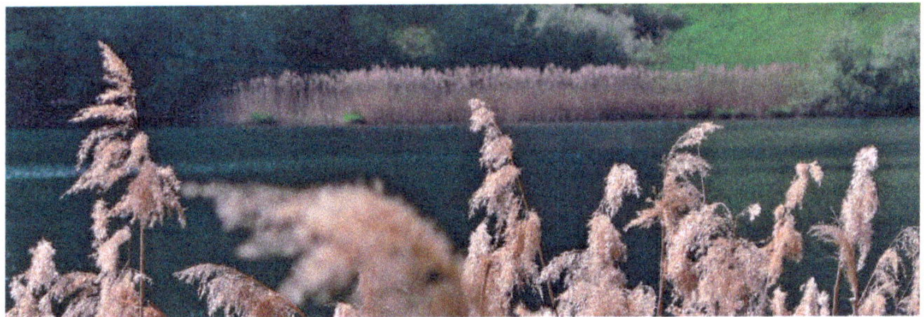

Vegetation and the corresponding environment are described by a large number of variables such as species scores, plant traits and environmental measurements. Multivariate comparison of sampling units is therefore a key issue in any analysis.

4.1 Resemblance in multivariate space

When talking about resemblance we address two kinds of measurement: *similarity*, where high values signify a high proportion of common features, and *distance*, where high values signify dissimilarity. As long as sampling units are described by one species or one site factor only, comparison is straightforward and the operational rules discussed in Section 3.1 on data types are valid. When more attributes exist, the technique is no longer trivial and several questions need to be clarified prior to data analysis:

- Are all attributes of the same type or is adjustment necessary?
- Do the attributes have the same weight or is some sort of adaptation advised?
- Are the attributes measured on the same scale or do scales have to be adjusted by appropriate transformation?

- Are some of the attributes correlated and therefore (partly) carrying identical information?

As a consequence of the multivariate nature of vegetation and site data, several attributes (or sampling units, when variables are compared) have to be taken into account simultaneously. The methods that do this belong to different approaches, three of which are being considered in this chapter: the geometric view, the statistical one (in the present case, measures of contingency) and the use of product moments (like covariance or correlation).

There exist more possible ways of multivariate comparison of sampling units or species, such as measures relying on information theory (Rényi 1961; Orlóci 1978). These methods focus on joint information, that is, information shared by any two sampling units, compared to the proprietary information of the units involved. I do not discuss information-based functions in the following sections as they are rarely used.

Considering a panoply of resemblance functions has a long tradition in vegetation analysis. But which one should be chosen in practice? Or, which one is the 'best'? The explanations given in this chapter are aimed at illustrating some main properties of resemblance functions and highlighting differences. But their 'performance' can only be assessed in the context of real-world applications, visually as well as quantitatively. In the last section, 4.7, references are given to various useful examples throughout this book.

4.2 Geometric approach

Multivariate similarity can easily be related to geometry, because geometry can handle high dimensionality. Geometric space may be one-dimensional (a straight line), two-dimensional (a surface) or three-dimensional (a volume). The dimensions can easily be extended to any number, say four or a hundred.

Operating in Euclidean geometric space implies some constraints. First, it is assumed that the dimensions (that is, the axes) are equally scaled. Secondly, the weight of all axes is the same. This also means that the attributes (and hence the axes) should be uncorrelated. The use of the multivariate Euclidean space is only justified if the attributes are measured on the same scale, as is the case when using cover percentage, for example. The principle is shown in the small example of Figure 4.1, with data given in panel (a). It is assumed that the pH values constitute the environmental, whereas the species scores form the biological space. In Figure 4.1(b) this two-dimensional biological space is shown, where the relevés are points in the scatter diagram. Whenever comprehensive species lists are used, the biological space may get extremely high-dimensional, with each species forming its own dimension. In Figure 4.1(c), however, it can be seen that a space may also be one-dimensional only. The relevés are still points, but now along a one-dimensional vector, pH, in this case.

Resemblance of any two sampling units in Euclidean space is most easily measured as a distance. If the distance is short then two relevés are similar. If the distance is long, the relevés involved diverge in many possible ways. Additional void species vectors do not add to the distance. This mainly welcome property distinguishes the geometric approach from various others.

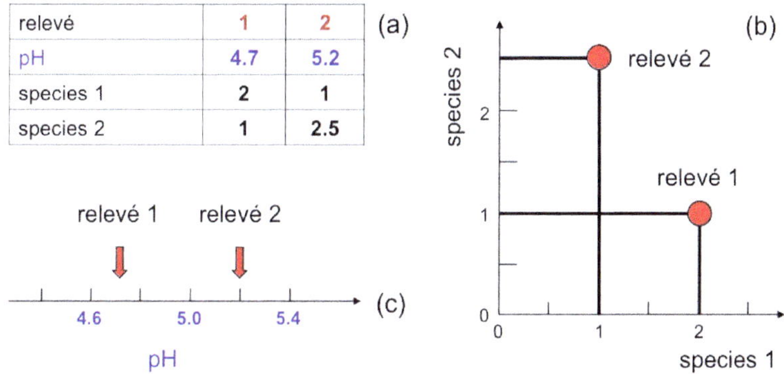

relevé	1	2
pH	4.7	5.2
species 1	2	1
species 2	1	2.5

Figure 4.1: Presentation of data in the Euclidean space. The data are shown in (a). In (b), the biological attributes are used to represent the relevés in two-dimensional (biological) space. (c) shows the one-dimensional environmental space.

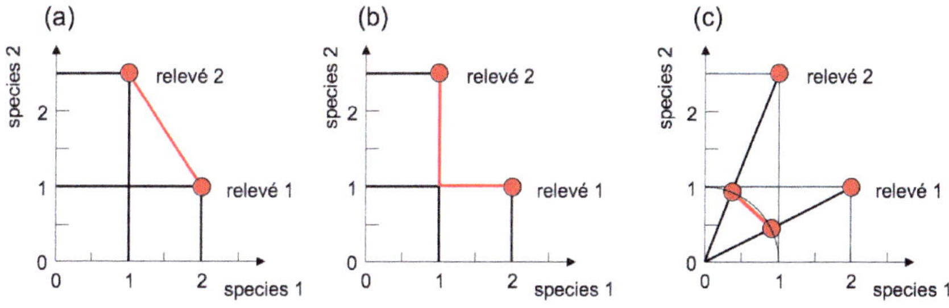

Figure 4.2: Three ways of measuring distance. (a) Euclidean distance. (b) Manhattan distance. (c) Chord distance.

There are different possibilities of calculating distance, as shown in Figure 4.2. A straightforward measure is *Euclidean distance*. The Euclidean distance between relevé 1 and relevé 2 is calculated as:

$$De_{1,2} = \sqrt{\sum_{j=1}^{p} (x_{1j} - x_{2j})^2} \tag{4.1}$$

In Figure 4.2(a), this is the direct connection between the corresponding data points. Equation (4.1) is written for p species and is therefore operational for any number of dimensions. The lower bound of $De_{1,2}$ is zero for identical relevés; the upper bound has no limit. As the number of dimensions (species) increases, the Euclidean distance tends to become larger.

An alternative measure is *Manhattan distance*, which is the sum of the differences of the scores calculated on all axes:

$$Dm_{1,2} = \sum_{j=1}^{p} |x_{1j} - x_{2j}| \tag{4.2}$$

The Manhattan distance [Equation (4.2)] has similar properties as the Euclidean, but it grows faster with increasing dimensionality, a property welcome when comparing species vectors [Figure 4.2(b) and Section 7.6.2]. Euclidean and Manhattan distance work perfectly as long as the number of species per relevé does not differ too much, because species missing in one relevé but present in the other always contribute to the apparent distance. The more species there are involved, the longer the distance will grow. Various methods presented below are attempts to compensate for this property.

Resemblance measures often differ by the intrinsic transformation involved. *Chord distance* is an example. It is identical to the Euclidean distance, but is calculated after normalizing the vectors [Figure 4.2(c)]. Embedding normalization into the calculation of Euclidean distance yields the corresponding formula:

$$Dc_{1,2} = \sqrt{2\left(1 - \frac{\sum_{j=1}^{p} x_{1j}x_{2j}}{\sqrt{\sum_{j=1}^{p} x_{1j}^2 \sum_{j=1}^{p} x_{2j}^2}}\right)} \tag{4.3}$$

Chord distance has a lower bound of zero (for identical relevés or species vectors). Unlike the previous measures, there is now a maximum value, the square root of two; that is, 1.414213. This is the case when relevés have no species in common. It is difficult to decide whether normalizing involved is ideal for applications: when transformation is really needed, many researchers prefer standardization (adjusting vector length and variance, Table 3.2) rather than normalization (adjusting vector length only). This issue will be further discussed in the context of the product moment measures (Section 4.4).

Related to Manhattan distance is *distance* [Equation (4.4)] (Lance and Williams 1966). This weights all distances by the species scores involved and as a result species with large scores get less weight than in Euclidean distance, a property many users seemingly appreciate:

$$Dca_{1,2} = \frac{1}{pp} \sum_{j=1}^{p} \left(\frac{|x_{1j} - x_{2j}|}{(x_{1j} + x_{2j})} \right) \tag{4.4}$$

Assuming one species to have score 1 and the other score 2, then the difference is divided by 3 and $Dca_{1,2} = 1/3$. But as the scores get 2 and 3 respectively, $Dca_{1,2} = 1/5$ only, a typical case of implicit transformation. In the original version of Lance and Williams (1966) division by pp (the number of non-zero species involved) was not yet included, but introduced later in the `vegan` package (see below). As a result, Canberra distance has a range from 0 to 1.

Something similar is done in *Bray–Curtis distance* [Equation (??)]. As the nominator cannot exceed the denominator the maximum value does not exceed 1:

$$Dbc_{1,2} = \frac{\sum_{j=1}^{p} |x_{1j} - x_{2j}|}{\sum_{j=1}^{p} (x_{1j} + x_{2j})} \tag{4.5}$$

As pointed out by Legendre and Legendre (2012, p. 311) this distance is no longer metric as it may violate the triangle inequality axiom. In the one-species case discussed above it is identical to Canberra distance, but no longer when two or more species are involved. Bray–Curtis is frequently used in conjunction with ordinations (Chapter 6) where the method has to coerce the similarity pattern into a metric graphic. Ordination methods that are able to do this are principal coordinates analysis (PCOA) and also nonmetric multidimensional scaling (NMDS). The popularity of Bray–Curtis has historical reasons, but also practical: if the user misses transforming a rank scale appropriately then Bray–Curtis will do it 'automatically'. In other words, it may yield the desired results even if the scores are not properly adjusted by scalar transformation (Section 3.4). Illustrations of this issue are found in Sections 5.8 and 6.8.

The resemblance measures most commonly used in vegetation ecology are available in the $\mathcal{R}$ package `vegan`. Function `vegdist()` applies these to data matrices, such as the two artificial relevés below (typed column by column):

```
two.rel<- matrix(c(0,1,0,1,1,1,1,0,1,0,0,0),nrow=2)
rownames(two.rel)<- c("r1","r2")
colnames(two.rel)<- c("s1","s2","s3","s4","s5","s6")
two.rel
```

```
   s1 s2 s3 s4 s5 s6
r1  0  0  1  1  1  0
r2  1  1  1  0  0  0
```

In this presence–absence type of data one species is common to both relevés, two are specific to either of the two and one is absent in both. Function `vegdist()` yields a lower triangular matrix of all resemblance values, including the diagonal elements if `diag=TRUE`:

```
de<-vegdist(two.rel,method="euclid",diag=TRUE)
de
```

```
   r1 r2
r1  0
r2  2  0
```

The sum of the squared differences in `two.rel` is 4 and the Euclidean distance is the square root of this. For the Manhattan distance we type:

```
dm<-vegdist(two.rel,method="manhattan",diag=TRUE)
dm
```

```
   r1 r2
r1  0
r2  4  0
```

The Manhattan distance is the sum of the differences of the two vectors. For chord distance $\mathcal{R}$ package **vegan** does not directly provide a function. But calculating this can be achieved by normalizing the relevés first as indicated in Table 3.2 and then using Euclidean distance. This is all done by combining functions `vegdist()` and `decostand()`:

```
dc<-vegdist(decostand(two.rel,"norm"),method="euclid",diag=TRUE)
dc
```

```
         r1       r2
r1 0.000000
r2 1.154701 0.000000
```

Because the maximum value chord distance takes on is 1.414, 1.154 can be considered a fairly large distance.

Directly available in **vegan** is the Canberra distance and also Bray–Curtis:

```
dca<-vegdist(two.rel,method="canberra",diag=TRUE)
dca
```

```
    r1  r2
r1 0.0
r2 0.8 0.0
```

```
dbc<-vegdist(two.rel,method="bray",diag=TRUE)
dbc
```

```
          r1        r2
r1 0.0000000
r2 0.6666667 0.0000000
```

Canberra and Bray–Curtis have range 0 to 1. When using presence–absence scores as in this example then Bray–Curtis distance is identical to one minus the Sørensen similarity index shown in Section 4.3.

4.3 Contingency measures

Contingency testing is a statistical approach, focusing on the joint occurrence of objects. In the case of vegetation relevés, these are common species. If there are many, one assumes that the relevés are similar. From a statistical point of view the question arises whether the number of common species is above, equal to or below expectation. Hence, everything depends on the meaning of 'expectation'. Contingency measures in their strict sense disregard abundance scores and rely on presence–absence only. Circumventing this restriction, however, is common practice in vegetation analysis and the way this is done is shown at the end of this section.

Table 4.1: Notations in contingency tables. a, b, c and d are frequency counts. A plus sign means presence and a minus sign means absence of species.

relevé 2 \ relevé 1	+	−	
+	a	b	$a+b$
−	c	d	$c+d$
	$a+c$	$b+d$	Σ

The standard setup for this type of measurement is the 2 by 2 contingency table (Table 4.1). This explains the elements termed a through d by which relevés are compared. For each species, common occurrence is counted in cell a. When a species occurs in one relevé only, it is counted in either cell b or cell c. If a species occurs in neither relevé 1 nor 2, it contributes to cell d.

The row and column sums yield useful numbers as well. The sum in row $a+b$ is the total number of species found in relevé 2; the sum in column $a+c$ is the same for relevé 1. The sum in row $c+d$ is the number of species that do not occur in relevé 2 and the sum in column $b+d$ the number of species that do not occur in relevé 1. The grand total Σ is the total number of species considered for calculations, including cell d, those occurring in neither relevé 1 nor 2. Consequently, contingency measures are affected by empty species vectors only in cases where variable d is included in the formula.

When combining a, b, c, and d from contingency tables, an almost unlimited number of coefficients can be devised. Many of these are listed in

Legendre and Legendre (2012, pp. 273–277). They differ in their properties and some are related to other types of resemblance measures. Three of them are shown in Table 4.2.

Table 4.2: Selected resemblance measures using the notations in Table 4.1.

Name	Formula	Distance measure	Property
Jaccard	$S_J = \frac{a}{a+b+c}$	$D_J = 1 - S_J$	Metric
Sørensen	$S_S = \frac{2a}{2a+b+c}$	$D_S = 1 - S_S$	Semimetric
Ochiai	$S_O = \frac{a}{\sqrt{(a+b)(c+d)}}$	$D_O = 1 - S_O$	Semimetric

The *Jaccard coefficient* S_J is the oldest, first published in 1901. It considers the number of common species and the total number of species present in either of the two relevés. The range is from zero (no species in common) to one (all species in common). When 50% of the species are common, $S_J = 0.50$.

The second is the *Sørensen coefficient* S_S (Sørensen 1948). It differs from the Jaccard coefficient in that common species have double weight. The range is also zero to one, but when 50% of the species are in common, $S_S = 0.667$. The derived distance measure (the complement) is semimetric, because it may happen that the distance configuration of three or more relevés cannot be presented in Euclidean space (that is, the triangle inequality is violated), limiting its application in some methods.

A third similarity measure is the *Ochiai coefficient*. As mentioned by Orlóci (1978) this is in fact a correlation coefficient (Section 4.4) because S_O is the cosine of the angle α shown in Figure 4.3.

For calculating contingency coefficients in $\mathcal{R}$ packages `vegan` and `labdsv` are used. In the examples below I use matrix `two.rel` from Section 4.2 again. For Jaccard distance we get:

```
dj<- vegdist(two.rel,method="jaccard",diag=TRUE)
dj
```

```
    r1  r2
r1 0.0
r2 0.8 0.0
```

Jaccard similarity, in this case, would be 0.2, the one complement of 0.8. Note that Jaccard in `vegdist()` also accepts scores different than presence–absence. Sørensen's coefficient is provided in `labdsv` with function `dsvdis()`:

```
ds<- dsvdis(two.rel,index="sorensen",diag=TRUE)
ds
```

```
          r1        r2
r1 0.0000000
r2 0.6666666 0.0000000
```

Compared with Jaccard this distance is somewhat smaller because the common occurrence, a, is counted twice. The Sørensen function in `dsvdis()` always transforms scores to presence–absence as does Ochiai's:

```
do<- dsvdis(two.rel,index="ochiai",diag=TRUE)
do
```

```
          r1        r2
r1 0.0000000
r2 0.6666667 0.0000000
```

For any other coefficients based on contingency, package **`vegan`** offers function `designdist()` where any favourite formula can be entered using the `method` parameter. Composing the Jaccard distance, for example, is done like this:

```
dj<- designdist(two.rel,method="1-(a/(a+b+c))",abcd=TRUE)
dj
```

Function `designdist()` supports a parameter, `abcd`, to invoke the terminology used in the present section. But it also supports an alternative, allowing the consideration of cover-abundance scores in all resemblance measures, the J, A, B and P parameters. Denoting the first relevé by x and the second by y, these are defined as follows:

$$
\begin{array}{lll}
J = \sum xy & \text{where} & a = J \\
A = \sum x^2 & \text{where} & b = A - J \\
B = \sum y^2 & \text{where} & c = B - J \\
P = a + b + c + d & \text{where} & d = P - A - B + J
\end{array}
$$

Function `designdist()` by default operates on this notation and for Jaccard distance an equivalent notation would be

```
dj<- designdist(two.rel,method="(A+B-2*J)/(A+B-J)",abcd=FALSE)
dj
```

This delivers exactly the same result. However, the authors of `designdist()` warn about long computation time and the increased risk of mistakes when manually typing its own distance function.

4.4 Product moments

Product moments are a group of measures expressing the degree to which vectors point into the same direction. This conforms with the basic concept of variance (the variance within one vector) and covariance (the variance shared by two vectors). Four related measures that differ in their implicit transformation only are listed in Table 4.3.

The *scalar product* is the vector product with no further transformation involved. If all scores are positive it ranges from zero towards infinity. The more attributes there are, the larger gets the scalar product.

The *centred scalar product* involves centring of the observational vectors; thus the mean of any vector will be zero. On average, half of the coefficients will be negative with no upper and lower bounds.

Covariance does the same thing as the centred scalar product, but in addition it offers a correction for the number of elements, n. It is used in analysis of variance. Note that $n-1$ corrects for the underestimation of variance in small samples of size n.

Table 4.3: Product moments. Types differ with regard to implicit data transformation.

Name	Formula	Transformation
Scalar product	$S_{jk} = \sum_{h=1}^{p} A_{hj} A_{hk}$	$A_{hj} = X_{hj}$
Centred scalar product	$S_{jk} = \sum_{h=1}^{p} A_{hj} A_{hk}$	$A_{hj} = X_{hj} - \overline{X}_h$
Covariance	$S_{jk} = \sum_{h=1}^{p} A_{hj} A_{hk}$	$A_{hj} = (X_{hj} - \overline{X}_h)/\sqrt{n-1}$
Correlation	$S_{jk} = \sum_{h=1}^{p} A_{hj} A_{hk}$	$A_{hj} = \frac{(X_{hj} - \overline{X}_h)}{\left(\sum_{e=1}^{n} X_{he} - \overline{X}_h\right)^{1/2}}$

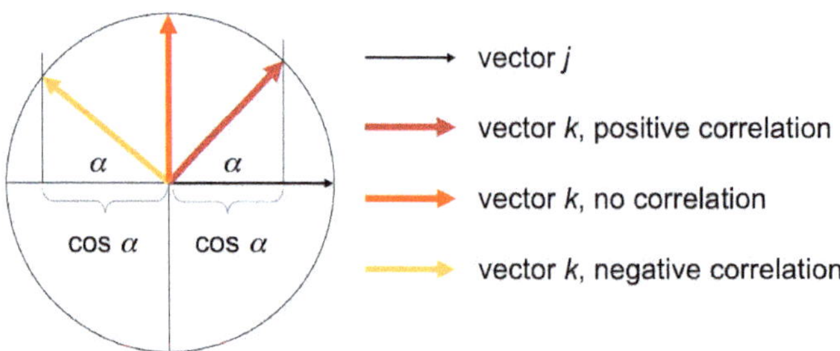

Figure 4.3: The correlation of vector j with vector k. The correlation coefficient is the cosine of the angle α between any two observational vectors j and k.

The *product moment correlation coefficient* (termed *correlation* in Table 4.3) standardizes the observational vectors implicitly. Their mean is zero and the standard deviation is equal to one. This has the practical advantage that there are fixed upper and lower bounds: $-1 \leq r \leq +1$. This is shown in Figure 4.3 in the form of a geometrical interpretation. When two vectors show the same trend, but in the opposite direction, correlation approaches cos $\alpha \approx -1$. When they are independent, r is zero. When they point in the same direction it approaches cos $\alpha \approx +1$.

There are methods for analysis which require a distance matrix. To profit from the properties of the correlation coefficient, r, transformation to *correlation as distance*, dr, is obtained as follows:

$$dr = (1 - r)/2 \tag{4.6}$$

This distance measure has fixed bounds $0 \leq dr \leq 1$. Thus, just as when using correlation coefficient directly, measurements taken at different scales become commensurable, and the variance is adjusted. If this is not desirable (because one expects important information from differences in variance, for instance) then another option should be considered. An example where this is suggested is the comparison of frequent species versus infrequent species. In Section 7.6.2 Manhattan distance is chosen for the purpose.

Calculating product moments does not require specialized functions. The *scalar product* is obtained directly when multiplying a matrix with its transpose, in mathematical terms:

$$S = AA' \tag{4.7}$$

In this S is the matrix product of A and its transpose A'. As $\mathcal{R}$ performs matrix operations in one step we get the square matrix of the *scalar product* as follows:

```
sp<- two.rel %*% t(two.rel)
sp
```

```
   r1 r2
r1  3  1
r2  1  3
```

Symbols $\% * \%$ are used in $\mathcal{R}$ for matrix multiplication and function `t()` transposes the matrix. The result is a two by two matrix in which score 3 is the sum of squared vectors and 1 is the scalar product of the two relevés. The dimension of the similarity matrix is the same as the number of relevés involved. For the *centred scalar product* we centre the data first by row as shown in Section 3.3. Then we take the matrix product:

```
ctwo.rel<- t(scale(t(two.rel),scale=FALSE))
csp<- ctwo.rel %*% t(ctwo.rel)
csp
```

```
     r1   r2
r1  1.5 -0.5
r2 -0.5  1.5
```

Score -0.5 is the centred scalar product of the two relevés involved. For *covariance* there is function `cov()` available in $\mathcal{R}$. Because this operates on columns by default we must use the transposed data to get covariance of rows:

```
covv<- cov(t(two.rel))
covv
```

```
    r1   r2
r1  0.3 -0.1
r2 -0.1  0.3
```

Score 0.3 is the *variance* of either of the relevés and -0.1 is the *covariance*. The name of the function used for *correlation* is `cor()`. This also operates on columns and our data have to be transposed in advance:

```
corr<- cor(t(two.rel))
corr
```

```
           r1         r2
r1  1.0000000 -0.3333333
r2 -0.3333333  1.0000000
```

As expected the diagonal elements are equal to 1.0 and the two relevés correlate by -0.33. To get correlation as distance [Equation (4.6)] this extends to:

```
dcorr<- (1-cor(t(two.rel)))/2
dcorr<- as.dist(dcorr,upper=TRUE,diag=TRUE)
dcorr
```

```
          r1        r2
r1 0.0000000 0.6666667
r2 0.6666667 0.0000000
```

$\mathcal{R}$ function `as.dist()` transforms `dcorr` into an object automatically recognized as a distance matrix in subsequent applications. At the same time default settings would display the lower triangular matrix, and `upper=TRUE` and `diag=TRUE` prints the full two by two matrix.

The centred scalar product, covariance and also correlation change when additional void species vectors are involved. This increasingly happens as data sets get larger. To see the issue we can add two more species with zero scores to our example and we get:

```
two.rel2<- matrix(c(0,1,0,1,1,1,1,0,1,0,0,0,0,0,0,0),nrow=2)
rownames(two.rel2)<- c("r1","r2")
colnames(two.rel2)<- c("s1","s2","s3","s4","s5","s6","s7","s8")
corr<- cor(t(two.rel2))
corr
```

```
            r1          r2
r1  1.00000000 -0.06666667
r2 -0.06666667  1.00000000
```

Correlation increases to $r = -0.0667$ compared with $r = -0.3333$ before. However, this does not hamper pattern recognition in classification (Chapter 5) and ordination (Chapter 6) because all relevés of a given data set are equally affected by this phenomenon.

4.5 The resemblance matrix

Whereas pairwise comparison of observational vectors (relevés, species, site factors, for instance) is useful for many purposes, assessing the pattern of an entire sample requires the computation of a resemblance matrix. This is done by comparing all possible pairs of sampling units, resulting in an $n*n$ matrix of resemblance coefficients. Such a matrix (Table 4.4) is generally symmetric and only the lower-left triangle (or the upper-right) has to be considered. Depending on the resemblance measure used, the diagonal elements, the self-similarity of the sampling units, may be of interest or not. When using Euclidean distance, for example, they are all zero; when using the correlation coefficient they all equal 1.0. When using covariance, however, they carry the variances of the sampling units and these usually vary.

Resemblance matrices may become very large. When computing the triangular matrix only, without the diagonals, the number of elements is $n*(n-1)/2$. This is far too much for immediate interpretation. The matrix therefore has to be processed further with the aim of pattern recognition, by cluster analysis (Chapter 5), component analysis (Chapter 6) or ranking (Section 6.7), for example.

A simple straightforward application of resemblance matrices is shown in Table 4.4, lower parts. The aim is to determine the *homogeneity* (or its complement, diversity) of a sample. From the data matrices in the upper part, the distance matrices are calculated and the mean Euclidean distance is computed. The mean distance is a measure of dissimilarity of the total set of relevés. Data matrix (a), with a relatively low average distance, in this view, is more homogeneous than data matrix (b).

Table 4.4: The average distance of a distance matrix of count data is a perfect measure of homogeneity (or its complement, diversity) of a sample. (a) High homogeneity (low mean distance); (b) low homogeneity (high mean distance).

(a) Data

	s1	s2	s3	s4
r1	10	20	4	6
r2	15	30	5	8
r3	18	25	15	12

Distance matrix, $\overline{d} = 13.1$

r1	0		
r2	11.4	0	
r3	15.7	12.2	0

(b) Data

	s1	s2	s3	s4
r1	4	11	2	5
r2	14	31	4	9
r3	20	41	24	17

Distance matrix, $\overline{d} = 29.8$

r1	0		
r2	22.8	0	
r3	42.2	24.4	0

To reproduce the result for data matrix (a) we first enter the data column by column, declaring this a `matrix()`:

```
data.a<- matrix(c(10,15,18,20,30,25,4,5,15,6,8,12),nrow=3)
data.a
```

```
     [,1] [,2] [,3] [,4]
[1,]   10   20    4    6
[2,]   15   30    5    8
[3,]   18   25   15   12
```

Using `vegdist()` for calculating the distance matrix yields the lower triangular matrix:

```
de<- vegdist(data.a,method="euclidean",diag=TRUE)
de
```

```
         1        2        3
1  0.00000
2 11.40175  0.00000
3 15.68439 12.24745  0.00000
```

```
mean(de)
```

```
[1] 13.11120
```

$\mathcal{R}$ function `mean()` automatically uses all elements in the lower triangular matrix `de` to yield the mean distance.

In cases where the relevés of a data set have a meaningful order then the corresponding resemblance matrix allows direct interpretation when displayed graphically. An example is Figure 10.4 where a time series is analysed. A similar form of presentation of resemblance matrices are heat maps, illustrated in the book of Borcard *et al.* (2011).

4.6 Assessing the quality of classifications

A more advanced application of graphically presenting resemblance matrices is in evaluating group patterns, Figure 4.4. This is a graph of the similarities within and between 71 forest vegetation types in Switzerland, published by Ellenberg and Klötzli (1972). The underlying historic data have been reconstructed from the original notes of the authors and the relevés found in the literature (Keller *et al.* 1998). From these 2533 relevés we know the corresponding classification used for definitions of the forest types. The coefficients in Figure 4.4 are not just pairwise similarities, but mean similarities within and between all relevés of the 71 groups involved. The diagonal elements are the mean similarities within the groups and thus a measure of homogeneity, as explained in Section 4.5.

Let us first look at some findings concerning the *diagonal* elements. There are examples of vegetation types exhibiting high internal homogeneity: the average similarity of relevés is high and therefore the corresponding

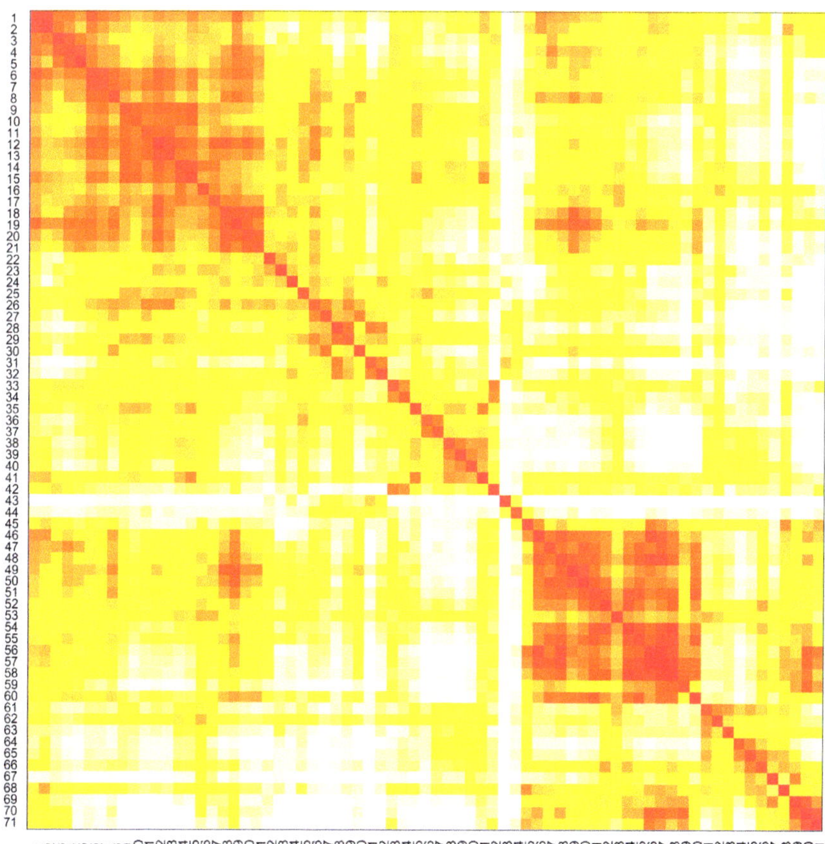

Figure 4.4: Similarities within and between the forest types of Switzerland according to Ellenberg and Klötzli (1972) based on the revision of Keller *et al.* (1998). Heat-colours used, red means high, light-yellow low similarity.

cell is red, when using so-called heat colours. Typical examples are forest types 49 (Equiseto-Abietetum), 56 (Sphagno-Piceetum typicum) and 70 (Rhododendro ferruginei-Pinetum montanae). The opposite is true for forest types 11 (Aro-Fagetum), 44 (Carici elongatae-Alnetum glutinosae) and 64 (Cytiso-Pinetum silvestris). When inspecting all diagonal elements it becomes clear that the internal homogeneity of the different vegetation types varies: dark red cells, indicating homogeneous groups, alternate with light-red cells, indicating heterogeneity. In practice this means that there are types that are easy to recognize in the field (homogeneous ones) and others that are difficult to recognize (heterogeneous ones).

The *off-diagonal* elements show which of the vegetation types are difficult to distinguish from others (dark-red cells) and which are easily differentiated (yellow cells). Forest types 1–21 form a block with dark off-diagonal symbols. These are beech (*Fagus sylvatica*) forests. Differences in species composition between these types are minor and careful inspection of the species lists is required for proper identification. A similar example is seen in spruce and fir forests (*Picea abies* and *Abies alba*), forest types 45–60. Interestingly, there are also certain forest types which bridge the two blocks when taking species composition into account (types 19 and 49). As can be seen from this example, a similarity matrix presented graphically can be an excellent tool for predicting problems in practical applications of classifications such as vegetation mapping.

Data to reproduce Figure 4.4 are located in two separate data frames as explained in Section 2.3.2, `EKv` holding vegetation data (2533 releveś) and `EKs` in which the original forest type membership is stored. Both data frames are implemented in `dave`. Typing `names(EKv)` would cause all the 1259 species names to be displayed. In data frame `EKs`, there is one column labelled `EK.Gesellschaftsnr`. This is the vector of 2533 vegetation type labels with range 1–71, the membership of releveś in the classification of Ellenberg and Klötzli (1972). Calling function `mxplot()` from package `dave` does all the operations needed to generate Figure 4.4:

```
o.mx<- mxplot(EKv,rmember=EKs$EK.Gesellschaftsnr,use="rows",y=1)
plot(o.mx)
```

The first parameter, `EKv`, addresses the entire data frame, whereas from `EKs` only the column specified after the dollar sign is taken. The list `o.mx` holds, among other results, the 71 by 71 matrix of mean correlations within and between vegetation types.

4.7 Which resemblance function?

Just as explained in the context of transformation in Section 3.5 there are no 'right' nor 'wrong' resemblance functions, although they will affect the final result and eventually interpretations as well. Again, there is no stand-alone criterion by which a choice could be justified. The performance of different functions is evaluated in view of a target, preferably the performance of an ecological model. The examples where evaluations of the kind are presented are the same as mentioned in Section 3.5.

The first instance is again found in Section 5.8 demonstrating that different classification methods yield different solutions, but depending on resemblance functions chosen and the transformation applied. Figure 5.8 visually demonstrates the effect of resemblance functions which sometimes lead to totally different solutions given the same combination of clustering method and transformation applied.

The same principle holds for ordination methods, Section 6.8. This is a visual comparison too (Figure 6.15), but there I argue that ordination should be sensitive to data transformation such that the result reflects different ways of looking at data. Some resemblance measures do this impressively well, others mask the effect of transformation.

The next example is a linear model expressing the interaction of vegetation and site factors. It is nonparametric multiple analysis of variance (NP-MANOVA) and it is presented in Section 7.5.2. The results suggest that when selecting pH as the explaining variable the best model is achieved when transforming cover-scale close to presence–absence (Table 7.8), all that simultaneously modified by the resemblance function.

Within the same chapter, in Section 7.5.3, the example from Section 5.8 is revisited. Here, the performance of clustering is measured quantitatively, displayed in Figure 7.9. The quality criterion is the proportion of variance that is explained by classification. In many cases the majority of transformations are almost equally successful, but specific distance functions, correlation as distance, for instance, appear to be highly recommended.

5 Classification

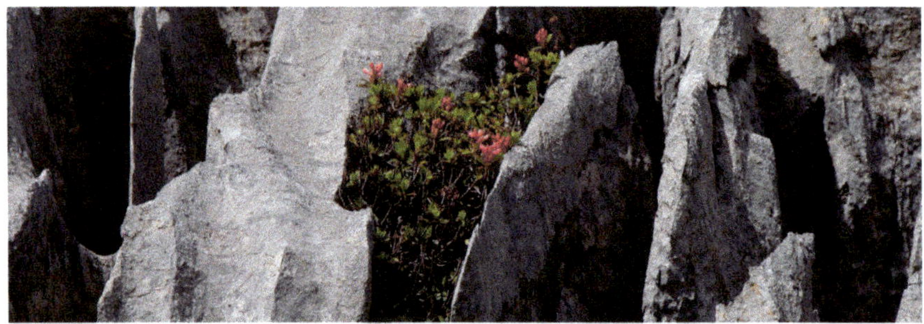

Classification aims to find groups of similar sampling units or descriptive variables. It is a means of data reduction, serving ecological model building and it helps in the communication of the state of vegetation among scientists and practitioners.

5.1 The legacy of vegetation classification

Unlike other analytical methods vegetation classification has a long history dating back to times when electronic data processing was unavailable. As a result countless descriptions and names of vegetation types are nowadays published in the literature and also in databases, many documented by field data, others residing on expert experience only (Peet and Roberts 2013). This is the legacy of long lasting traditions in vegetation science, one that sometimes conflicts with what is described in this chapter: using formal classification as a tool for pattern recognition and subsequent model building. However, the main problem is not the age of a classification, but the way in which the data are perceived. In numerical classification we intend to screen a sample for patterns (Section 2.3) and we assume that this sample is representative for a sampling environment (wetlands, for instance, or forests). This implies that a standardized method of sampling was used, plot locations chosen at random, etc. Numerical classification methods would then reveal vegetation types existing in the real world. Numerically derived types are probably similar (but rarely identical) to types described in the past. Because vegetation types do not seem to be very stable constructs, we may also want to revise these eventually, a topic addressed in Section 5.7 and in Chapter 12.

Big classification systems have their strengths and weaknesses. They are built upon a consistent approach, such as the system of Braun-Blanquet (1932) where millions of relevés exist and all use more or less the same cover-abundance scale (Section 3.4). Yet this system tends to grow uncontrolled as any author is allowed to assess his own vegetation type – hoping that in the end the system would become comprehensive. Better managed is the more recent system of the United States national vegetation classification (Jennings *et al.* 2009). This consists of a predefined set of vegetation types (also called associations) and extensions follow established rules. Also there are efforts to keep archived data available for future use (Wiser 2016). A recent and consistent system seems to be the British National Vegetation Classification programme (Rodwell 2006). So far, this has been developed and led by a single person (John Rodwell).

Classifications devised in the framework of big systems are important tools of communication and indispensable for various applications, such as drawing vegetation maps, assessing endangered vegetation types for legal protection or as references in monitoring projects (Mucina 1997). In the present chapter, however, we are exclusively striving for results intended for further use in the analytical context, such as environmental analysis (Chapter 7) or vegetation modelling (Chapter 9).

5.2 Group structures

'The aim of classification is to obtain groups of objects (samples, species) that are internally homogeneous and distinct from other groups' (Lepš and Šmilauer 2003). Working with a small number of groups rather than a large number of relevés and species is the main practical advantage of classification, and because species combinations tend to reoccur at different locations, classification is justified from the theoretical point of view as well. A group structure is a kind of pattern and therefore classification can be considered a tool for pattern recognition. As shown in Figure 5.1, however, these patterns are easily hampered. Figure 5.1(b) illustrates a case where two groups are formed by dividing a perfectly continuous point cloud. Additional investigation is needed to distinguish this from the case in Figure 5.1(a) where groups reflect an obvious pattern and therefore are termed *natural*, in contrast to *artificial* in Figure 5.1(b). It can be seen from the intermediate data points in Figure 5.1(a) that this distinction is not always trivial. Just a few additional data points may connect two natural groups, which then become artificial. And yet another issue is the shape of the group. In ecology it frequently happens that groups form 'internal' gradients as illustrated in Figure 5.1(c). Groups, in fact, can have any shape; Gan *et al.* (2007) show many more examples.

Classification also reduces the dimensionality of data, just as ordination does (Chapter 6). The maximum number of dimensions of a multivariate data set, m, is the smaller of either the number of relevés n or the number of species p; that is, $m = min(n, p)$. After forming relevé and also species groups, the number of dimensions will not exceed the smaller number of either the relevé groups or the species groups. Whereas in Figure 5.1 the entire point cloud is presented in two dimensions, in the univariate case it is one-dimensional. Forming groups when only one or two dimensions exist is usually a simple task to be carried out visually. Mathematical methods forming groups are needed when the number of dimensions is high, a property typical for vegetation data.

When a species or a relevé is assigned to one group only, then the pattern is discrete. However, classification also involves a continuous concept known as fuzzy classification (Roberts 1986; Oldeland *et al.* 2010; Everitt *et al.* 2011). In fuzzy classification all relevés and species have a degree of belonging to all groups, measured on a continuous scale from 0 to 1, an analogue to ordination where coordinates are continuous. In this case classification and ordination converge (see Section 5.6).

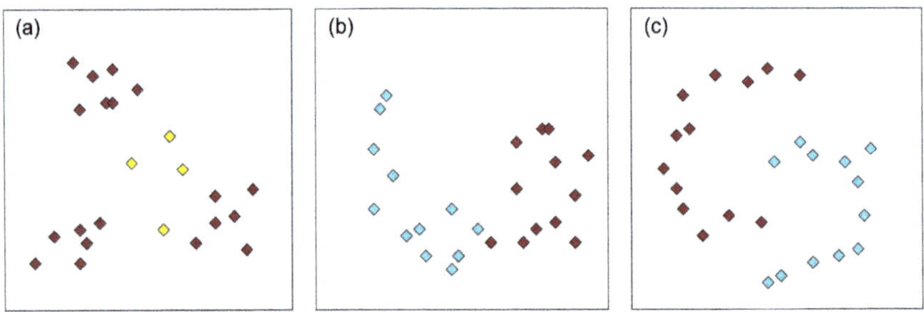

Figure 5.1: Two-dimensional group structures. (a) Natural groups and intermediate, yellow points. (b) Continuous structure with artificial division into two groups. (c) Two gradient-shaped groups.

The issue of finding the best – or at least a good – group structure is by no means trivial. First, one has to find an appropriate number of groups, m, in the range of:

$$1 \leq m \leq n$$

under the assumption that a group consists of at least one data point (i.e. sampling unit) and n is the sample size. Secondly, a clustering technique has to be chosen from among the many available methods to group the sample accordingly.

Why are there so many methods for clustering and not just one? The reason is that no single method works perfectly, even though the ultimate solution exists in theory: it is the enumeration of all possible assignments of sampling units to groups followed by the selection of the 'best' one. As noted by Anderberg (1973), the number of combinations $\mathcal{S}_n^{(m)}$ is:

$$\mathcal{S}_n^{(m)} = \frac{1}{m!} \sum k = 0^m (-1)^{m-k} \binom{m}{k} k^n. \tag{5.1}$$

Even for the very moderate task of assigning 25 sampling units to five groups this yields:

$$\mathcal{S}_{25}^{(5)} = 2\ 436\ 684\ 974\ 110\ 751$$

combinations. This number is so vast that the procedure is out of reach of today's computers. The many classification methods invented are attempts to approximate the ideal solution provided this exists. The books of Gan *et al.* (2007) and Everitt *et al.* (2011) offer insight into the still evolving field of classification. They also classify methods, for example, according to the following scheme:

Heuristic	*vs*	Formal
Agglomerative	*vs*	Divisive
Hierarchical	*vs*	Nonhierarchical
Deterministical	*vs*	Stochastical
Crisp	*vs*	Fuzzy

A heuristic algorithm implies two elements. First, assumptions are made concerning the initial group structure. Second, the initial configuration is improved through a formal, sequential reallocation procedure. Anderberg (1973) discusses these methods in the context of nonhierarchical clustering: 'Such algorithms begin with an initial point and then generate a sequence of moves from one point to another, each giving an improvement of the objective function, until a local optimum is found' (p. 156). It is worthwhile to note that the optima found depend on the initial assumptions and even on the order in which the data is processed. For very large data sets, processing the sampling units sequentially may be the only feasible method, considering the constraints of storage and computing power. However, heuristic methods should be avoided in favour of formal agglomerative and divisive methods whenever possible. The majority of methods of this chapter are agglomerative, with DIANA being the only divisive method included.

For ecological applications it is often sufficient to distinguish groups without considering hierarchy. However, hierarchy has the advantage that it reflects similarity relationships between the groups, which allows us to change the number of resulting groups by altering the hierarchical level. Hierarchy is of course needed when an analysis is based on hierarchy theory (Allen and Starr 1982).

5.3 Agglomerative clustering

This is the most popular approach used in numerical classification. Agglomerative clustering generates a hierarchy which is usually displayed in the form of dendrograms; that is, resemblance trees depicting the similarity of individuals and groups, as shown in Figure 5.2. On the horizontal axis, sampling units 1 through 7 are lined up, connected by arches. The height of any arch measures the dissimilarity (distance) between the corresponding sampling units – or groups of sampling units. However, the order of the sampling units and hence the arches is not given by the algorithms. The configuration is allowed to spin around all vertical axes (Figure 5.2). Thus, the vicinity of data points along the x-axis has no specific meaning and cannot be used for interpretation.

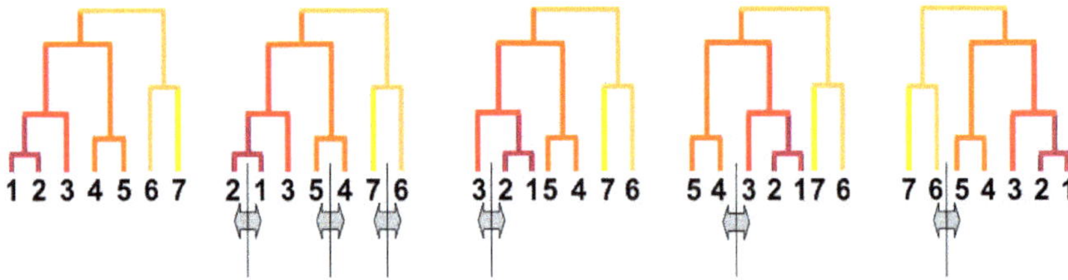

Figure 5.2: A dendrogram as output from an agglomerative hierarchical clustering method. All branches may be in spin and therefore vicinity does not signify similarity.

5.3.1 Linkage clustering

The process of agglomerative hierarchic clustering is demonstrated for three methods: single, complete and average linkage clustering. Whereas single- and complete linkage clustering are unambiguous terms, average linkage clustering is used for an entire class of methods. The one presented below is also called centroid clustering; three more are listed in Section 5.3.2.

All three methods use the same rule for the comparison of any two sampling units. The resemblances (distance or similarity) are directly taken from the resemblance matrix of the sample, but methods differ in the way they consider larger groups [Figure 5.3(a)]. *Single linkage* always uses the distance (or similarity) of the closest members of any two groups. *Complete linkage* refers to the two most distant members of any two groups, thus distance between groups is much larger. *Average linkage* measures the

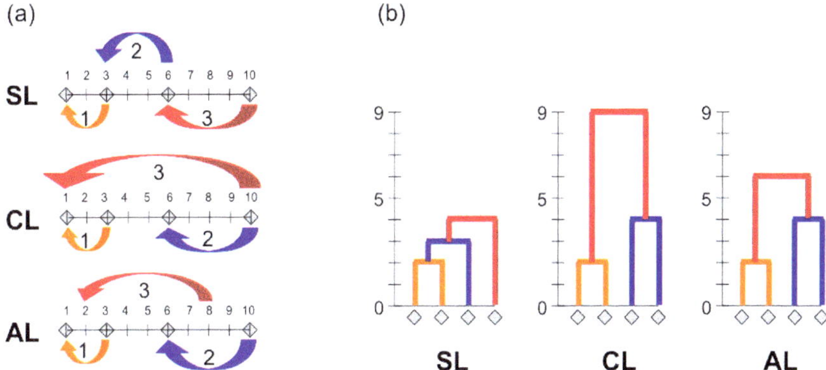

Figure 5.3: Comparing single (SL), complete (CL) and average linkage (AL) clustering. (a) Group definitions applied in a one-dimensional example. (b) Results displayed as dendrograms.

similarity between the centroids of the groups; in the one-dimensional example of Figure 5.3 this is the centre of two (or more) sampling units involved. The definitions imply that the distance between groups is generally larger in complete linkage than in single linkage, and intermediate in average linkage.

The example in Figure 5.3 is one-dimensional. The dissimilarity of any two points is therefore just their distance on a straight line. When arranging the sampling units from left to right, the distance matrix is:

$$\mathbf{D} = \begin{matrix} 0 & & & \\ 2 & 0 & & \\ 5 & 3 & 0 & \\ 9 & 7 & 4 & 0 \end{matrix} \tag{5.2}$$

In the first step (numbered arrows in Figure 5.3) all methods do the same. They find the first two points to be the most similar, separated by two units on the x-axis. This yields a first arch in the dendrogram, which is two units in height. In the second step the next closest neighbours are searched for. In single linkage this is the third point with a distance of three units from the group formed before. For complete linkage the third point is five units apart from the newly formed group. The next fusion is therefore composed of points three and four being only four units apart. The corresponding arch has height four. The same holds for average linkage. However, there are now two equivalent solutions as the distance between the first two points and the third point is also four units. In the third and final step, single linkage adds the fourth point to the previous cluster. This is done at height four of the arch, according to the distance of the third and the fourth points. In complete linkage, the two two-member groups are fused at arch level nine, the distance between the most remote points. The same is done in average linkage, but the height of the arch reflects the distance between the centres of the two groups involved.

Although this is a tiny example, the shape of the resulting dendrograms is typical for the methods. In single linkage, the chaining effect can be seen. The dendrogram formed by complete linkage is more balanced and typically much higher. The average linkage dendrogram has an intermediate shape.

Reproducing Figure 5.3 in $\mathcal{R}$ begins with input of scores along the x-axis and from this distance matrix `mde` is derived using function `vegdist()`:

```
x<- matrix(c(1,3,6,10),nrow=4)
colnames(x)<- c("s1")
rownames(x)<- c("r1","r2","r3","r4")
mde <- vegdist(x,method="euclidean",diag=TRUE)
mde
```

```
   r1 r2 r3 r4
r1  0
r2  2  0
r3  5  3  0
r4  9  7  4  0
```

Function `hclust()` available in package `stats` provides all the methods used in this example:

```
o.hcls<- hclust(mde,method="single")
o.dendro<- as.dendrogram(o.hcls)
plot(o.dendro, horiz = FALSE)
```

Function `as.dendrogram()` changes the dendrogram into a different format (called 'class' in $\mathcal{R}$) to deliver the graph in Figure 5.3. Just as shown in Figure 5.2 there is a good chance that the order of relevés is changed in some way. For the second dendrogram in Figure 5.3 method `"complete"` is chosen in function `hclust()` and for the third this is `"centroid"`. Again, this third dendrogram may reflect the alternative solution mentioned above and therefore differ from that in Figure 5.3.

5.3.2 Average linkage clustering revisited

In many applications an intermediate solution with properties inherited from single linkage and complete linkage clustering is preferred as a standard. The term 'average linkage clustering' has been used for various different methods (Sneath and Sokal 1973; Everitt *et al.* 2011), such as those listed in Table 5.1. They have in common that the definition of group similarity is based on all members of a group and not just one as in single and complete linkage clustering. It is in this regard that they are also related to minimum-variance clustering (Section 5.3.3).

The four methods are discussed in detail in Legendre and Legendre (2012), where a small numerical example is given for illustration and comparison. In the following I use the abbreviations given in Table 5.1. The methods are distinguished by two alternative criteria. UPGMA and WPGMA use the average resemblance of all group members as a criterion for between-group resemblance. In UPGMC and WPGMC, a group centroid is established: in geometrical terms this is the centre of gravity of any one group. Between-group resemblance is thus the distance or similarity of any two centroids.

The second criterion of distinction concerns weighting. 'U' (UPGMA, UPGMC) signifies 'unweighted', which, however, may be somewhat misleading. When computing average resemblance as well as centroids, group sizes are taken into account. 'Unweighted' means that the weight of the original set of resemblances is retained. 'W' (WPGMA, WPGMC) signifies 'weighted'. This means that groups of different size get the same weight when fused. In this case, the weight is the inverse of group size.

The relationship between the four methods is shown in Table 5.1, an extension of Table 8.2 in Legendre and Legendre (2012), p. 353. Method options offering these in function `hclust()` stem from Borcard *et al.* (2011, p. 59). Reversals may occur in the dendrograms: subsequent fusions may take place at lower distance levels than the previous. Dendrograms of this kind are both difficult to draw and difficult to interpret (Section 5.5).

All four methods are available in function `hclust()`, for instance as method `"average"` for UPGMA:

```
x<- matrix(c(1,3,6,10),nrow=4)
colnames(x)<- c("s1")
rownames(x)<- c("r1","r2","r3","r4")
```

Table 5.1: Properties of four average linkage clustering methods and parameters used in function `hclust()` (adapted from Legendre and Legendre 2012, p. 353, and Borcard *et al.* 2011, p. 59).

Properties	Consider the average similarities or distances of all members of a cluster as candidates for further fusions	Consider the centroid of all members of a cluster as candidates for further fusions
Give equal weight to the original resemblances (weight of groups proportional to group size)	**UPGMA** (unweighted arithmetic average clustering), `"average"`	**UPGMC** (unweighted centroid clustering), `"centroid"`
Give equal weight to any two branches of the dendrogram (weight of groups identical irrespective of size)	**WPGMA** (weighted arithmetic average clustering), `"mcquitty"`	**WPGMC** (weighted centroid clustering), `"median"`

```
mde <- vegdist(x,method="euclidean")
o.hcls <-hclust(mde,method="average")
plot(as.dendrogram(o.hcls))
```

For the remaining methods one uses "`centroid`" for UPGMC, "`mcquitty`" for WPGMA and "`median`" for WPGMC.

5.3.3 Minimum-variance clustering

This method deserves special attention as it proved to yield classifications having high predictive power in many ecological models (see, for example, Sections 5.8 and 7.5.3). Ward (1963) called it 'Minimum-variance clustering' and Orlóci (1967) 'Sum of squares clustering'. Ward's method is described in detail in Legendre and Legendre (2012, p. 360 ff) in a way compatible with the software in the **stats** package of $\mathcal{R}$. This is why Ward's, rather than Orlóci's method is described here. In any case, the objective of minimum-variance clustering is to merge groups such that the increase of within-group variance is minimized. Since all this happens in Euclidean space, the elements involved in clustering can be illustrated as shown in Figure 5.4.

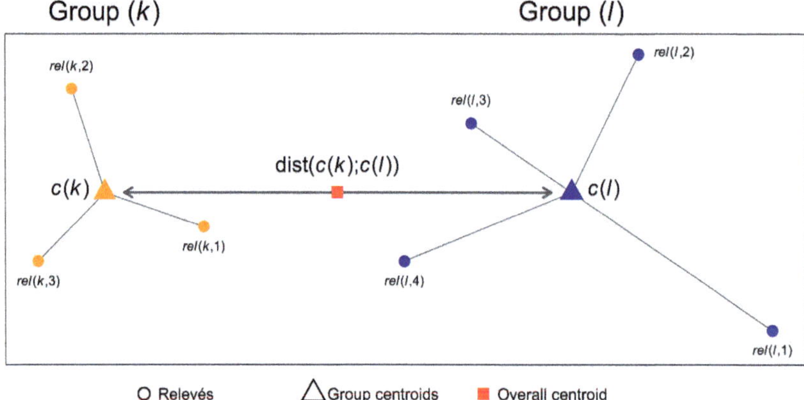

Figure 5.4: Variance within and between groups in minimum-variance clustering. The variance within groups is the sum of the squared distances between relevés and group centroids (light lines). $c(k)$ and $c(l)$ are group centroids and $dist[c(k); c(l)]$ the distance between these. Adapted from Anderson (2001).

The variance within any group, g, is derived from the original data set as follows:

$$Q_g = \sum_{i=1}^{p} \sum_{j=1}^{n_g} (x_{ij} - \overline{x}_i)^2 \tag{5.3}$$

where x_{ij} is the score of species i in relevé j, n_g is the size of group g and p is the number of species. The mechanism to be applied in clustering is rather simple: for all possible pairwise combinations of groups (whether encompassing a single or several sampling units) the resulting sum of squares have to be computed as shown above (Equation 5.3). The pair in which the increase of the sum of squares is lowest is then chosen for fusion.

Such an algorithm would be rather time consuming due to the many calculations to be discarded once an ideal fusion has been found. Most descriptions (for example, Legendre and Legendre 2012, p. 360 ff; Orlóci 1978, p. 204 ff) make use of the fact that multivariate sum of squares are far more efficiently derived and modified based on a squared Euclidean distance matrix, D^2:

$$Q_g = \frac{1}{n_g} \sum_{i<j} d_{ij}^2 \tag{5.4}$$

This is the same quantity as in Equation 5.3. It is the sum of all elements of a triangular squared Euclidean distance matrix of group g divided by the number of group members involved. This relationship is illustrated by example in Figure 7.8 and it is used to derive the small example below.

To explain the principle I use the univariate example from Figure 5.3. Clustering starts with the squared Euclidean distance matrix of the sampling units, that is, the squared matrix from Equation 5.2:

$$\mathbf{D^2} = \begin{array}{lllll} r1 & 0 & & & \\ r2 & 4 & 0 & & \\ r3 & 25 & 9 & 0 & \\ r4 & 81 & 49 & 16 & 0 \end{array} \tag{5.5}$$

In the realm of squared Euclidean distance, the candidates for fusion are $r1$ and $r2$ with $d_{r1,r2}^2 = 4$. These form the new cluster and we could update the matrix of squared distances using their centroid. But this would end up in centroid clustering. For minimum-variance clustering the size of the groups involved has to be considered. The appropriate formula to update the squared distances, leading to the final shape of the dendrogram generated by method `"ward.D2"` in $\mathcal{R}$ package `hclust()`, is found in Legendre and Legendre (2012), p. 365:

$$d_{kl,g}^2 = \frac{n_k + n_g}{n_k + n_l + n_g} d_{k,g}^2 + \frac{n_l + n_g}{n_k + n_l + n_g} d_{l,g}^2 - \frac{n_g}{n_k + n_l + n_g} d_{k,g}^2 \tag{5.6}$$

In this k and l are the recently merged groups and all the remaining (to which the distances have to be updated) are denoted by g. For example, to update the distance of the new cluster to $r3$ we get $d_{r12,r3}^2 = (\frac{2}{3} * 25) + (\frac{2}{3} * 9) - (\frac{1}{3} * 4) = 21.33$. And this is the fully updated matrix:

$$\mathbf{D^2} = \begin{array}{lllll} r1, r2 & 0 & & & \\ - & - & - & & \\ r3 & 21.33 & - & 0 & \\ r4 & 85.33 & - & 16 & 0 \end{array} \tag{5.7}$$

The next fusion concerns $r3$ and $r4$ with a squared distance of 16. After updating we get:

$$\mathbf{D^2} = \begin{array}{lllll} r1, r2 & 0 & & & \\ - & - & - & & \\ r3, 4 & 72.0 & - & 0 & \\ - & - & - & - & - \end{array} \tag{5.8}$$

Orlóci (1978) takes the increase in within sum of squares for drawing the y-axis. Legendre and Legendre (2012) in their example use the squared minimum distances for the same, 4, 16 and 72 in the example above. In method "ward.D2" of function hclust() the square root is taken, 2, 4, and 8.49 respectively:

```
x<- matrix(c(1,3,6,10),nrow=4)
colnames(x)<- c("s1") ; rownames(x)<- c("r1","r2","r3","r4")
mde <- vegdist(x,method="euclidean")
o.hcls <-hclust(mde,method="ward.D2")
plot(as.dendrogram(o.hcls))
```

Minimum-variance clustering is capable of distinguishing groups of different size, shape and density; that is, groups with high internal variance versus groups with low variance. This is a situation often encountered in vegetation data. As pointed out by Legendre and Legendre (2012), minimum-variance clustering is often also considered one of the many variants of centroid clustering offering yet another intermediate solution between the extremes represented by single and complete linkage clustering.

5.4 Divisive clustering

In divisive clustering the classification process starts with the entire sample forming the initial group. Subsequently, meaningful divisions are sought resulting in a steadily growing number of groups. The process is the opposite of agglomerative clustering (Section 5.3). Divisive clustering has played an important role at the outset of numerical ecology (Anderberg 1973; Everitt *et al.* 2011, p. 84 ff) when many methods were monothetic (that is, considering one single variable only at a time) and therefore manageable by less performing computers. On the downside misclassification occurred frequently in multivariate data. An improved old method still in use nowadays is TWINSPAN (Hill 1979b). Rather than selecting a single species at a time for a first division, TWINSPAN uses the first axis of correspondence analysis when striving for a division. Because ordination axes account for a fraction of total pattern only, misclassifications still happen, although recent implementations try to correct this by post-processing. The weaknesses of TWINSPAN are discussed in various publications (Belbin and McDonald 1993; Dufrêne and Legendre 1997; Legendre and Legendre 2012; Lötter *et al.* 2013) all suggesting that its use should be avoided.

One of the very few and probably the best implementation of polythetic divisive clustering available to date is `diana()` in package `stats`, DIvisive ANAlysis clustering, introduced by Kaufman and Rousseeuw (1990). Splitting off new groups always proceeds in two steps. First, the most isolated sampling unit in any one group is identified and declared a new (temporal) group, called a 'splinter group'. Then all members of the entire sample are screened to eventually join this splinter group in cases where the mean distance to this is lowest (Everitt *et al.* 2011, p. 86 ff). Our example in Figure 5.3 is sufficiently small for demonstrating this by longhand calculation. Unlike in Equation 5.2 we inspect the full square distance matrix, D:

$$\mathbf{D} = \begin{matrix} 0 & 2 & 5 & 9 \\ 2 & 0 & 3 & 7 \\ 5 & 3 & 0 & 4 \\ 9 & 7 & 4 & 0 \end{matrix} \tag{5.9}$$

We first compute the average distance from each of the four sampling units to the respective remaining three. This is the mean of the rows (or columns) in D, divided by three. We get $d_{i.} = \{5.3; 4.0; 4.0; 6.7\}$ for any row i. Sampling unit number 4 is the most isolated forming the centre of the new 'splinter group'. Then, the re-assignment process starts. For this we calculate the mean distance of each individual to the remaining (A) and to the new splinter group (B) as shown in Table 5.2.

Table 5.2: Reassigning plots to splinter groups in divisive clustering, function `diana()` (Everitt *et al.* 2011).

Division	Plot no.	Mean dist. to {1, 2, 3} (A)	Mean dist. to {4} (B)	A-B
1	1	7/3	9	-20/3
1	2	5/3	7	-16/3
1	3	8/3	4	-4/3
Division	Plot no.	Mean dist. to {1, 2} (A)	Mean dist. to {3} (B)	A-B
2	1	1.0	5	-4.0
2	2	1.0	3	-2.0

In `diana()` the largest mean distance of any sampling unit to number 4 is taken as the level of division when plotting the resulting dendrogram, that is, $d = 9.0$. As all differences (A-B) in the upper part of Table 5.2 are negative, no individual is moved to group with member {4} and we proceed with splitting group {1, 2, 3}. According to Equation 5.9 the mean distance of the three elements to the remaining two is $d_{i.} = \{3.5; 2.5; 4.0\}$ and we now see that number 3 will split off.

As shown in the lower portion of Table 5.2, the level of division is the largest mean distance to sampling unit 3 and 5.0 will be used for plotting the dendrogram. Again, all differences (A-B) are negative and none of the members {1; 2} is moved to the group with member {3}. The last and only division remaining concerns group {1; 2} and according to Equation (5.9) the level at which this takes place is 2.0, the distance between the two sampling units.

In summary, the levels where divisions take place in our example are $h = \{9.0; 5.0; 2.0\}$. The resulting dendrogram resembles the same of single linkage clustering in Figure 5.3 with just the two right-hand branches being somewhat higher. The use of **diana()** in $\mathcal{R}$ is easy because this function generates an object of class 'diana' that can be transformed to already used classes 'hclust' and 'dendrogram'.

```
dm<- dist(c(1,3,6,10),method="euclid")
o.diana<- diana(dm)
plot(as.dendrogram(as.hclust(o.diana)))
```

How does divisive clustering differ from agglomerative and is it any better at all? As shown in Sections 5.5 and 5.8 the results are rather promising in comparison with agglomerative clustering. But the success when implementing it into an ecological context will tell us more about this (Chapters 7 and 9).

5.5 Forming groups

A dendrogram does not directly yield groups, but it offers unique flexibility in doing so. When cutting horizontally, for example, we divide the sample into groups. The dissimilarity of the resulting groups will be as large as the level at which cutting takes place, as is illustrated in Figure 5.5. Cutting the dendrogram just below the uppermost arc will divide the sample into two groups. When moving down to the next arc, one new group is formed (unless two or more arcs are found to be at exactly the same level). The procedure automatically ends when all groups consist of one sampling unit only.

But how many groups should be formed, and how large and how homogeneous should they be? Considering the huge number of potential solutions to this problem [Equation (5.1)], the decision has to be based on the purpose of classification. This also means that a criterion has to be found which is independent of the data yielding the classification. As an example, a site factor can be used for testing the explanatory power of classified vegetation samples (see Section 7.2.1 and 12.4). Only in rare cases will the structure found in clustering be unambiguous.

The result of cutting a dendrogram not only depends on the number of groups chosen, but also on the clustering method. Figure 5.5 shows how the shape of the dendrogram is primarily dependent on the method rather than the similarity pattern of the relevés involved. To get an impression of the latter, an ordination diagram (Chapter 6) is much better suited than a dendrogram. Single linkage clustering frequently exhibits a chaining effect getting even more pronounced when applied to larger data sets. Its strength is in detecting group shapes shown in Figure 5.1(c). Even in the very small `nveg` example, complete linkage, minimum-variance clustering and divisive clustering confirm their tendency to form groups of rather balanced size, whereas average linkage clustering represents an intermediate solution. Centroid clustering [Figure 5.5(e)], a member of the average linkage family, shows inflections mentioned in Section 5.3.2.

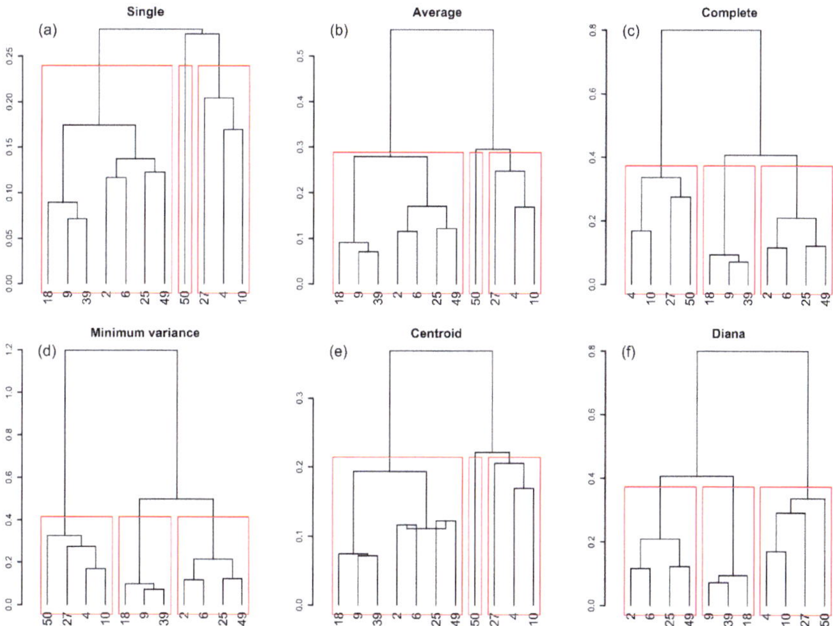

Figure 5.5: Cutting dendrograms derived by six different methods into three groups. (a) Single linkage; (b) average linkage; (c) complete linkage; (d) minimum variance clustering; (e) centroid clustering and (f) divisive clustering (diana).

When group number and size are chosen there is sometimes an opportunity to test these for significance based on an independent site factor. In Section 7.2 the use of analysis of variance for measuring the predictive power of a classification will be raised. The test criterion, the F-value, helps us to compare alternative solutions for group number and size. It is defined as:

$$F = \frac{Var_{betweengroups}}{Var_{withingroups}} \tag{5.10}$$

The significance of any one F-value can be checked in the F-table of a statistical textbook. The F-value has two degrees of freedom, $df1$ and $df2$, where:

$$\begin{aligned} df1 &= m - 1 \quad \text{and} \\ df2 &= n - m \end{aligned} \tag{5.11}$$

in which m is the number of groups and n is the sample size. Upon inspection of F-tables it becomes clear that n must be sufficiently large: much larger than m, the number of groups. However, if the number of groups is too low, the predicting power of the classification is also poor. Two groups ($df1 = 1$), from the point of view of the analysis of variance, are less favourable than a number around five ($df1 = 4$) promising better results. Furthermore, the number of sampling units per group should also not drop below, say, five.

The dendrograms in Figure 5.5 provide classifications of the 11 relevés in data frame `nveg` (Appendix 14). Function `hclust()` processes a matrix of correlations used as distance [Equation (4.6)]. Cutting is done by function `cutree()` in package `stats`, delivering an output where the labels of the resulting three groups are assigned to relevé numbers within the data set. Figure 5.5(a) is obtained as follows:

```
ddr<- as.dist((1-cor(t(nveg)))/2)
o.hclr<- hclust(ddr,method="single")
plot(as.dendrogram(o.hclr))
o.grel<- cutree(o.hclr,k=3)
```

For Figure 5.5(b) the method parameter is changed to `"average"`, for (c) to `"complete"`, for Figure 5.5(d) to `"ward.D2"`, for Figure 5.5(e) to `"centroid"`, respectively. We now draw the boxes and inspect the assignment of groups:

```
rect.hclust(o.hclr,3,border="red")
o.grel
```

```
2  4  6  9 10 18 25 27 39 49 50
1  2  1  1  2  1  1  2  1  1  3
```

As shown in Section 5.4 the graph in Figure 5.5(e) requires somewhat different instructions. A call of function `diana()` is required and the resulting object has to be transformed for access by function `rect.clust()`:

```
o.diana<- diana(ddr)
o.hcl<- as.hclust(o.diana)
o.grel<- cutree(o.hcl, k = 3)
plot(as.dendrogram(o.hcl))
rect.hclust(o.hcl,3,border="red")
```

The assignment of relevés to groups is again found in object `o.grel` and it is used in all further applications where a group structure is involved.

5.6 Silhouette plot and fuzzy representation

Dendrograms are limited in presenting within and between group similarity in populations. In the preceding sections they are mainly used to illustrate the progress of clustering. The fusion levels along the y-axis show the distance at which assignment of any one individual, or group of individuals, has taken place. This reports on the heterogeneity of groups, that is, which members are well fitting a group and which are somewhat different to typical group members. Figure 5.5 suggests that the group with members 18, 9 and 39 is the most homogeneous according to all four methods used. But it does not answer the question of how well a sampling unit would fit into an alternative group. This is what a silhouette plot (Rousseeuw 1987) does explicitly (Figure 5.6).

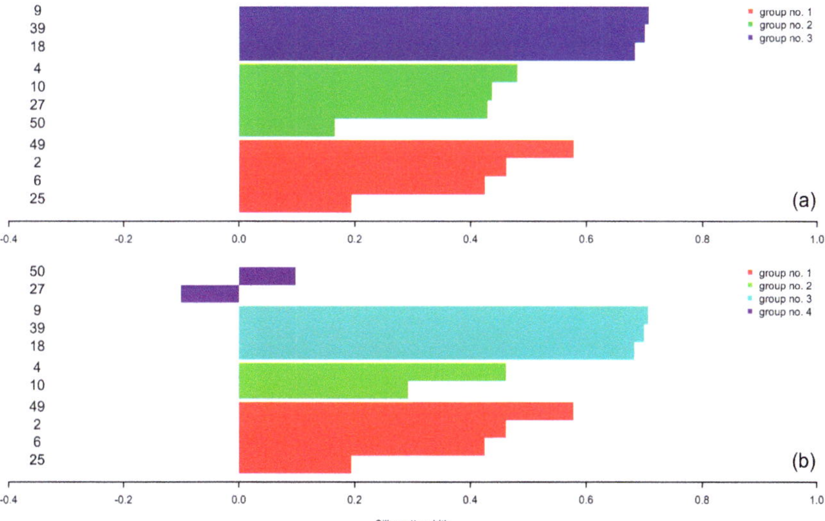

Figure 5.6: (a) Silhouette plot of classification resulting from clustering of object `nveg` shown in Figure 5.5(b) (complete linkage clustering). (b) The same data and methods used, but four groups formed instead of three groups.

Silhouette width, $s(i)$, is based on the mean distance of any one object to all other objects of the same group, $a(i)$, and to the mean distance to any one object of the nearest group it does *not* belong to, $b(i)$. According

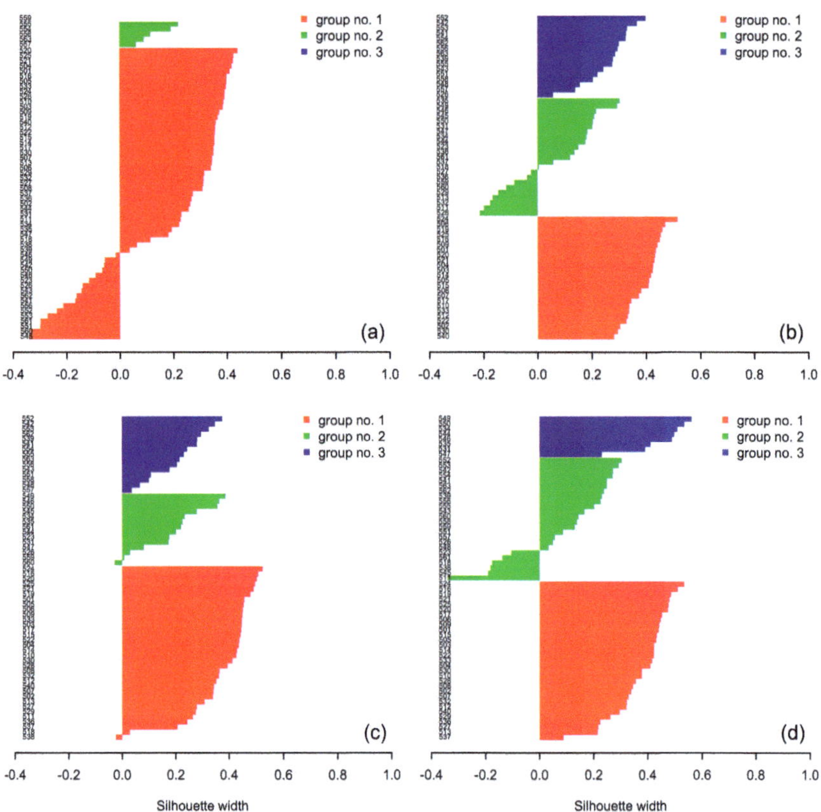

Figure 5.7: Silhouette plot of four clustering solutions, object sveg. (a) Single linkage clustering. (b) Complete linkage clustering. (c) Average linkage clustering (UPGMA). (d) Ward's minimum-variance clustering.

to Kaufman and Rousseeuw (1990) the definition accords with:

$$s(i) = \frac{b(i) - a(i)}{max(a(i), b(i))}; i = 1, ..., n \tag{5.12}$$

In this, n is the number of elements included in clustering. It can easily be seen that a very distinct assignment with small distance to other members of a cluster results in an $s(i)$ value approaching 1. In the case where $a(i)$ and $b(i)$ are identical, the silhouette width becomes zero. Negative values indicate misclassifications. Because sample size of object **nveg** is only $n = 11$ we can easily determine silhouette width of relevé number 9 by longhand calculation, for example. Printing the corresponding row (or column) of distance matrix *ddr* (Section 5.5) shows the distances of relevé 9 to all others:

```
round(as.matrix(ddr,nrow=9)["9",],digits=3)
```

```
    2     4     6     9    10    18    25    27    39    49    50
0.321 0.509 0.249 0.000 0.404 0.089 0.174 0.429 0.071 0.353 0.280
```

For the mean similarity to the group it belongs to, that is group 1, we get $a(i) = (0.089 + 0.071)/2) = 0.080$, including distance to relevés 18 and 39, respectively. Mean distance to the second nearest group, that is group number 3, is $b(i) = (0.321 + 0.249 + 0.174 + 0.353)/4 = 0.274$, including distances to relevés 2, 6, 25 and 49. According to Equation (5.12) the silhouette width for relevé 9 is $s(9) = (0.274 - 0.080)/0.274 = 0.708$.

The silhouette plot is computed by function `davesil()` in package `dave`. This is merely a so-called 'wrapper' of function `silhouette()` in package `cluster`. The plot method this function provides labels the bars with the original relevé numbers. Showing computation from scratch we first cluster object `nveg` from Section 5.5:

```
ddr<- as.dist((1-cor(t(nveg)))/2)
o.hclr<- hclust(ddr,method="complete")
o.relgr<- cutree(o.hclr,k=3)
```

Function `davesil()` accesses the distance matrix `ddr` previously generated for clustering and also the output of functions `hclust()` and `cutree()`:

```
o.davesil<- davesil(ddr,o.hclr,o.relgr)
plot(o.davesil)
```

This generates Figure 5.6(a). The silhouette plot confirms the solution to be meaningful. In Figure 5.6(b) the same data and method are used as in Figure 5.6(a), but four groups are chosen in function `cutree()` instead of three groups. It can easily be seen that the fourth group lacks consistency and the result would hardly serve any purpose. Silhouette plots in fact help to identify bad solutions when cutting dendrograms. In critical cases the plot method offers a parameter `range` to plot just part of the graph, bars 1 to 9 only in this case:

```
plot(o.davesil,range=c(1,9))
```

The strength of silhouette plots comes from the evaluation of all units individually. But as can be seen from Equation (5.12) it does this by considering the nearest neighbour distance criterion. However, some of the more advanced clustering methods, like Ward's, identify groups with variable internal variance, density and diameter. When subjecting the results to silhouette plot then parts of the groups emerge as misclassifications, although the graphs just express properties of the respective clustering methods. This is shown in Figure 5.7 using vegetation data `sveg` with sample size 63. Figure

5.7(a), for instance, illustrates the chaining effect occurring in single linkage analysis and the frequently rather uneven group sizes it yields. Because the shape of groups can be far from round and different in range, negative silhouette widths occur. The same happens with complete linkage analysis, Figure 5.7(b), but the method devises more balanced group sizes. Average linkage analysis, that is UPGMA in terms of Table 5.1, Figure 5.7(c), provides the optimal solution – based on the criteria valid in silhouette plot. Ward's method, Figure 5.7(d), offers a good balance for group size, but different group shapes mimic outliers.

In all clustering methods mentioned so far the results are 'crisp' classifications, where relevés belong to one group only. In fuzzy classification, units get a degree of belonging to all groups (Roberts 1986; Marsili-Libelli 1989). In this view crisp classification is just a special case of fuzzy classification with one degree of belonging being 1 and all remaining being 0. Silhouette plot can be seen as a step towards fuzzy classification in that it encapsulates the degree of belonging (the silhouette width) of all units to the group they belong and also to the most similar to which they do not belong. As shown for example by Everitt *et al.* (2011), there are various methods available to offer full fuzzy clustering. While these are not considered here, fuzzy representation of vegetation types also results from ecological modelling. Real world examples of fuzzy maps are shown in Section 9.8 and Section 13.3.3.

5.7 Revising classifications

The clustering methods devised in the previous sections generate classifications from scratch. In vegetation ecology, however, established classification systems are widespread and used in landscape management, biodiversity conservation, forestry and agriculture, for instance. To serve these purposes methods are requested for gently revising existing groups, testing these for consistency and allowing improvements if required.

In the present section I explain the simple case where the classification of samples shall be revised. A more complex task would be the revision of a classification system, one derived from many samples or an entire database (De Cáceres *et al.* 2015). Hence, I only address the question if a given classification can be improved to better reflect the similarity pattern. This is where the k-means method comes into play (Anderberg 1973; Everitt *et al.* 2011; Legendre and Legendre 2012), although there exist various alternatives (see Roberts 2015 for a comparison). The k-means algorithms as they are usually described offer a divisive clustering method (Section 5.4). In the present context, however, the revising part of the algorithm serves the purpose of improving already existing relevé groups.

The k-means method tries to minimize the sum of within-group variances just as does minimum-variance clustering (Section 5.3.3; Ward 1963; Everitt *et al.* 2011, p. 124 ff) through iteration:

1. Compute group centroids using the temporarily valid classification. This may consider the entire relevé sample or a part of it only.
2. Sequentially remove relevé by relevé from the sample. Reassign it to the group whose centroid is nearest.
3. Recompute the centroids whenever a sampling unit has changed group membership. Proceed with step 1 if at least one relevé has changed group membership in step 2.
4. If no more relevés change group membership then stop and adopt the present classification.

This description reveals that k-means re-allocation (and also k-means clustering) operates on an initial classification. Most algorithms try to avoid getting trapped in a local, sub-optimal solution, for instance, by varying this classification.

The following example demonstrates reallocation by k-means (Table 5.3). In this it is obvious that the first two relevés should form a first, the third relevé a second group. The groups we assess before reallocation intentionally represent a bad solution and it is hoped that the k-means algorithm would improve on this. The two group centroids are the following:

Group 1 {3.0; 1.0}
Group 2 {1.5; 1.5}

The centroid of group 1 is identical to relevé 1, whereas the same of group 2 consists of the mean species scores of relevés 2 and 3 respectively. Our example reveals a complication happening whenever a group has one member only: in cases where we would remove relevé 1 from group 1 (as suggested in a normal case) we would lose one centroid. Therefore such centroids are left unchanged.

Starting the process with the first relevé we find the squared distance of relevé 1 to group 1 to be $d_{r1,g1} = 0.0^2 + 0.0^2 = 0.0$. The same of relevé 1 to group 2 is $d_{r1,g2} = 1.5^2 + 0.5^2 = 2.5$. Hence, relevé 1 will remain a member of group 1.

The assignment of relevé 2 is slightly different as this contributes to group centroid 2. To find the best solution it is wise to recalculate the centroids first without considering relevé 2:

Group 1 {3.0; 1.0}
Group 2 {1.0; 2.0}

Clearly, an assignment to group centroid 1 (which is presently the same as relevé 1) is better than an assignment to group centroid 2 (presently identical to relevé 3). Based on this finding we recalculate the centroids, noting that relevé 3 is now the only member of group centroid 2:

Group 1 {2.5; 1.0}

Group 2 {1.0; 2.0}

If we would repeat the entire reassignment process starting with the revised classification ('Group after re-allocation' in Table 5.3) we could observe that group memberships are no longer changed. Therefore the revision stops at this point.

Table 5.3: Data set illustrating the k-means algorithm.

Species	1	2	Group before re-allocation	Group after re-allocation
Relevé 1	3	1	1	1
Relevé 2	2	1	2	1
Relevé 3	1	2	2	2

$\mathcal{R}$ function **kmeans()** does not handle initial classifications directly, but group centroids have to be provided instead. Centroids are easily obtained by function **aggregate()**. For the purpose of illustration I use the still fairly small **nveg** and **nsit** data sets again. In the first step I generate an initial classification on a distance matrix, **dr**, of pH-values, rather than vegetation data:

```
dr<- vegdist(nsit[,"PH"],method="euclid")
o.clr<- hclust(dr,method="complete")
o.grel<- cutree(o.clr,k=3)
```

The classification is now in object **o.grel** (a list) which is used for computing the centroids **cent** based on vegetation data **nveg**:

```
cent<- aggregate(nveg,list(o.grel),mean)
o.kmeans<- kmeans(nveg^0.5, cent[,-1])
```

Upon running function **kmeans()** the first column of the centroids, **cent**, must be skipped as it contains the row names. Before displaying the classification before and after revision the elements of **o.grel** get the row names of **nveg** to adjust the formats:

```
o.grel<- as.array(o.grel) ; row.names(o.grel)<- row.names(nveg)
o.grel ; o.kmeans$cluster
```

```
2  4  6  9 10 18 25 27 39 49 50
1  2  1  3  2  2  1  2  1  1  3
2  4  6  9 10 18 25 27 39 49 50
1  2  1  3  2  3  1  2  3  1  3
```

The result reveals that relevé numbers 18 and 39 have changed group membership. An even better overview of what happens in the course of revision reveals a contingency table for which function `table()` is used:

```
c.t<- table(o.grel,o.kmeans$cluster)
c.t[c.t == 0]<- "." ; c.t
```

```
o.grel 1 2 3
     1 4 . 1
     2 . 3 1
     3 . . 2
```

In the diagonal of `c.t` we find the number of sampling units not changing the group. The off-diagonal elements count the number of changes and they show to which group they were moved. The replacement of zeros by dots is made for convenience.

5.8 Which classification method?

Here is an attempt to give hints to the use of classification methods. It is a case study in which the same data set is analysed by various methods allowing the results to be compared visually. As shown in Figure 5.2 dendrograms are not really suited for this purpose as they can be drawn in many different ways. Silhouette plots (Section 5.6), used here, are certainly easier to be compared.

The data set chosen is the Schlaenggli vegetation data (object `sveg`) with sample size $n = 63$. This size allows for the results to be fitted on a single page while details still remain visible. The graphs can be evaluated in many different ways, depending on the objective of the analysis. But what constitutes a 'good' classification? I concentrate on three criteria. The first is how balanced group sizes are. Groups of similar size promise to reflect overall composition better than unbalanced solutions where, for example, one group dominates the entire sample. Another criterion is whether a clustering method responds to changes in data transformation (a favourable property) or if it suppresses this (the less favourable case). Finally, silhouette plots also show the homogeneity of groups.

Vegetation classification has tremendous practical relevance as outlined in more detail in Chapter 12.1 and discussed in the papers of Dengler *et al.* (2011) or De Cáceres *et al.* (2015). The comparisons shown below concern classical clustering methods only, excluding preferential solutions and revisions of existing classifications (Section 5.7).

The clustering method is not the only source of variation of classifications, but data transformation and the effect of the resemblance measures must simultaneously be taken into account. The outcome of transforming cover-abundance scores as explained in Section 3.4 is apparent when inspecting the rows of Figure 5.8. The original vegetation data `sveg` consists of a

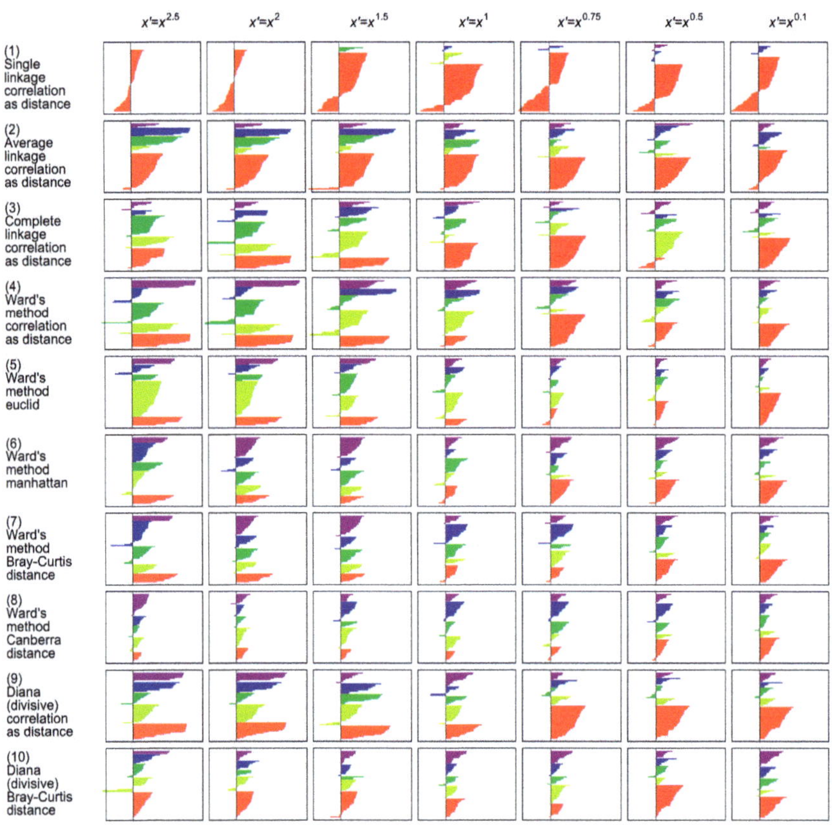

Figure 5.8: Comparison of classification methods. Data transformation and the choice of the resemblance function are varied simultaneously. The individual graphs are silhouette plots of classifications, data set used is sveg.

rank scale with steps 0 to 6. When applying transformation $x = x^y$, any $y > 1.0$ tends to approximate cover percentage, whereas $y < 1.0$ changes scores into the direction of presence–absence (Table 3.4). When $x' = x^{2.5}$ then the scale mimics cover percentage.

A selection of clustering methods is chosen for comparison. Rows one through four in Figure 5.8 present the results of the four clustering methods, single linkage (1), average linkage (2), complete linkage (3), and as one more 'intermediate' solution, Ward's method (4). In all these cases, correlation as distance (Equation 4.6) is taken as a resemblance function. The conclusion these cases allow are rather striking:

- The proper choice of the method is crucial, the outcome of classification varies tremendously.

- Data transformation is essential, the graphs in the left hand column (quantitative approach) differ widely from the same in the right hand column (qualitative approach).
- As expected, in the context of vegetation description by species scores, single linkage probably does not provide what most users would like to see and may therefore be avoided.
- Probably the best balanced group sizes are achieved when using Ward's method.

It is for this reason that for the next block, rows five through eight, only Ward's method is taken, but the distance measures are now varied with Euclidean distance (5), Manhattan distance (6), Bray–Curtis distance (7) and Canberra distance (8). This complements the above conclusions as follows:

- The choice of the distance measure is as influential as is the choice of the clustering method: distance strongly affects the resulting classifications.
- When concentrating on the quantitative view (right hand column) then the Bray–Curtis distance (7) provides a well-balanced group structure – notably very much like correlation as distance (4).
- Bray–Curtis (7), but even more so Canberra distance (8), deliver well-balanced group sizes. Unfortunately they do hardly respond on data transformation. Their intrinsic adjustments (Equation 4.4 and Equation 4.5) allow a qualitative view of the data only, no matter what the user intends by the initial data transformation applied.

The last block extends the evaluation to the divisive method DIANA (Section 5.4). This is based on correlation as distance (9) and Bray–Curtis distance (10). The following can be seen:

- DIANA differs from all agglomerative methods. Especially the solutions of the quantitative data (left hand column) deliver a very well-balanced group structure.
- In combination with Bray–Curtis the same drawbacks occur as noted for row 7. Bray–Curtis downscales the effect of cover scores into the direction of presence–absence.
- In summary DIANA, despite rarely being used, has the potential to perform well in further applications.

In all examples of Figure 5.8 the group structures vary considerably depending on the combination of methods chosen. This is typical for a data set describing a continuous gradient where no natural group limits exist. As soon as discontinuities occur in the data most methods will reliably detect these.

Figure 5.8 allows a visual comparison only. It is in Figure 7.9 where the resulting classifications are quantitatively evaluated. An alternative way of comparing clustering methods is found in the paper of Lengyel and Podani (2015).

6 Ordination

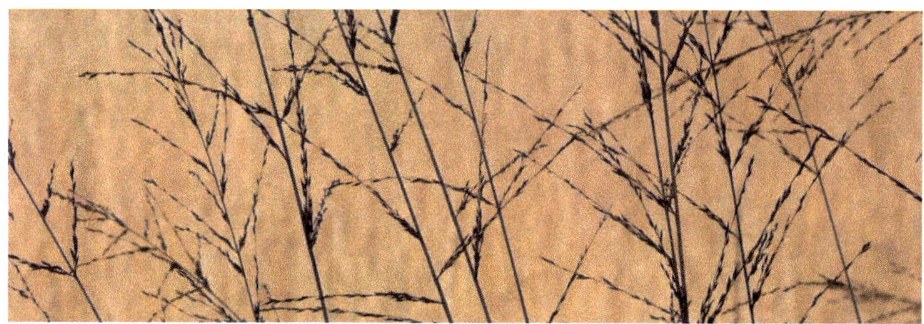

Ordination is a means to visualize the similarity pattern of a sample, typically in two- or three-dimensional plots where the individual observations are points. Despite operating in reduced data space, ordination diagrams reveal dominant patterns like groups, gradients and outliers.

6.1 Why ordination?

Ordination is a graphical representation of the similarity of sampling units or attributes in resemblance space. An example of a two-dimensional ordination is the one in Figure 4.1(b), where the axes are two plant species and the data points relevés. This graph displays the similarity of the two relevés involved, a trivial case as it presents the full configuration given in the raw data without having to improve insight into the system. But ordination is a tool for analysing and visualizing complex data sets with a high number of sampling units and many attributes involved.

Recognizing patterns in large multivariate data sets means operating in a resemblance space of very high dimension. When considering four data points, for example, the configuration of resemblance is three-dimensional. Although more than three dimensions are difficult to display graphically, in vegetation ecology large data sets often include hundreds of species, requiring hundreds of dimensions to be considered. The aim of ordination is to reduce this number, to derive a graph that can be plotted or inspected as a

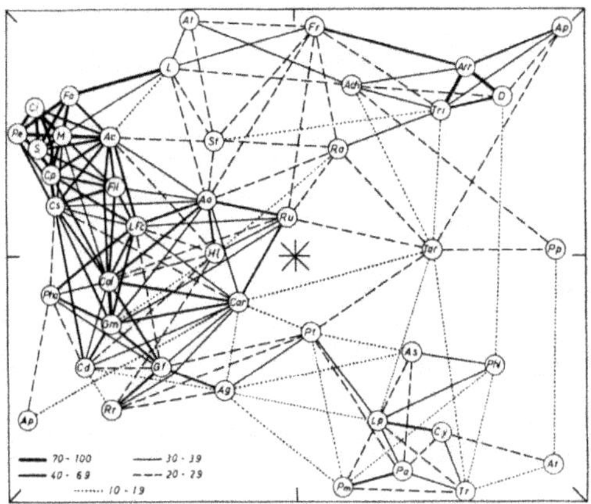

Figure 6.1: Three-dimensional representation of similarity relationships (Mueller-Dombois and Ellenberg 1974).

two- or three-dimensional and eventually rotating point cloud. This is why ordination has always played a key role in vegetation ecology, and Figure 6.1 is a historical example of this effort, illustrating the attempt to display all the important interactions found in a multidimensional configuration.

In the meantime ordination methods have evolved and new ones have found their way to plant ecologists. As von Wehrden *et al.* (2009) noticed, 'ongoing discussions and the diversity of available methods have led to considerable uncertainty among researchers and students'. The present chapter intends to reconcile this situation in various ways. First, ordination is strictly conceived as an aid in visualizing high dimensional similarity patterns. Second, as explained below, all methods follow the same strategy: reducing dimensionality while maximizing resolution. Third, I postpone using ordination as an ecological model, allocating the latter to Chapters 7 and 9, including constrained ordination: the tremendous progress in the field of modelling no longer justifies statistical testing in reduced ordination space (see, for example, Elith *et al.* 2006 and Venables and Ripley 2010).

Most methods for gaining the desired insight into the similarity patterns of multivariate data sets roughly proceed along the steps illustrated in Figure 6.2:

- Centre the attributes in a data matrix to shift the origin of the new coordinate system into the centre of the point cloud. This first step serves computational convenience only and it does not alter the point pattern.

- Rotate the point cloud such that the maximum possible variance is found along the first axis.

- Continue rotating the point cloud, while keeping the first axis fixed, and maximize the remaining variance on the second axis.
- Continue this operation until all the axes are processed.
- Represent the result graphically by omitting higher dimensions.

This is basically what principal component analysis (PCA) does, the 'mother' of all ordination methods. All others mainly differ from this in the intrinsic transformation applied to the data, the resemblance measure used and the way the point cloud is rotated. An illustration of the variation of frequently used ordination methods is given at the end of this chapter (Section 6.8).

6.2 Principal component analysis

6.2.1 Operational steps

Principal component analysis (PCA) is nothing else but a mathematical solution to the objectives of ordination listed in Section 6.1. It strictly relies on *linear correlation* of attributes, operates in the orthogonal Euclidean (that is, geometric) space and therein strives for useful projections of point clouds. Because it is based on the concept of variance partitioning, the result it finds can always be reproduced – even when using different computers and software. Orthogonality (that is, uncorrelated axes) also means that the variance carried by the axes is *additive*. Whatever projection is chosen, the variance explained by the graph is equal to the sum of the explanatory power of the axes involved.

To explain the operations involved in PCA I use the small example given in Figure 6.2(a) that mimics a short gradient of six relevés described by two species. Although this is a trivial case as the similarity of the raw data can be presented in two dimensions, it allows the steps involved in PCA to be reproduced graphically [Figure 6.2(b)]. The species scores used in Figure 6.2(a) are entered in $\mathcal{R}$ as a matrix (using functions `matrix()` and `c()`), typing:

```
raw<- matrix(c(1,2,2.5,2.5,1,0.5,0,1,2,4,3,1),nrow=6)
colnames(raw)<- c("s1","s2")
rownames(raw)<- c("r1","r2","r3","r4","r5","r6")
raw
```

```
    s1 s2
r1 1.0  0
r2 2.0  1
r3 2.5  2
r4 2.5  4
r5 1.0  3
r6 0.5  1
```

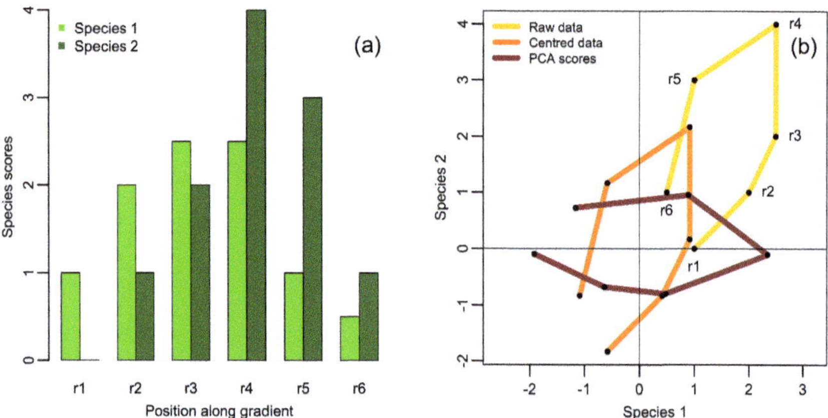

Figure 6.2: Operations in PCA ordination. (a) Raw scores of two species along a gradient of six relevés. (b) Relevés plotted by raw scores of two species, by centred scores of the same species and by two PCA score vectors. Note that PCA does not change point configurations.

These are now centred by columns using function `scale()`:

```
cent<- scale(raw,scale=FALSE)
cent
```

```
           s1         s2
r1 -0.5833333 -1.8333333
r2  0.4166667 -0.8333333
r3  0.9166667  0.1666667
r4  0.9166667  2.1666667
r5 -0.5833333  1.1666667
r6 -1.0833333 -0.8333333
...
```

Setting `scale=FALSE` suppresses any further transformation after centring. For convenience we denote the now centred data by X, still a matrix of six relevés by two species.

PCA sores are the result of a rotating process. Rotation is achieved by a matrix multiplication of the centred scores by square matrix α of dimension p, where p is the number of species. This matrix of *eigenvectors* α is obtained from the data by eigenanalysis (Batschelet 1975), yielding the desired properties of the final result – orthogonality of axes – and maximizing variance on the first axis. Eigenanalysis uses the scalar product, covariance or correlation matrix of the species, and as a result it not only finds the eigenvectors, but also the eigenvalues, λ, the variance of the newly found axes. In $\mathcal{R}$ this is done by function `eigen()`:

```
o.eig<- eigen(t(cent) %*% cent)
```

```
$values
[1] 11.982346  2.559321

$vectors
          [,1]       [,2]
[1,] 0.3491944 -0.9370503
[2,] 0.9370503  0.3491944
```

What is printed under `$values` are the eigenvalues, λ, ordered by size where the first element, $\lambda_1 = 11.982$, is the variance explained by the first, and $\lambda_2 = 2.559$ is the same of the second axis. This means that roughly 82% of the total variance concentrates on the first axis. Under `$vectors` are the eigenvectors, α. The elements of the eigenvector matrix are correlation coefficients between the original attributes (the species) and the new ordination axes. Element 0.349 signifies that the first species has a positive correlation of $r = 0.349$ with the first ordination axis. The correlation with the second axis is $r = -0.937$. The second species correlates with the first axis by $r = 0.937$ and with the second axis by $r = 0.349$. Hence, the eigenvectors are a useful tool for interpreting the ordination axes, because they express the contribution of the species to the axes.

To get the final ordination scores centred data X is multiplied by the eigenvectors, α:

$$Y = X\alpha \tag{6.1}$$

In terms of $\mathcal{R}$ this is:

```
Y <- cent %*% o.eig$vectors
Y
```

```
         [,1]        [,2]
r1 -1.9216223 -0.09357697
r2 -0.6353776 -0.68143293
r3  0.4762699 -0.80076373
r4  2.3503705 -0.10237502
r5  0.8895287  0.95400610
r6 -1.1591692  0.72414255
```

This is the final result used in Figure 6.2(b). There are various functions available in $\mathcal{R}$ for doing PCA conveniently in one step, like function `pca()` in package `labdsv` and it even provides a plot method:

```
o.pca <- pca(cent)
plot(o.pca)
lines(o.pca$scores)
```

Function `lines()` draws the dark lines in Figure 6.2(b). Further details about the result are available through the summary method of function `pca()`:

```
summary(o.pca)

Importance of components:
                            [,1]      [,2]
Standard deviation     1.5480534 0.7154468
Proportion of Variance 0.8240009 0.1759991
Cumulative Proportion  0.8240009 1.0000000
```

This gives the standard deviations of the first and the second axis and it confirms the contributions of the axes to total variance obtained above from function `eigen()`.

6.2.2 Interpretation by example

In real world applications the result of PCA deserves careful interpretation, as illustrated in Figure 6.3, using data frame `sveg` of 63 sampling units (relevés) and 119 attributes (species). The environmental factors are not analysed at this point (but will be so in Section 6.6, for example), hence the focus is on the interpretation of vegetation data only.

Just as in the previous section the results as provided by function `pca()` are presented. Computation including a transformation (the fourth root of the species scores taken) is achieved in a single step as follows:

```
o.pca <- pca(sveg^0.25,cor=TRUE)
o.s<- summary(o.pca) ; o.s[,1:5]

                            [,1]       [,2]       [,3]      [,4]       [,5]
Standard deviation     4.9534260 3.10432989 2.68773866 2.0888545 1.98836927
Proportion of Variance 0.2061885 0.08098205 0.06070537 0.0366665 0.03322363
Cumulative Proportion  0.2061885 0.28717053 0.34787590 0.3845424 0.41776603
```

The output of the summary is restricted to the first five axes reporting the importance of the corresponding eigenvalues. Note that the full output would encompass 63 non-zero elements. What is denoted by 'Proportion of Variance' is the relative contribution of the axes (and the related eigenvalues) to the explaining power of the ordination. Here, the x-axis in Figure 6.3(a) explains 20.6% of the variance and the y-axis 8.09%. The entire ordination shown in Figure 6.3(a) uses the scores of the first two axes of PCA as coordinates, summing up to 28.7%. But the third axis, explaining another 6%, is used as well for the symbol size. The total therefore is 34.8%.

Is 34.8% explained variance good or poor for a three-dimensional ordination? Peres-Neto *et al.* (2005) have written a review on papers discussing the issue of 'nontrivial axes'. The authors suggest a randomization test to identify the number of relevant axes. In practice, however, it may be safe to screen for patterns beyond this number. The proportion of explained variance depends on the type of data analysed and of course on the number of axes considered for viewing. For the present sample size, experience suggests that 28.7% in two dimensions usually reveals the dominating pattern, which in this case is a classical horseshoe (see Section 6.5), indicating that a (nonlinear) vegetation gradient exists. From many more examples it can be inferred that data sets of several hundreds of relevés usually deliver a first eigenvalue explaining around 10% of the total variance or even less. As a rule the explaining power of the first axis will decrease as the sample size is increasing.

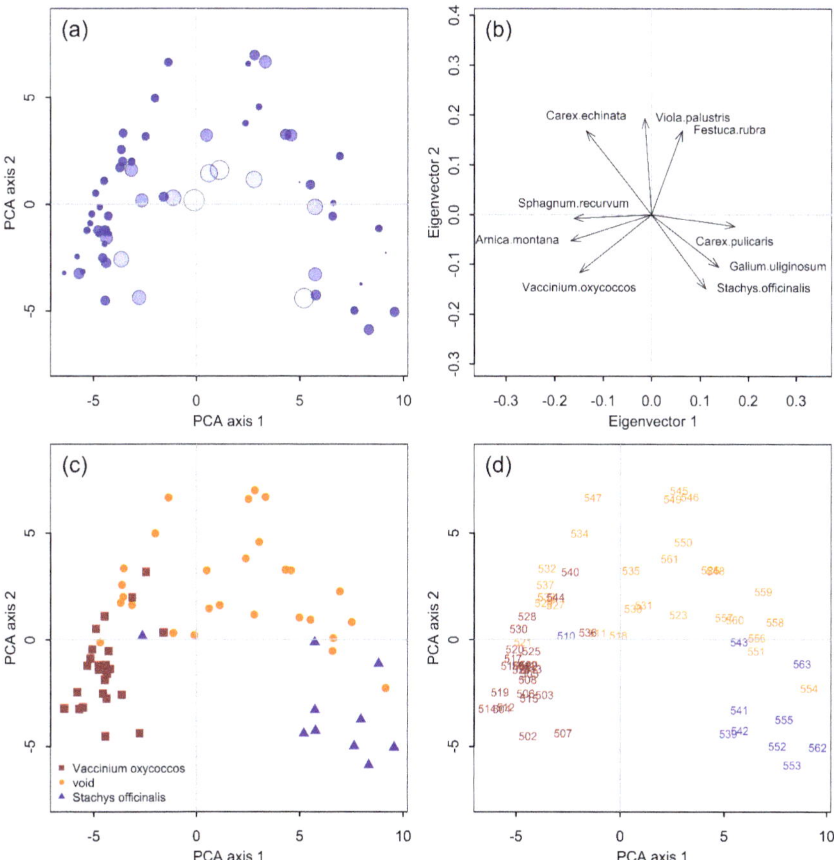

Figure 6.3: Interpretation of PCA results using real world data (data frame sveg). Panel (a) is the ordination of relevé scores in two dimensions with symbol size and colour intensity proportional to the third axis. The eigenvectors in panel (b) are used to draw arrows pointing into the direction of the centres of species occurrence. An alternative is including the distribution of selected species in the ordination, panel (c). To identify the individual relevés their numbers can be plotted instead of symbols, panel (d).

The point cloud simply shows the similarity space, with the only complication arising from the as yet hidden higher dimensions. If any two data points are in close neighbourhood then the corresponding relevés are similar. However, they may still be separated in higher dimensions.

Explicit plotting rather than using the more convenient plot methods (`plot(o.pca)`, in this case) offers enhanced flexibility. For panel (a) we first re-scale the third axis to range from 0 to 1 and use this for adjusting the size (cex) of circles (`pch=1`) proportionally:

```
z<- o.pca$scores[,3] ; z<- (z-(min(z)))/(max(z)-(min(z)))
plot(o.pca$scores[,1],o.pca$scores[,2],pch=1,cex=z*2.0,asp=1)
```

The factor 2.0 enlarges the symbols for convenience. An important parameter for drawing ordinations is `asp=1`, causing identical scaling of the x- and y-axes, hence avoiding distortion of the strictly metric graph. Function `abline()` draws the grey zero lines, when typing `abline(h=0, v=0, col="gray")`. In this `h` is responsible for the horizontal and `v` for the vertical line.

For proper interpretation of PCA results, the eigenvectors (sometimes called component coefficients) have to be considered as well. They are the correlations of the species with the ordination axes. Due to the very high dimensionality of the resemblance space, most correlations are rather low, with none even reaching $r = 0.2$. Geometrically, correlation coefficients are cosines of vectors (Figure 4.3) and they can therefore be used to draw species vectors [Figure 6.3(b)]. The species arrows point into the direction of their centres of occurrence. Selecting just a few species for display will avoid a proliferation of symbols in the graph.

The detailed interpretation of Figure 6.3(a) and (b) proceeds as follows: the relevés (data points) in the lower-left quadrant of the ordination are characterized by high abundance of *Vaccinium oxycoccos*, a species typical of peat bogs. In the relevés of the lower-right quadrant *Stachys officinalis* and *Galium uliginosum* occur frequently. The horseshoe-shaped point cloud reveals a gradient from the lower right (acidic conditions) towards the lower left (neutral conditions). Analysing the relationship to pH, however, requires other methods such as constrained ordination (Section 7.4).

The coordinates for drawing the eigenvectors in Figure 6.3(b) are found in `o.pca$loadings` where `loadings` is yet another term for eigenvectors. We restrict these to a list, `s.sp`, of nine species, displayed in Figure 6.3(b), with their index stored in `s.sp`:

```
s.sp<- c(3,11,23,31,39,46,72,77,96)
snames<- names(sveg[,s.sp])
snames
```

```
[1] "Vaccinium.oxycoccos" "Carex.echinata"      "Arnica.montana"
[4] "Festuca.rubra"       "Carex.pulicaris"     "Sphagnum.recurvum"
[7] "Viola.palustris"     "Galium.uliginosum"   "Stachys.officinalis"
```

The names of the selected species are taken from object `sveg` and stored in vector `snames` with the aid of function `names()`. Again, the coordinate system is drawn first with function `plot()`, then nine arrows are generated from the origin of the coordinate system ($x = 0, y = 0$) to the coordinates of the eigenvectors. Finally we add the species names using function `text()`:

```
x<- o.pca$loadings[,1] ; y<- o.pca$loadings[,2]
plot(x,y,type="n",asp=1)
arrows(0,0,x[s.sp], y[s.sp],length=0.08)
text(x[s.sp],y[s.sp],snames,pos=1,cex=0.6)
```

Parameter `cex=0.6` reduces font size to about half of the standard and `length=0.08` reduces the size of the arrowheads. Argument `pos` allows the species names to be positioned below (1), to the left (2), on top (3) or to the right of the arrow head (4) to eventually avoid overlap.

One more means to improve interpretability of ordination is to use a different symbol (`pch=sym`) or colour (`col=co`) for relevés in which a typical species occurs, like *Vaccinium oxycoccos* and *Stachys officinalis* in Figure 6.3(c). The most elegant way is to use function `ifelse()` with three parameters involved: the first is a logical expression (does *Vaccinium.oxycoccos* occur, for instance), the second the value if the expression is true (symbols 15, 16, and 17 for squares, circles and triangles respectively) and the third if the expression is false. Below, grey tones of different density are assigned to the symbols:

```
sym<- ifelse(sign(sveg[,"Vaccinium.oxycoccos"])==1,15,16)
sym<- ifelse(sign(sveg[,"Stachys.officinalis"])==1,17,sym)
co<- ifelse(sym==15,gray(0.2),gray(0.7))
co<- ifelse(sym==17,gray(0.4),co)
x<-  o.pca$scores[,1] ;  y<- o.pca$scores[,2]
plot(x,y,pch=sym,cex=1.0,col=co,asp=1)
```

Any colour can be assigned to the symbols instead of the grey scale. The same variable `co` is now also used in Figure 6.3(d) illustrating the identification of data points by plotting the relevé names (or numbers) instead of symbols:

```
plot(x,y,type="n",asp=1,cex.axis=0.8,cex.lab=0.8)
text(x,y,rownames(sveg),col=co)
```

Finally, a good practice is to view all possible projections of the high-dimensional point cloud, as shown in Figure 6.4 for the first five axes. The `graphics` package in $\mathcal{R}$ offers function `pairs()` to do this routinely:

```
pairs(o.pca$scores[,1:5],pch=sym,col=co,cex=0.5)
```

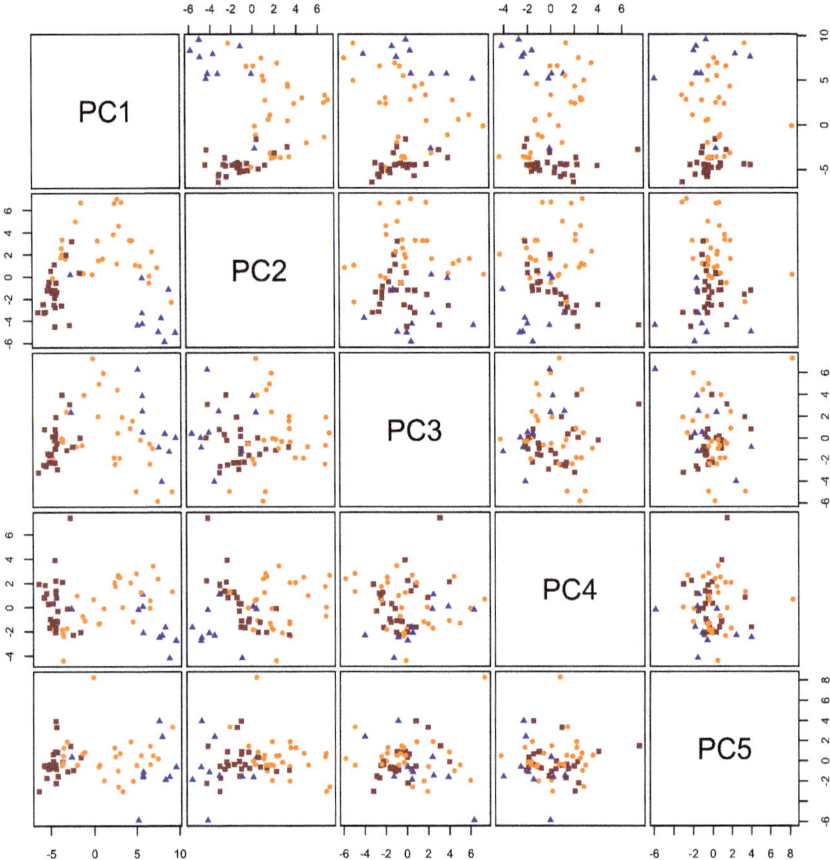

Figure 6.4: Projection of five-dimensional PCA ordination by function `pairs()`. See Figure 6.3(c) for a legend.

In this `cex=0.5` chooses small symbols for plotting the scattergrams. One could consider many more dimensions, such as 10 instead of 5, but the individual graphs would get really small inside the resulting 10 by 10 matrix of ordinations.

Figure 6.4 confirms the distribution of scores of higher dimensions, four and five for example, to be almost normal. Typically, a few sampling units may appear slightly isolated, although not really qualifying as outliers (see Section 12.3.1 for identification of these). As an example, in all graphs where the fourth dimension is involved there is one such observation. It is the uppermost data point in the ordinations of row four, and the one to the right in ordinations of column four. In the `base` package of $\mathcal{R}$ there is function `which.max()` to identify the sampling unit with the highest score, inside the fourth dimension in this case:

```
which.max(o.pca$scores[,4])
```

```
507
  7
```

This data point is number 7 and the corresponding plot name is 507. The two extremes observed in the fifth dimension can be assessed in the same way too. These are 518 and 539, respectively; the latter found by function `which.min()`.

6.3 Principal coordinates analysis

This method offers an alternative solution to the PCA problem, first published by Gower (1966). It is not only known as *principal coordinates analysis* (PCOA) but also as *principal axis analysis* or *metric multidimensional scaling*. In PCOA, when the relevés are ordinated, the eigenvalues and eigenvectors are derived from a resemblance matrix of the relevés. This differs from PCA, where the species-similarity matrix is used to derive the coordinates of the relevés and vice versa. In PCOA there are no eigenvectors available to support the interpretation of the ordination, but correlation of axes with species can be calculated subsequently.

The outset of PCOA is a distance matrix of relevés. If this is Euclidean, then the final coordinates are the same as when doing a PCA based on the scalar product. In all other cases the PCOA solution will differ from the same of PCA and it will even work if the distance measure used is non-metric. It then happens that 'the representation of objects in the reduced space of the first few principal coordinates forms the best possible Euclidean approximation of the original distances' (Legendre and Legendre 2012, p. 499).

Denoting the distance matrix of relevés by D, then the first step is to calculate a similarity matrix, S, with elements:

$$s_{ij} = -\frac{1}{2}d_{ij}^2 \tag{6.2}$$

These are interpreted as direction cosine and to move the origin of the coordinate system into the centre of the ordination they have to be adjusted for range. The new matrix, A, has the elements:

$$a_{ij} = s_{ij} - \overline{s}_i - \overline{s}_j + \overline{s} \tag{6.3}$$

where $\overline{s}_i$ and $\overline{s}_j$ are the row and column means of S and $\overline{s}$ is the same of the grand total. Equation (6.3) is sometimes referred to as Gower's doubly centring. From A the eigenvalues, $\lambda_1, \ldots, \lambda_n$, and the corresponding eigenvectors, $\beta_1, \ldots, \beta_n$, are found. The eigenvectors are adjusted to the eigenvalues to satisfy the condition:

$$\beta_{1i}^2 + \ldots + \beta_{pi}^2 = \lambda_i^2 \tag{6.4}$$

This can be verified using the example from Section 6.2.1 with six relevés by two species:

```
raw<- matrix(c(1,2,2.5,2.5,1,0.5,0,1,2,4,3,1),nrow=6)
colnames(raw)<- c("s1","s2")
rownames(raw)<- c("r1","r2","r3","r4","r5","r6")
```

Similarity matrix A is derived via Euclidean distance and Equation (6.2) as follows:

```
de<- dist(raw,method="euclid",diag=TRUE,upper=TRUE)
(A<- as.matrix(-0.5*de^2))
```

```
       r1     r2     r3     r4     r5     r6
r1  0.000 -1.000 -3.125 -9.125 -4.500 -0.625
r2 -1.000  0.000 -0.625 -4.625 -2.500 -1.125
r3 -3.125 -0.625  0.000 -2.000 -1.625 -2.500
r4 -9.125 -4.625 -2.000  0.000 -1.625 -6.500
r5 -4.500 -2.500 -1.625 -1.625  0.000 -2.125
r6 -0.625 -1.125 -2.500 -6.500 -2.125  0.000
```

We now apply Equation (6.3) to get the doubly centred matrix A:

```
S<- A
 mrow<- apply(S,1,mean)
 mcol<- apply(S,2,mean)
 nd<- nrow(S); A<- S
 for (i in 1:nd) for (j in 1:nd) {
    A[i,j]<- S[i,j]-mcol[i]-mrow[j]+mean(S)
 }
 round(A,digits=2)
```

```
      r1    r2    r3    r4    r5    r6
r1  3.70  1.28 -0.84 -4.51 -1.80  2.16
r2  1.28  0.87  0.24 -1.42 -1.22  0.24
r3 -0.84  0.24  0.87  1.20 -0.34 -1.13
r4 -4.51 -1.42  1.20  5.53  1.99 -2.80
r5 -1.80 -1.22 -0.34  1.99  1.70 -0.34
r6  2.16  0.24 -1.13 -2.80 -0.34  1.87
```

Eigenanalysis now yields the squared eigenvalues and eigenvectors:

```
o.e<- eigen(A)
o.ev<- t(o.e$vectors[,1:2])*o.e$values[1:2]^0.5
round(o.e$values,digits=3) ; o.ev
```

```
[1] 11.982  2.559  0.000  0.000  0.000  0.000
            [,1]       [,2]       [,3]      [,4]      [,5]       [,6]
[1,] -1.92162229 -0.6353776  0.4762699  2.350371 0.8895287 -1.1591692
[2,] -0.09357697 -0.6814329 -0.8007637 -0.102375 0.9540061  0.7241426
```

The eigenvalues are identical to the same in the PCA-example given in Section 6.2.1 and the adjusted eigenvectors are the ordination coordinates used in Figure 6.2.

But PCOA is used beyond the capabilities of PCA such as for the analysis of distance matrices that are not entirely Euclidean. Figure 6.5(a) is an example for this. Data frame `sveg` includes cover-abundance scores changed into a rank scale that are further transformed according to $x' = x^{0.25}$. In $\mathcal{R}$ one can use function `pco()` in package `labdsv` and apply this, for instance, to a Bray–Curtis distance matrix. Ordination scores are used as x- and y-coordinates, the first two parameters in function `plot()`. Parameter `k=3` limits the dimensions to be calculated to three:

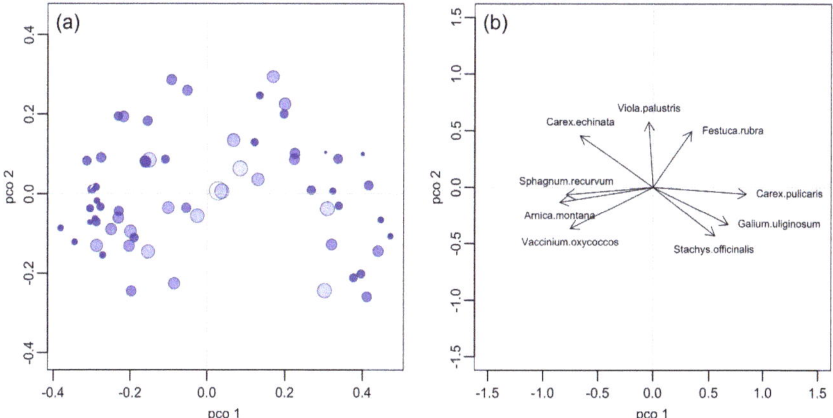

Figure 6.5: PCOA ordination using the 'Schlaenggli' data set of 63 relevés, the same as in Figure 6.3(a). Bray–Curtis distance is used. Panel (a): plot of relevés, symbol size and colour intensity are proportional to the third axis. Panel (b): correlation of selected species with ordination axes.

```
o.pco<- pco(vegdist(sveg^0.25,method="bray"),k=3)
x<- o.pco$points[,1] ; y<- o.pco$points[,2]
z<- o.pco$points[,3] ; z<- (z-(min(z)))/max(z-(min(z)))
plot(x,y,asp=1,pch=1,cex=z*2.0)
```

As already demonstrated in Figure 6.3(a) the third axis is used for symbol size, `cex`, which is further enlarged by factor 2.0 for convenience. Whenever using PCOA, inspection of the eigenvalues is recommended. These are given in the output list as vector `o.pco$eig`. Because the Bray–Curtis distance is semi-metric, not only positive eigenvalues result (from the metric part of resemblance space), but also some negative ones (for the nonmetric part of resemblance space). A simple way to inspect the sum of the positive as well as of the negative ones proceeds as follows:

```
pos.eig<- sum(o.pco$eig[o.pco$eig > 0])
neg.eig<- sum(o.pco$eig[o.pco$eig < 0])
pos.eig ; neg.eig
```

```
[1] 10.20894
[1] -1.380561
```

Then, the percentage of variation that the first two axes account for is calculated:

```
eig1<- o.pco$eig[1]/pos.eig
eig2<- o.pco$eig[2]/pos.eig
eig1*100 ; eig2*100
```

```
[1] 42.64565
[1] 11.39147
```

As a surrogate for the missing eigenvectors the correlations between the species vectors and the ordination scores from PCOA can be calculated subsequently to draw Figure 6.5(b) in analogy to Figure 6.3(b). After selecting the same nine species the correlations of the species vectors in `sveg` and the ordination scores in `o.pco$points` build the new matrix `alpha`. Function `cor()` is used for this. Then the arrows (function `arrows()`) are drawn and the species names included by function `text()`:

```
s.sp<- c(3,11,23,31,39,46,72,77,96)
alpha<- cor(x=sveg^0.25,y=o.pco$points)
plot(alpha[,1],alpha[,2],type="n",asp=1)
arrows(0,0,alpha[s.sp,1],alpha[s.sp,2],length=0.08)
text(alpha[s.sp,1],alpha[s.sp,2],snames,pos=1,cex=0.6)
```

Alternatively function `pcobiplot()` in the `dave` package delivers graphs (a) and (b) in Figure 6.5 as well in one step. Obviously this ordination differs somewhat from the PCA solution. This is because Bray–Curtis distance is used instead of covariance. The two ordinations would be identical when doing PCOA with Euclidean distance.

PCOA is the ideal tool to compare the effect of the many resemblance functions developed for the analysis of vegetation data. An example is shown in Figure 6.6. Superimposed on the PCOA ordinations (circles in different colours, reflecting group pattern) is always the PCA solution (small dots). As can be seen from the axes annotations, the scales differ because the resemblance measures have a different range. To adjust the PCA solution to

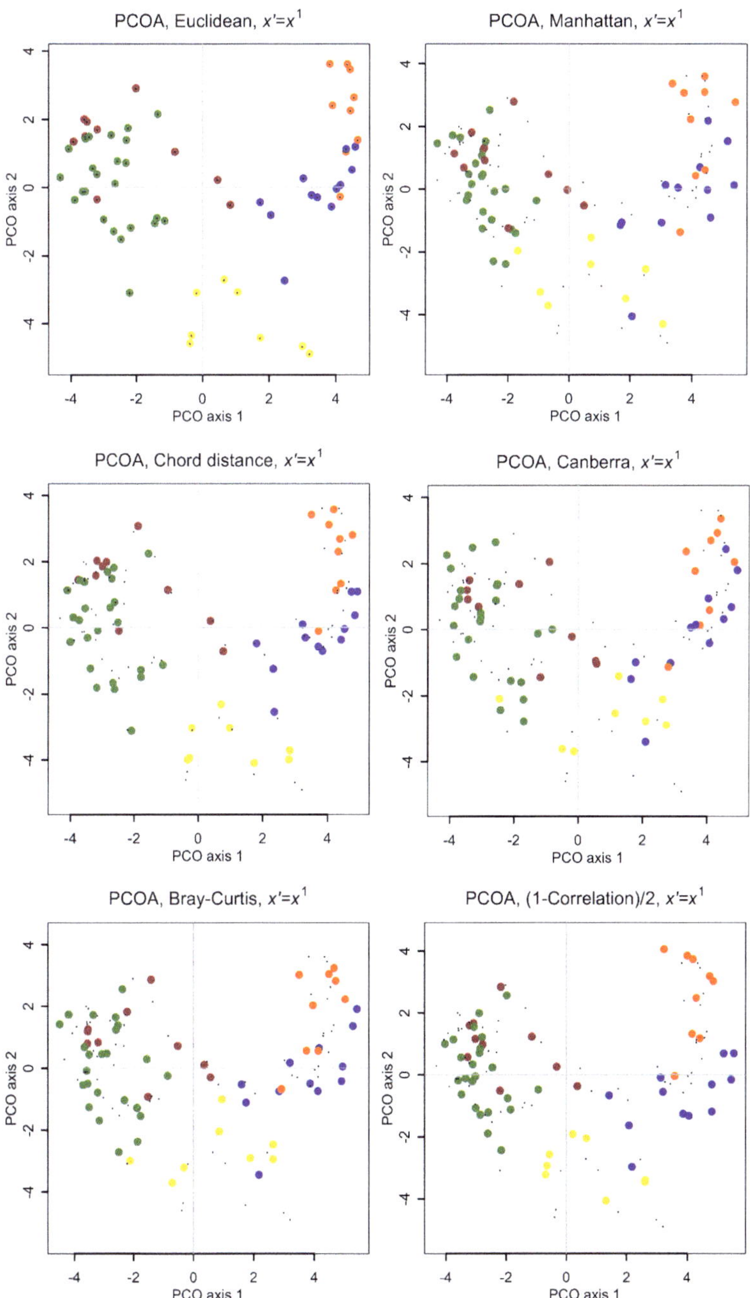

Figure 6.6: PCOA ordinations with six different resemblance measures. Small dots are PCA ordination adjusted to the PCOA solutions by Procrustes analysis for comparison. Data frame sveg used. Colours illustrate group-wise resemblance pattern to facilitate interpretation.

the different ordinations Procrustes analysis is used (Peres-Neto and Jackson 2001). This minimizes the difference between any two ordinations by rotating the axes as needed and then rescaling these by a constant factor.

Figure 6.6 first of all confirms that PCOA is identical to PCA when Euclidean distance is used as a resemblance measure (top left graph). Euclidean and Manhattan distances represent the full, unaltered geometric resemblance space. Chord, Canberra, Bray–Curtis and Correlation distances are all surprisingly similar. These have in common that they normalize the relevé vectors intrinsically, although in a different way (Chapter 4).

To try this with any other vegetation data set such as `nveg`, `wetveg` or `ws200`, for instance, function `pcovar()` can be run. While in Figure 6.6 the original cover–abundance scale of `sveg` is used, this can also be varied by specifying the exponent y of the function $x' = x^y$. Hence, Figure 6.6 is generated as follows:

```
o.pcovar<- pcovar(sveg,y=1)
plot(o.pcovar,reversals=c(0,0,0,0,0,0))
```

Ordinations will not always be oriented as seen in Figure 6.6; they may appear reversed. This is because in eigenanalysis the sign of axes is set arbitrarily. When plotting with `pcovar()` this can be changed by adapting parameter `reversals`. For example, to mirror the top middle graph where Manhattan distance is used, we set `reversals=c(0,1,0,0,0,0)`.

6.4 Correspondence analysis

Correspondence analysis (CA) is distinct from PCA and PCOA due to the intrinsic assumptions and the corresponding transformations applied. In CA the data matrix, F, is assumed to be a contingency table; that is, a table consisting of counts. Some of the many alternative names for CA reflect this fact:

- Contingency table analysis (Fisher 1940)
- Analyse factorielle des correspondances (Benzécri 1969)
- Reciprocal averaging (Hill 1973)
- Reciprocal ordering (Orlóci 1978)
- Dual scaling (Nishisato 1980)

The elements of F, f_{hj}, are frequency counts. These are not used directly, but deviations from expectations, u_{hj}, instead:

$$u_{hj} = \frac{f_{hj}}{\sqrt{f_{h.}f_{.j}}} - \frac{\sqrt{f_{h.}f_{.j}}}{f_{..}} \tag{6.5}$$

In this $f_{h.}$ is the sum of all frequencies in row h, $f_{.j}$ the same in column j and $f_{..}$ the grand total, the sum of all elements in matrix F. Like in PCA, the calculations are then based on a matrix of product moments, $S = UU'$, computed for p attributes with a characteristic element:

$$s_{hi} = \sum_{j=1}^{s} u_{hj} u_{ij} \tag{6.6}$$

From this similarity matrix the non-zero eigenvalues, $\lambda_1, \ldots, \lambda_t$, and the associated eigenvectors, $\alpha_1, \ldots, \alpha_t$, are extracted. The eigenvalues have the form of correlation coefficients: the mth eigenvalue is the square of the mth canonical correlation. The eigenvector matrix, A, after the adjustment shown below, is used as ordination scores for the attributes.

As explained in Legendre and Legendre (2012, p. 469), there are different ways to scale the scores. For ecological applications it is most convenient to choose an adjustment that allows the joint plotting of row (relevé) and column (species) scores. The species scores X are directly derived from the eigenvectors by weighting with the square root of the inverse of the marginal totals:

$$x_{hm} = \frac{(\alpha_{hm} - \overline{\alpha}_m)}{\sqrt{\sum_{h=1}^{p} (\alpha_{hm} - \overline{\alpha}_m)^2}} \sqrt{\frac{f_{..}}{f_{h.}}} \tag{6.7}$$

This formula also involves standardization of the eigenvectors to fulfil the following conditions:

$$\sum_{h=1}^{p} \alpha_{hm}^2 = 1 \quad \text{and} \quad \sum_{h=1}^{p} \sqrt{f_{h.}} \alpha_{hm} = 0. \tag{6.8}$$

To compute the relevé scores, Y, one could transpose the data matrix F and repeat all the computations. If so, one would observe that the resulting eigenvalues were exactly the same. But there is a direct way to derive the relevé scores, Y, from the species scores:

$$y_{jm} = \sum_{h=1}^{p} \frac{f_{hj} x_{hm}}{f_{.j} R_m} \tag{6.9}$$

where R_m is the mth canonical correlation. When analysing large data sets, carrying out the analysis on the smaller similarity matrix of either relevés or species will save computation time.

The difference between PCA (or PCOA) and CA is in the underlying assumption about the nature of data and the intrinsic transformation in CA.

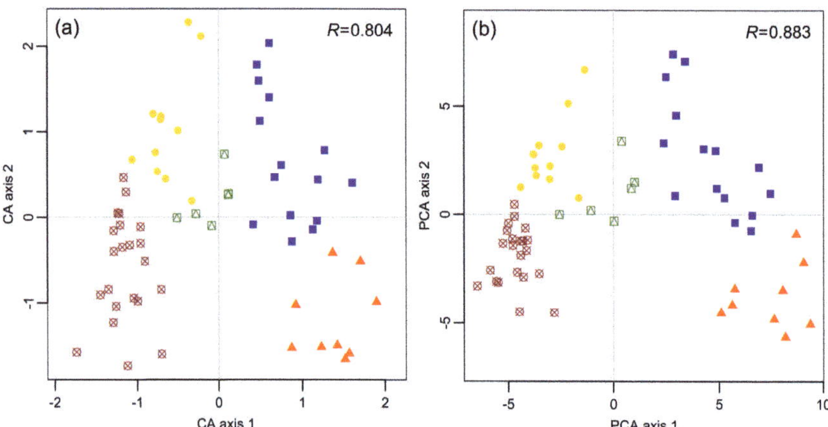

Figure 6.7: Comparison of CA (a) and PCA (b). Data points are identically classified. 'Schlaenggli' data set (sveg) used (Appendix 14).

As outlined by Legendre and Gallagher (2001), CA resides on χ^2 transformation of data:

$$f'_{hj} = \frac{f_{hj}\sqrt{f_{..}}}{f_{h.}\sqrt{f_{.j}}} \tag{6.10}$$

CA ordination displays so-called χ^2 distance and this is the Euclidean distance of scores when transformed according to Equation (6.10). This transformation is embedded in function **decostand()** of the **vegan** package and a small example with a matrix of dimension 2 illustrates the effect achieved through this:

```
before<- matrix(c(3,1,1,0),nrow=2)
after<- decostand(before,method="chi.square")
before ;  round(after,digits=2)
```

```
     [,1] [,2]
[1,]    3    1
[2,]    1    0

     [,1] [,2]
[1,] 0.84 0.56
[2,] 1.12 0.00
...
```

Notably, proportions of scores change in a way not always expected by users. The first and largest element after transformation is only the second largest and initially identical scores with counts 1 differ afterwards.

Using the identical data set, a CA and a PCA ordination are computed and displayed in Figure 6.7 for comparison of methods. Superimposed is a classification of relevés. As in many other cases, it can be observed that the gradient displayed in CA is V-shaped, whereas in PCA it tends to be U-shaped. The order of the relevés along the main gradient is roughly the same and I therefore conclude that both methods reveal the underlying pattern. The R values in the graph stem from analysis of similarities (ANOSIM; Clarke 1993) and they express the power of the two ordination axes, respectively, in resolving the classification used for symbol plotting. As expected from visual inspection they are almost identical. ANOSIM in this case is not an ecological analysis but only used to evaluate the graphs and the R values would likely turn out differently when changing the classification or including three or more axes.

CA has some unexpected properties to be kept in mind when using it. First, it is more sensitive to outliers (see Section 12.3.1) than other ordination methods. Unlike in PCA and PCOA, the range of the coordinates increases with higher dimension (and smaller eigenvalue). It is good practice to restrict adjustment of scales of the ordination axes to the ones used for plotting.

There exist various packages for calculating CA in $\mathcal{R}$. In Figure 6.7 function `cca()` from package `vegan` is applied to `sveg` and this offers a plot method:

```
o.ca<- cca(sveg^0.5)
plot(o.ca)
```

This yields a simultaneous plot of releveś and species (see also Section 6.6). All results are stored in object `o.ca` from where the coordinates of the first two axes of releveś are isolated as follows:

```
x<- o.ca$CA$u[,1] ; y<- o.ca$CA$u[,2]
plot(x,y)
```

The scores of the species are found in `o.ca$CA$v` and the 62 non-zero eigenvalues in `o.ca$CA$eig`. The proportion of variance the first two eigenvalues account for is:

```
o.ca$CA$eig[1:2]/sum(o.ca$CA$eig)
```

```
      CA1       CA2
0.1938717 0.0784178
```

This is 27%, a rather high value for an ordinary vegetation data set.

6.5 Heuristic ordination

6.5.1 The horseshoe or arch effect

The fact that relevés originating from an environmental gradient form a curve-shaped ordination (Figure 2.3, for example) continues to cause confusion. In many papers and textbooks this was considered a deficiency of the ordination method used, and much emphasis was put on finding 'better' methods. However, the origin of the problem can easily be understood when inspecting original data. Biological parameters used as axes, such as species performance scores, usually correlate nonlinearly. A plot of performance against an environmental variable typically results in a bell-shaped (Gaussian) response curve. Using two species as axes in a two-dimensional graph yields the arch or horseshoe demonstrated in Figure 6.8(b). As a result, the extreme points of a gradient are located in close proximity. This happens because near the extremes of the gradient any two species involved may encounter unfavourable site conditions responding to this with low scores. Introducing a third variable may alter this, but still not in the desired manner [Figure 6.8(a) and (b)]: the configurations then form spirals in three dimensions. In fact, there is no remedy to avoid this phenomenon, perfectly understood and described by van Groenewoud (1965) a long time ago.

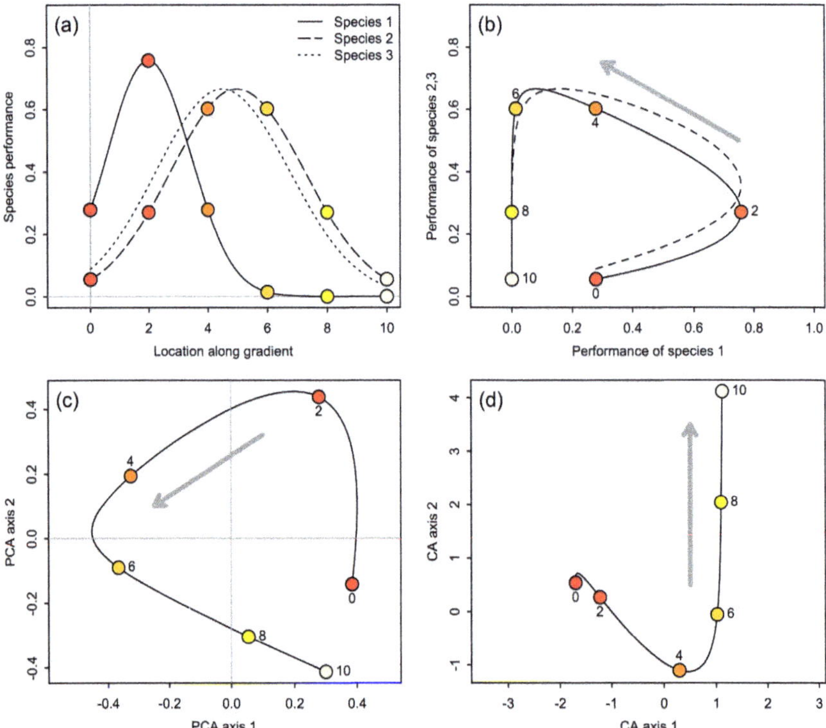

Figure 6.8: Origin of the arch effect: performance of three species along a hypothetical environmental gradient (a). A species-by-species plot delivers an arch-shaped projection (b). PCA finds the same arch, but rotated (c). Transformation inherent in CA delivers an entirely different response (d).

Plant ecologists widely agree that the horseshoe is caused by nonlinear correlation of species, but there is disagreement about the merit of it. Protagonists see the bent pattern as an expression of a natural process, whereas opponents are striving for the ultimate nonlinear ordination method. Unfortunately nonlinear models require an assumption about the nature of the underlying mathematical functions and this must be provided by the analyst [see, for example, the comment by Crawley (2005) in the context of nonlinear regression]. Although all methods described so far do reveal horseshoes when applied to vegetation gradients [Figure 6.8(c) and (d), for example], the shape and strength of the patterns vary.

The methods I explain and demonstrate below with examples are popular or at least widely used. They illustrate the difficulty of finding better ordination methods or improving on the existing. All deviate in various ways from linear methods, such as PCA. In the course of analysis the first instance to apply an alternative strategy is on the level of the resemblance matrix, trying to eliminate effects of nonlinearity (as in flexible shortest path adjustment, FSPA). The second instance is in the search of configuration of data points, where the eigenvalue procedure is replaced by an alternative approach not relying on linearity (nonmetric multidimensional scaling, NMDS). And finally it is also possible to just alter the final order of points according to a preconceived strategy (detrended correspondence analysis, DCA).

All methods presented in the following sections are 'adaptive' aimed at improving on already existing ordinations or ordination methods. This means that any resulting ordination should be compared with its related linear counterpart. For example, DCA should be compared with ordinary CA, NMDS with PCA, from which the initial configuration is usually taken, and FSPA with its underlying PCOA ordination. Furthermore, these methods are flexible by nature, offering options for altering the process by various means. In the following I am only presenting one or very few among an almost infinite number of solutions, and readers should be aware that testing robustness of the results is highly recommended when choosing among methods.

6.5.2 Flexible shortest path adjustment

FSPA is a nice example of the hidden origin of many methods in data analysis. It was used in a paper in *Ecology* by Bradfield and Kenkel (1987) citing Floyd (1962), but like many other straightforward ideas, it has since been reinvented for different applications. This can be seen in two papers appearing in *Science* (Roweis and Saul 2000; Tenenbaum *et al.* 2000), where further examples illustrate the idea, a rather straightforward one: ecological distances should not be measured directly in Euclidean space, but along the underlying gradient, that is, along the horseshoe. The first step in the analysis is to calculate an ordinary distance matrix. Because long distances are assumed to be underestimated in Euclidean space they are erased and replaced by the shortest sum of distances via intermediate data points, hence the name shortest path. In the new matrix the longest distances appear now increased, whereby the metric property of the resemblance space is lost (that is, the triangle inequality axiom is violated). Deriving an ordination from this matrix requires a method that can handle nonmetric space and this is either NMDS (see later) or PCOA used in the present example.

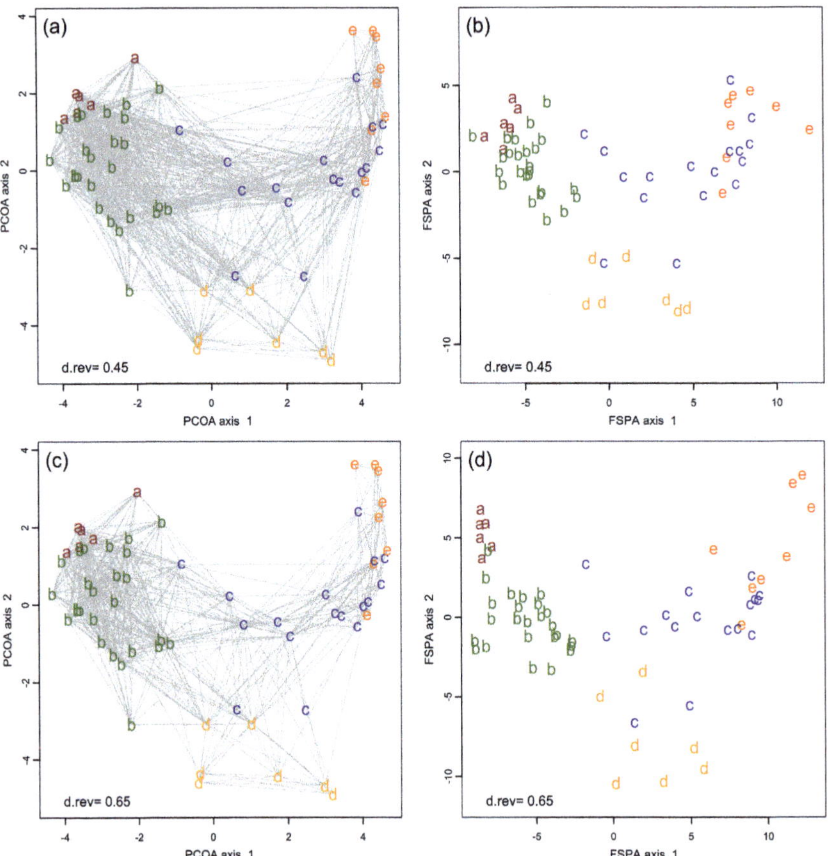

Figure 6.9: Comparing PCOA [graphs (a) and (c)] with two FSPA solutions [graphs (b) and (d)]. Groups of data points originate from minimum-variance cluster analysis. `d.rev` is the proportion of distances revised within FSPA, 45% and 65% respectively in the present examples. Grey lines in the left hand graphs are the distances FSPA uses for updating the distance matrix.

The way FSPA works is presented in Figure 6.9. Graphs (a) and (c) show the PCOA ordinations based on Euclidean distance. The grey lines represent approved distances after erasing all long connecting lines. Revised distances sometimes encompass various steps across intermediate data points. As many as 45% of the original distances are recalculated in panel (a) and 65% in panel (c) (parameter `d.rev`). The effect of FSPA is more pronounced the more distances are replaced. However, there is an upper limit beyond which the point cloud becomes disconnected resulting in a breakdown of the calculations. Figure 6.9 has been generated using $\mathcal{R}$ function `fspa()` in the `dave` package. To access an ordinary vegetation data matrix, `sveg` in this case, within the call of the function we type:

```
o.fspa<- fspa(sveg,method="euclidean",d.rev=0.45,n.groups=5)
plot(o.fspa)
```

The plot method generates the graphs (a) and (b) in Figure 6.9. The distance matrix used for Figure 6.11(a) is found in `o.fspa$dmat.before` and for Figure 6.11(b) in `o.fspa$dmat.after`. The groups formed (parameter `n.groups`) serves the interpretation of the results only and they are not part of the ordination method.

FSPA usually delivers nice ordinations, but the results change considerably when altering the number of distances to be revised. This is best seen when experimenting with parameter `d.rev` done in graphs (c) and (d). Generally the larger `d.rev` is chosen the larger is the proportion of nonmetric variation. The latter cannot be shown visually and it is lost for further ecological interpretation. Using $\mathcal{R}$ function `sum()` one can sum up the positive as well as the negative eigenvalues stored in `o.fspa$eig` as follows:

```
pos.eig<- sum(o.fspa$eig[o.fspa$eig > 0])
neg.eig<- sum(o.fspa$eig[o.fspa$eig < 0])
pos.eig ; neg.eig
```

```
[1] 6192.069
[1] -2028.359
```

To circumvent the use of eigenanalysis the modified distance matrix may also be subjected to NMDS (Section 6.5.3), typing:

```
o.mds<- isoMDS(o.fspa$dmat.after)
plot(o.mds$points)
```

The plots obtained reveal a pattern quite similar to those in Figure 6.9, graphs (b) and (d).

6.5.3 Nonmetric multidimensional scaling

NMDS dates back to the work of Gower (1966) and an advanced algorithm to solve the iterative procedure, now available in $\mathcal{R}$, is found in Ripley (1996). NMDS starts with an initial ordination, which can be random, but often a

PCA solution is taken. In subsequent steps data points in reduced ordination space (two when using defaults in functions used below) are shifted by small amounts and the distances in the new ordination compared with the same in the original distance matrix. This comparison yields the Shepard diagram (Shepard 1962), shown in Figure 6.10(d). The divergence of ordination and original distances is measured as stress, S^2, given as a percentage in Figure Figure 6.10(d). From this goodness of fit, R^2, is obtained as the one-complement:

$$R^2 = 1 - S^2 \tag{6.11}$$

In the present example the squared correlation is $1 - 0.1523 = 0.8477$, an almost perfect agreement. However, NMDS is not using this to evaluate goodness of fit, but deviation from the red line in Figure 6.10(d) is taken instead. This is monotonous, but allowed to be nonlinear as expressed by the name of the method.

Comparison with PCOA (and also PCA, because Euclidean distance is used) and the outcome of $\mathcal{R}$ function `isoMDS()` (the original contained in the `MASS` package) as well as function `metaMDS()` (a more elaborated version found in package `vegan`, also relying on `isoMDS()`) is shown in Figure 6.10 using 'Schlaenggli' data (data frame `sveg`). The similarity between PCOA and the two NMDS solutions is striking. The algorithm implemented in `isoMDS()` may end up in a local optimum. To avoid this, `metaMDS()` makes several attempts and selects the solution with the best fit. Both functions use random numbers and whenever repeating the analysis the results will differ. When comparison of results is intended one should therefore fix the seed in $\mathcal{R}$ using function `set.seed()`, typing `set.seed(57)`, for example.

Both of the two $\mathcal{R}$ functions mentioned above accept a distance matrix as input, the Euclidean `mde` in the following calls:

```
mde <- vegdist(sveg,method="euclidean")
set.seed(57)
o.imds<- isoMDS(mde,k=2)
set.seed(57)
o.mmds<- metaMDS(mde,k=2)
```

To generate Figure 6.10(b) and (c), respectively (without group symbols), plotting is done like this:

```
plot(o.imds$points)
plot(o.mmds$points)
```

Without setting the seed it is unlikely that the ordinations are identical to those in Figure 6.10. The Shepard diagram too will not be exactly the same:

```
stressplot(o.imds,mde)
```

Function `stressplot()` is located in `vegan` and similar functionalities are available in package `MASS`.

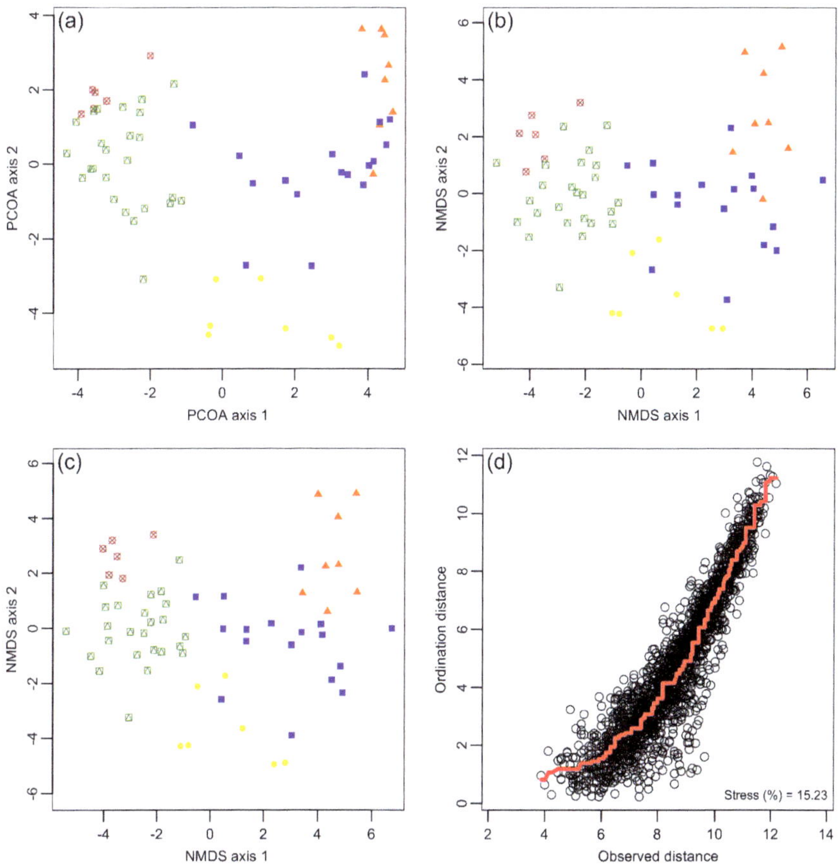

Figure 6.10: Comparison of PCOA and NMDS. (a) PCOA (Euclidean distance used). (b) NMDS generated by isoMDS(). (c) NMDS generated by metaMDS(). (d) Shepard diagram based on isoMDS().

6.5.4 Detrended correspondence analysis

The third heuristic method is DCA, devised by Hill and Gauch (1980) with the corresponding computer program DECORANA (Hill 1979a). It became popular because it implements corrections for properties of CA, the horseshoe effect and the fact that data points near the extremes of the axes may get squeezed. DCA in its original mode splits the axes into segments and shifts the points within these to become centred. This is shown graphically in the book of Legendre and Legendre (2012), p. 482 ff, where it is also criticized. For example, Wartenberg *et al.* (1987) argue that the horseshoe effect is an inherent property of the distance pattern of gradients and not a mathematical artefact, just as discussed in Section 6.5. DCA has recently been implemented in various program packages and von Wehrden *et al.* (2009) in their literature overview mention that results vary depending on the computer program used.

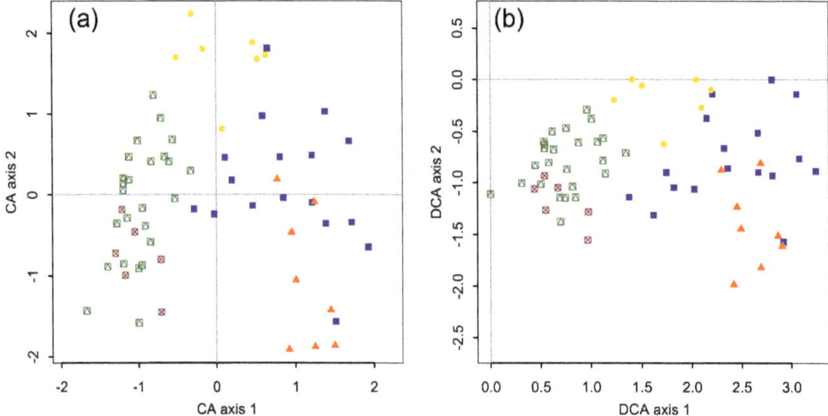

Figure 6.11: Comparison of CA (a) and DCA (b). For (b) shifting by 10 segments is used.

A comparison of CA and DCA is shown in Figure 6.11 confirming the striking effect of detrending: the horseshoe-shaped point cloud is coerced to become almost straight. Taking into account that the outset of the ordination in Figure 6.11(b) is the one in Figure 6.11(a) raises the question of justification of the process, its control and interpretation.

Whereas the call of CA is shown in the explanations to Figure 6.7 the same for DCA by function `decorana()` in the **vegan** package is slightly different:

```
o.dca<- decorana(sveg,mk=10)
plot(o.dca$rproj,asp=1)
```

The default value for the number of segments, `mk`, is 26. Function `decorana()` apparently uses an improved method for smoothing the detrending process and the result varies only little when changing the number of segments.

6.6 How to interpret ordinations

Which ordination method should be preferred is a permanent issue among plant ecologists. As Podani and Miklós (2002) show, the mechanisms behind ordination are fairly well understood, but tradition and taste play a crucial role in the choice of method too (von Wehrden *et al.* 2009). Clearly, ordination in reduced resemblance space is not suited for ecological model testing because there is no strict control over variation included in what is displayed.

In this section the focus is on interpretation only. From Figure 6.8 we conclude, for example, that the appearance of a horseshoe likely suggests the existence of a vegetation gradient. Other properties of ordination methods have to be taken into account as well. There is an agreement that CA, for example, reflects the order of a gradient along the first axis whereas in PCA inflections occur (Figure 6.7), a fact graphically illustrated in Legendre and Legendre (2012) and considered in practical context in Section 7.6.

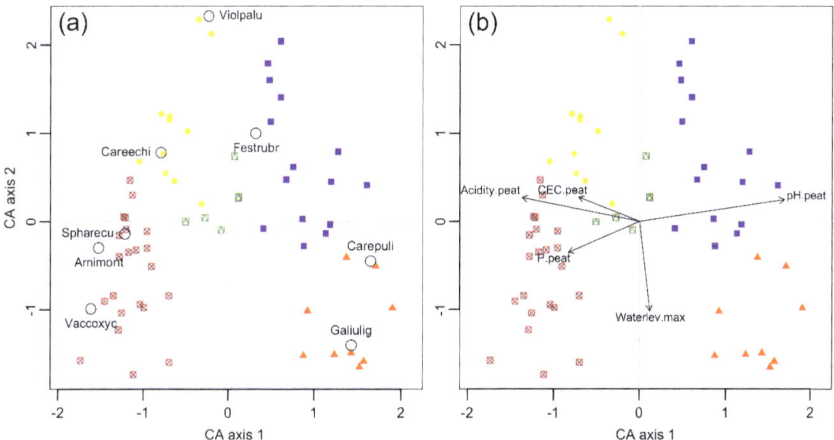

Figure 6.12: Interpretations of CA by scores of selected species (a) and by correlating environmental factors (b).

Interpretation of ordination is very much a visual issue and the emphasis concentrates on generating an easy to interpret graph of high dimensional data. A first way is using biplots of relevés and species simultaneously. CA, for example, delivers properly scaled coordinates pointing to the centre of occurrence of species (Section 6.4). Hence, one can add species scores to the ordination in Figure 6.7(a). As shown in this example relevé scores are located in `o.ca$CA$v` of the output list of CA and for species they are in `o.ca$CA$u`. Plotting all data points of the 119 species present in this data matrix including the names would result in an extremely crowded graph. We therefore use a restricted selection of species considered sufficient for interpretation, as in Figure 6.3, storing the column numbers of these:

```
s.sp<- c(3,11,23,31,39,46,72,77,96)
snames<- colnames(sveg[,s.sp])
```

The names of the selected species are now in vector `snames`, because function `names()` extracts these from the column names of `sveg`. First we compute CA and plot the ordination as shown in Section 6.4:

```
o.ca<- cca(sveg^0.5)
plot(o.ca$CA$u,asp=1,pch=16)
```

We now add data points and names of selected species, without closing the plot window:

```
sx<- o.ca$CA$v[s.sp,1]
sy<- o.ca$CA$v[s.sp,2]
points(sx,sy,pch=1,cex=1.2)
snames <- make.cepnames(snames)
text(sx,sy,snames,pos=c(1,2,2,1,3,2,4,3,1),cex=0.8)
```

This generates Figure 6.12(a) with (black) circles for species by setting `pch=16`. Reducing over-plotting of symbols and text requires a good deal of editing in $\mathcal{R}$. Function `make.cepnames()` in package `vegan` generates abbreviations of the species names to further save space. Reducing font size of species names in `text()` to `cex=0.8` or even less, supports the effort. A more tedious, but helpful task is to place species names individually either below the symbol (`pos=1`), to the left (2), on top (3) or to the right (4).

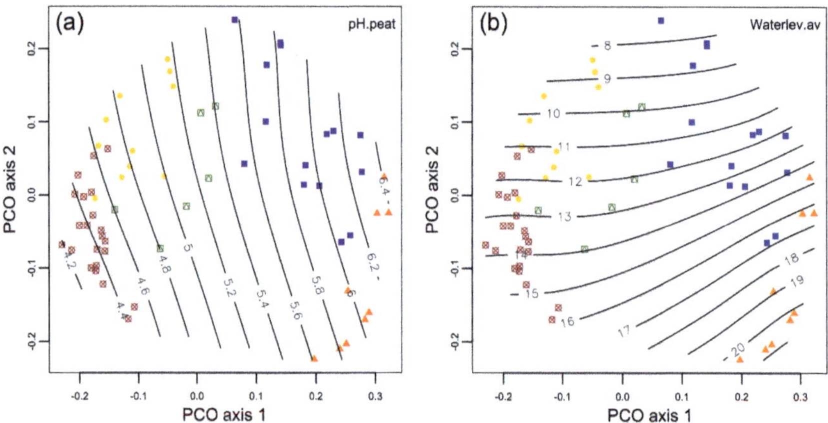

Figure 6.13: Surface fitting to interpret a PCOA ordination. (a) pH (measured in peat). (b) Average water level (measured in vertical tubes).

Just as when correlating species in PCA (Figure 6.3) the same can be done with site factors using function `envfit()`. This requires access to the data set of environmental measurements, `ssit` in this case. There is also a selection made in `sit5`, this time using the explicit names of the site factors chosen [the same could have been done for Figure 6.12(a), typing the species names instead of their corresponding column number]:

```
sit5<- c("pH.peat","Acidity.peat","CEC.peat","P.peat","Waterlev.max")
```

```
o.ev<- envfit(o.ca,ssit[,sit5])
plot(o.ca$CA$u,asp=1)
plot(o.ev,add=TRUE)
```

The first argument of `envfit()` is the output list of function `cca()` and the second is the data matrix of site factors, reduced to the five chosen. Plotting (`plot(o.ev)`) completes the ordination in Figure 6.12(b). Again, adding arrows and text requires the plot window with the CA ordination to be kept open. Just as in Figure 6.12(a), font size may be reduced and the names of the site factors positioned individually. Arrows now point into the direction of highest values of site factors and arrow length is proportional to correlation. Significance of the latter can be inspected when typing `o.ev`.

The above method is restricted to linear correlations. To see the relation between ordination and a site factor one could annotate the data points with measurements. A better method is to fit a nonlinear surface to the two-dimensional ordination. There are various methods for doing this (see, for example, Venables and Ripley 2010), some of these inherited from spatial statistics. One of the more suitable for the purpose is function `ordisurf()` found in the `vegan` package. According to the descriptions it uses generalized additive models (GAMs, Section 9.4) to fit a smooth surface. In Figure 6.13 pH measured in peat samples and average water level are fitted to a PCOA ordination. `pH.peat`, Figure 6.13(a), results in an almost linear solution whereas `Waterlev.av` in Figure 6.13(b) is slightly nonlinear and almost perpendicular to pH.

Function `ordisurf()` adds trend surfaces to ordination graphs. Figure 6.13 uses PCOA with correlation used as distance in distance matrix `vdm`:

```
vdm<- as.dist((1 - cor(t(sveg)))/2)
o.pco<- pco(vdm)
plot(o.pco$points)
```

As usual site factors are taken from a separate object (`ssit`) and `ordisurf()` requires the ordination output, `o.pco`, and one site factor, `pH.peat` for example, for adding the trend surface:

```
ordisurf(o.pco,ssit$pH.peat,add=TRUE)
```

The two surfaces in Figure 6.13(a) and (b) are superimposed if `ordisurf()` is called twice without plotting a new graph in the mean time. Alternatively there is a similar function in package `labdsv` called `surf()` which does almost the same, but sometimes a little more flexibly than `ordisurf()`. It is easy to check for the difference as both functions use the same parameters.

6.7 Ranking by orthogonal components

6.7.1 RANK method

Ranking by orthogonal components is not an ordination method, but it illustrates how variation in data sets can be partitioned and redundancy erased. This section starts by introducing the RANK algorithm, that is, decomposition of a correlation or covariance matrix by orthogonal components. In contrast to PCA, the axes excerpted are not complex linear combinations, but selected variables of the raw data sets. As we shall see below, the purpose of this is not only a didactic one but there are various practical applications.

The ranking procedure proposed here is striving for independent components within matrices of sum of squares, covariances or correlations (Orlóci 1973, 1978) and in this sense it is closely related to PCA. The variables chosen should explain as much of the total variance as possible and should be independent (that is, not correlated) as far as possible. The fact that the variables are chosen from among the original attributes (or sampling units) and not generated by linear combination makes ranking somewhat less efficient than PCA, but also easier to interpret. The description presented below is for ranking attributes (Wildi and Orlóci 1996):

1. The data, X, is centred within the attributes in order to obtain a new matrix, A, with elements:

$$A_{hj} = \frac{X_{hj} - \overline{X}_h}{Q_h} \tag{6.12}$$

   where $\overline{X}_h$ and Q_h are the mean and a factor of adjustment, respectively, in attribute h. X_{hj} is the value of attribute h in relevé j. For Q_h I refer to Table 4.3, where it can be seen that this choice affects the type of the similarity coefficient, S.

2. Cross-products, $S = AA'$, are computed. A characteristic element is given by:

$$S_{hi} = \sum_{j=1}^{n} A_{hj} A_{ij} \tag{6.13}$$

   where n is the number of relevés. If the attributes are standardized, then S is a correlation matrix.

3. Dispersions and highest values are calculated:

$$SS = max\Big(\sum_{h=1}^{p} \frac{S_{hi}^2}{S_{hh}}\Big) \qquad h = 1, \ldots, p \tag{6.14}$$

where p is the number of attributes. The quantity, SS, is a measure of redundancy in the sample of p attributes with respect to attribute m. Rank 1 is declared for attribute m associated with SS.

4. Residuals are computed:

$$S_{hi} := S_{hi} - Y_{hm}Y_{im} \text{ for any } h, i = 1, \ldots, p \tag{6.15}$$

in which:

$$Y_{hm} = \frac{S_{hm}}{\sqrt{S_{mm}}} \text{ and } Y_{im} = \frac{S_{im}}{\sqrt{S_{mm}}} \tag{6.16}$$

5. Computation of a new value for SS from the elements of the residual, S, and declaration of rank 2 for the corresponding attributes. Then repeat steps 3 and 4 as many times as necessary until all attributes are ranked.

An interactive version of this has been proposed in Wildi (1984), concerning step 3: instead of selecting the variable with the highest value for SS, the choice is left to the user. This can reduce the efficiency of the procedure but it allows attributes to be omitted that do not seem feasible for the application in mind, such as species that are difficult to identify.

The RANK algorithm is most efficient when applied to data sets with a very high number of attributes. This can be seen from the following example, demonstrating the mechanism involved and the interpretation. The data matrix is shown in Table 6.1, where the results are also summarized.

Table 6.1: Data set and results illustrating the RANK algorithm.

Relevé	1	2	3	4	Rank no.	Explained variance (%)
Species 1	2	2	1		4	0.0
Species 2	2	1	1		2	17.0
Species 3			1	1	1	78.5
Species 4			2	1	3	4.5

The procedure starts by calculating the resemblance matrix of species, R. In the example this implies standardization of the vectors (Equation 6.12) and the computation of the cross-product (Equation 6.13). We get a correlation matrix:

$$\mathbf{R} = \begin{pmatrix} 1.0 & 0.85 & -0.91 & -0.64 \\ 0.85 & 1.0 & -0.71 & -0.34 \\ -0.91 & -0.71 & 1.0 & 0.91 \\ -0.64 & -0.43 & 0.91 & 1.0 \end{pmatrix}$$

As can be expected from the data, species 1 and 2 as well as 3 and 4 are highly correlated ($r = 0.85$ and 0.91, respectively). The dispersions (correlations) explained by the individual attributes are found in the respective rows or columns of the correlation matrix R (Equation 6.14). The correlation coefficient simplifies matters as S_{hh}, the diagonal elements, are equal to 1. The variances the attributes account for are:

$$
\begin{aligned}
SS_1 &= \tfrac{1}{1}\left[(1.0)^2 + (0.85)^2 + (-0.91)^2 + (-0.64)^2\right] = 2.95\\
SS_2 &= \tfrac{1}{1}\left[(0.85)^2 + (1.0)^2 + (-0.71)^2 + (-0.43)^2\right] = 2.41\\
SS_3 &= \tfrac{1}{1}\left[(-0.91)^2 + (-0.71)^2 + (1.0)^2 + (0.91)^2\right] = 3.14\\
SS_4 &= \tfrac{1}{1}\left[(-0.64)^2 + (-0.43)^2 + (0.91)^2 + (1.0)^2\right] = 2.41
\end{aligned}
$$

Species 3 has the highest explanation power and will get rank no. 1. It is important to note that the other species achieve high values as well. Taking species 1 instead of species 3, for example, would reduce the efficiency only moderately. This situation is typical for high-dimensional vegetation data.

The correlation matrix is now reduced by the fraction of variance explained by species 3 according to Equation 6.15. For the first column this is:

$$
\begin{aligned}
r'_{11} &= 1.0 - (-0.91 * -0.91) &= 0.17\\
r'_{12} &= 0.85 - (-0.71 * -0.91) &= 0.20\\
r'_{13} &= -0.91 - (1.0 * -0.91) &= 0.0\\
r'_{14} &= -0.64 - (-0.91 * 0.91) &= 0.19
\end{aligned}
\tag{6.17}
$$

In the new, reduced matrix, R' the rows and columns related to species 3 are now all zero:

$$
\mathbf{R}' = \begin{pmatrix} 0.17 & 0.20 & 0.0 & 0.19\\ 0.20 & 0.50 & 0.0 & 0.21\\ 0.0 & 0.0 & 0.0 & 0.0\\ 0.19 & 0.21 & 0.0 & 0.17 \end{pmatrix}
$$

The procedure according to Equation 6.14 is now applied to matrix R'. As can be seen in Table 6.1 the variance explained decreases rapidly, underpinning the efficiency of the method. It even turns out that species number 1 no longer contributes to the total variance, indicating that the dimension of the total resemblance matrix is equal to 3 only.

The above results are reproduced in $\mathcal{R}$ by function `orank()` in package `dave`. We first create the data frame of raw data, `r.test`, as in Table 6.1 where this is presented in traditional organization with species as rows and releveś as columns. For standard use in $\mathcal{R}$ we enter the elements row by row to get the arrangement shown in Section 2.3.2:

```
r.test<- matrix(c(2,2,1,0,2,1,1,0,0,0,1,1,0,0,2,1),nrow=4)
rownames(r.test)<- c("r1","r2","r3","r4")
colnames(r.test)<- c("s1","s2","s3","s4")
r.test<- as.data.frame(r.test)
r.test
```

```
   s1 s2 s3 s4
r1  2  2  0  0
r2  2  1  0  0
r3  1  1  1  2
r4  0  0  1  1
```

The first parameter in function `orank()` is the data frame of raw data, the second causes the species to be ranked (rather than the relevés), the third limits the number of ranks to be computed to 3. $y = 1$ means that no further transformation is applied to the scores. Provision is made for entering an x- and y-axis used for plotting the results:

```
x.axis=NULL ; y.axis=NULL
o.orank<- orank(r.test,use="columns",rlimit=3,y=1,x.axis,y.axis)
```

The result is written to the $\mathcal{R}$ console (or given in more detail in `o.orank`):

```
Call:
orank.default(veg = r.test, use = "columns", rlimit = 3, y = 1,
    x.axis = x.axis, y.axis = y.axis)

RANK
Variable              rank no.    var.       var.%   cum. var.%
s3                       1        3.14      78.409       78.409
s2                       2        0.68      17.045       95.455
s1                       3        0.18       4.545      100.000
```

Unlike in Table 6.1 the last rank is assigned to species 1 rather than species 4. What happened? The SS value of species 1 and 4 was the same when computing the last rank. Function `orank()` for simplicity's sake takes the first possible choice leaving no more variance to species 4. Hence, the two solutions are equivalent.

The list `o.orank` can be used to reduce a vegetation data frame by retaining the highest ranked species only:

```
r.test.red<- r.test[,o.orank$var.names]
dim(r.test.red)
```

```
[1] 4 3
```

Because the first index of `r.test` is omitted, all rows are taken, but only the three columns are included, as specified in `o.orank$var.names`, resulting in reduced dimension of the new data frame `r.test.red`.

6.7.2 A sampling design based on RANK (example)

RANK proved to be useful as well in selecting typical plots for permanent observation. As plots chosen accordingly account for maximum covariation in the data, they are considered ‘typical’. In order to design an efficient plan for permanent plot research in a specific area (Wildi 1990) the following steps are taken:

1. Complete an initial investigation of the area. The sampling intensity should be sufficiently high to give an accurate account of vegetation types and gradients.

2. Select a reduced number of (typical) plots using RANK. Survey and analyse these periodically.

3. As soon as a considerable trend occurs in the selected plots, re-survey the entire sample.

The example below aims to demonstrate the efficiency of the method. The criterion for measuring efficiency is the proportion of variance explained by the plots chosen for permanent survey. The ‘Schlaenggli’ data set (data frame `sveg`) is a typical example of a moderate-sized sample.

The result of ranking is shown in Table 6.2, derived from a correlation matrix of 63 sampling units. There are only 5 out of the 63 needed to account for 50% of the total variation. Increasing the sub-sample to 10, as shown in Table 6.2, increases the explained variance to 62%. Obviously, the contribution of further plots does not add much to the representativeness of the survey: 24 plots are needed for 80% of the variance explained and all 63 for 100%.

Ranking shown in Table 6.2 is also obtained from function `orank()`, but this time setting parameter `use="rows"`:

```
x.axis=NULL ; y.axis=NULL
o.orank<- orank(sveg,use="rows",rlimit=24,y=0.25,x.axis,y.axis)
```

By setting `y=0.25` species scores are scaled close to presence–absence, almost neglecting cover scores.

Output list `o.orank` now offers the chance to reduce the vegetation data frame to the relevés accounting for a maximum of explained variance in a given sample. First, we reduce this to the rows selected by the RANK algorithm:

```
sveg.red<- sveg[o.orank$var.names,]
```

There is a risk that some of the species vectors are empty. We therefore apply the recipe explained in Section 7.6.3 to eventually remove these:

```
f.s<-apply(sign(sveg.red),1,sum)
sveg.red<- sveg.red[,f.s > 0]
```

In the present case, none of the species vectors is empty and the dimensions of the data frame `sveg.red` is still `dim(sveg.red)` equal to 24 and 119.

Table 6.2: Ranking relevés of the 'Schlaenggli' data set.

Rank no.	Plot no.	Explained variance (%)	Cumulative variance (%)	pH peat
1	520	26.47	26.47	4.6
2	560	10.57	37.04	6.2
3	527	5.51	42.54	4.6
4	546	4.18	46.72	5.5
5	553	4.06	50.78	6.2
...	...	...	...	...
10	557	1.85	62.82	6.3
...	...	...	...	...
24	523	0.95	80.02	4.7

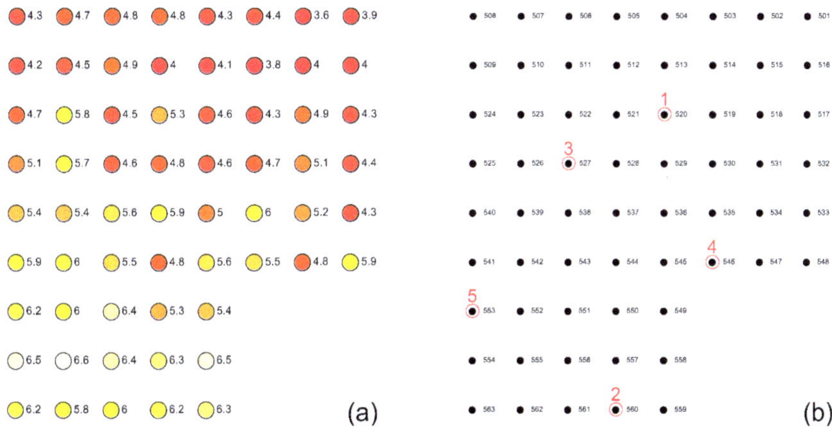

Figure 6.14: Relevés chosen by RANK for permanent investigation. (a) pH measured in peat. (b) Plot numbers and first five ranks of plots.

As the plots ranked are as independent as possible, RANK takes these from contrasting locations within the investigation area. In Figure 6.14(b) the first five plots listed in Table 6.2 are shown within the sampling plan. The fact that they represent contrasting composition directly translates to spatial location: they are dispersed across the sampling area. Due to spatial

Table 6.3: Ranking species of the 'Schlaenggli' data set.

Rank no.	Species	Explained variance (%)	Cumulative variance (%)
1	*Carex pulicaris*	15.38	15.38
2	*Vaccinium oxycoccos*	5.08	20.46
3	*Drosera rotundifolia*	4.98	25.44
4	*Aster bellidiastrum*	4.27	29.71
5	*Dactylorhiza fistulosa*	2.83	32.54
6	*Rhytidiadelphus squarrosus*	2.74	35.27
7	*Cirsium oleraceum*	2.64	37.92
8	*Rhinanthus minor*	2.55	40.46
9	*Melampyrum pratense*	2.49	42.95
10	*Hylocomium splendens*	2.31	45.25
...	...	...	...
20	*Sphagnum subsecundum*	1.58	64.66
...	...	...	...
42	*Acer pseudoplatanus*	0.79	90.17

autocorrelation (Section 7.3.3) it is rather unlikely that neighbouring (and hence similar) plots occur within the first few ranks. The independence of the sampling units can also be observed in the pH values, Figure 6.14(a). The orthogonality of ranks is reflected in large pH steps, from 4.6 up to 6.2 and down again to 4.6 (but probably with other factors causing a difference to rank 1). I emphasize that the analysis is entirely based on species composition, and pH just shows the effect in terms of measured site conditions.

Ranking of plots as in Figure 6.14(b) is possible in any vegetation survey where spatial coordinates are known. In the present example the latter are taken from `ssit` and assigned to parameters `x.axis` and `y.axis`, respectively:

```
x.axis=ssit$x.axis ; y.axis=ssit$y.axis
o.orank<- orank(sveg,use="rows",rlimit=5,y=0.25,x.axis,y.axis)
plot(o.orank)
```

In this `rlimit` is again the number of ranks now restricted to 5 and `y=0.25` transforms the species scores. The specification of axes addresses their source, `ssit`. The plot method also works when no axes are specified as in the previous examples. In this case ranks are displayed within a two-dimensional PCOA ordination of vegetation data.

The same algorithm can also be used to select a meaningful set of indicator species. Generally, species correlations are lower than relevé correlations (Section 3.2). As in the 'Schlaenggli' data set, it frequently happens that the number of species exceeds the number of relevés. The species ranked first will therefore account for less variance than the relevé ranked first. The result of this analysis is listed in Table 6.3. The 10 first ranks account for about 45% of total variance; 42 out of 119 are needed for 90%; 100% is reached with 63 species. This is the dimensionality of the similarity space as given by the number of relevés.

Table 6.3 is obtained in the same way as Table 6.2, only the orientation of the data matrix is changed by setting `use="columns"`.

6.8 Which ordination method?

The user of ordination profits from the fairly large number of techniques offering much flexibility in the choice of the method. On the downside, a consistent theoretical background for a generally valid evaluation of methods is still missing. Austin (2013a) in his review paper admits that 'ordination is a powerful technique for summarizing complex multivariate ecological data, yet confusion about appropriate methods is still apparent'. In many of the papers comparing methods, simulated data with known properties (unimodal species response, for example) are used, like in the paper of Swan (1970) who is frequently cited for describing the horseshoe effect for the first time (Section 6.5) or Minchin (1987) working with two-dimensional distribution functions. In the present section I decided to use a real-world data set to investigate the variability of ordination results, specifically when varying the transformation of species performance scores, the use of different resemblance functions and the use of the ordination technique in its strict sense. This idea goes back to the suggestions of Noy-Meir and Whittaker (1977): 'Comparison of ordination techniques have tended to confound three factors: the methodological algorithm, the resemblance measure employed and the standardization used.' Below, these three factors are varied and applied to a data set for illustrational purpose.

Using the 'Schlaenggli' vegetation data (object `sveg`) the effect of transforming cover-abundance scores as explained in Section 3.4 is investigated in first place. Untransformed vegetation data `sveg` consists of a rank scale (steps 0 to 6). When applying transformation $x = x^y$, any $y > 1.0$ tends to approximate cover percentage, whereas $y < 1.0$ changes scores in the direction of presence–absence (Table 3.4). In second place the resemblance measure is varied where feasible (Chapter 4). This is most easily done in principal coordinates analysis (PCOA, Section 6.3) accepting any distance function. Principal components analysis (PCA, Section 6.2), on the other hand, allows the scalar product, covariance or correlation only to be used. Correlation, as shown in the sixth row of Table 4.3, implicitly standardizes the data vectors. An opposite case is correspondence analysis (CA, Section 6.4) where similarity used is part of the method and no variation is possible in this regard. In third place only the ordination technique is varied to hopefully capture the most relevant cases.

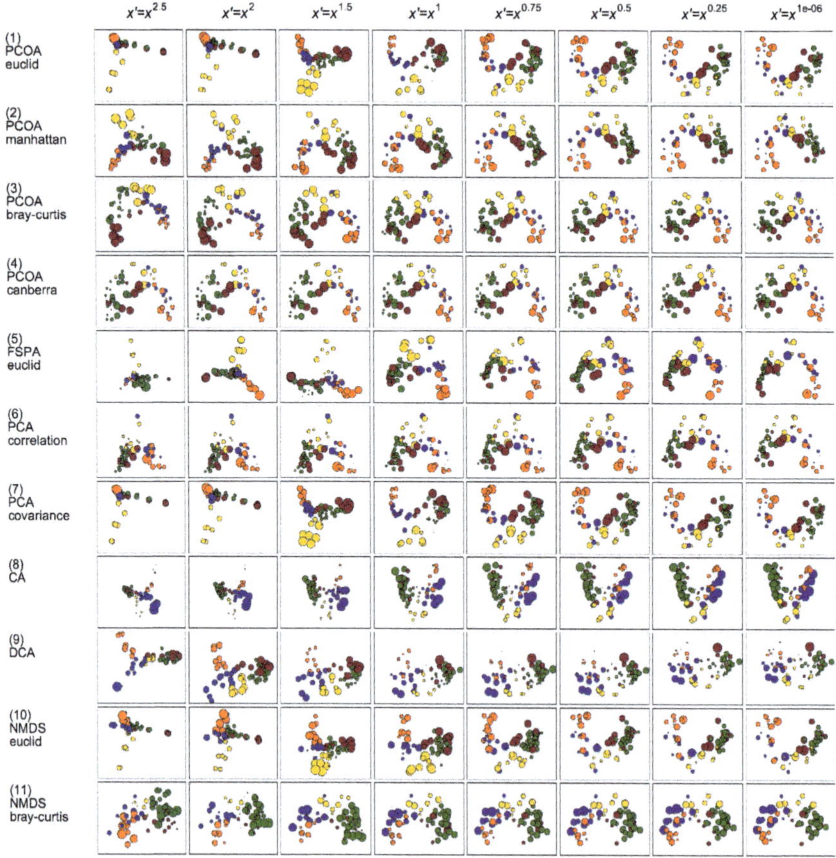

Figure 6.15: Ordinations of the sveg vegetation data set. Transformation of the cover-abundance scale, the choice of the resemblance measure and ordination technique are varied.

To understand what happens when transforming abundance scores it is essential to recognize their extremely skewed frequency distribution in most vegetation samples, including `sveg`:

```
table(as.vector(as.matrix(sveg)))
```

Function `table()` displays all occurring scores in a first, the corresponding frequencies in a second row:

```
   0    1   2   3   4   5   6
5212 1939 263  44  14  16   9
```

This pattern with dominating zeros, a high number of low scores and very few high scores, is typical for almost all vegetation data sets. As explained in Section 6.5 it is the main culprit for causing the horseshoe-shaped pattern of the majority of ordinations. Keeping this in mind the results given in Figure 6.15 are useful for answering various questions:

- *Is scalar transformation reflected in the ordination pattern?* The answer varies with the method used. In row one where PCOA is applied to Euclidean distance, the effect is strong. On the left side, high cover-abundance values are squared and therefore get high weight in the ordination. On the right side the scores approach presence–absence and all non-zero entries equally affect the final shape of the scatter plot. The same applies to the second row with PCOA applied to a matrix of Manhattan distances. In rows three and four, however, we observe that transformation of species scores causes minor changes in ordinations. Obviously, Bray–Curtis distance as well as Canberra distance level out inequalities of scores as can be inferred from Equations 4.4 and 4.5. Exactly the same can be observed in nonmetric multidimensional scaling (NMDS, Section 6.5.3), the two last rows in Figure 6.15. When using Euclidean distance, transformation is strongly affecting the ordination; in the case of Bray–Curtis transformation it is almost invisible.
- *What is the role of the resemblance measure?* Similar to the effects of Bray–Curtis and Canberra distances we see the effect of standardization involved in the correlation coefficient (standard in PCA) in row six when compared to row seven. Standardization mitigates the effect of data transformation considerably, whereas PCA based on the scalar product is identical to PCOA using Euclidean distance (Section 6.3).
- *What is the role of the ordination technique?* As seen in rows eight and nine CA and DCA also respond to transformation of scores. This cannot be further varied as the use of the resemblance measure is part of the ordination technique. In conclusion it is evident that mentioning the ordination technique only in publications is insufficient, especially when addressing the more flexible methods, that is, PCOA, PCA and NMDS.

  An example of a heuristic method is FSPA, row 5. The initial configurations used for these ordinations are the same as in row 1 of which 35% of the distances are subsequently revised. Whereas cover percentage delivers erratic patterns, presence–absence scores end up in a classical horseshoe. Heuristic too is DCA, row 9. This is the only method really suppressing the horseshoe effect and the final result is unique when compared to all others, including CA.

There are some further observations contradicting current paradigms. One concerns the use of Bray–Curtis distance. This is considered the best choice by many (e.g. Minchin 1987; Austin 2013a). The often observed stability resides on the fact that results remain almost unchanged irrespective of the scale used for measuring cover-abundance. But it also means that an inappropriate transformation still delivers a pleasing result, because scores are internally transformed close to presence–absence.

Conversely, the often-mentioned inappropriateness of PCA has its origin in the fact that this method is very flexible and inadequate transformations generate unfavourable results. Furthermore, using the correlation coefficient in PCA is default in many computer programs and alternatives offering improved flexibility are sometimes missing.

Also obvious from Figure 6.15 is the fact that PCOA and NMDS yield very similar results. And in fact some computer programs use PCOA as standard initial configuration for NMDS. The gains, if visible, can therefore be minor and they come at the expense of reduced transparency of the technique.

7 Ecological patterns

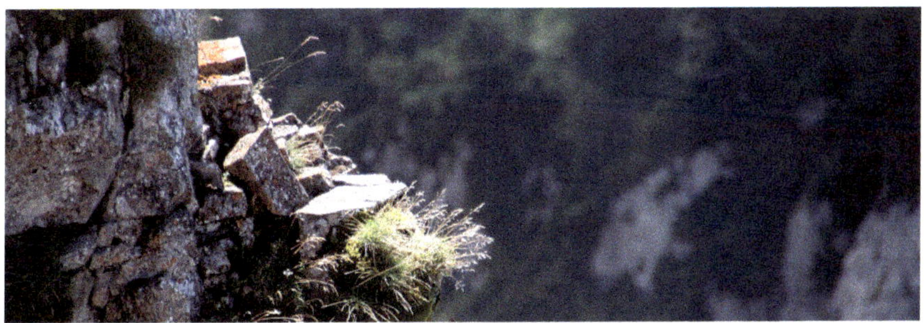

The similarity of patterns found in biological, environmental and geographical space is a hallmark of ecological interactions. Recognition of joint patterns requires vegetation data, ecological measurements and spatial coordinates to be amalgamated within the framework of analysis.

7.1 Pattern and ecological response

The identification of patterns is a first step in finding rules governing systems. Patterns can be found in biological space, but also in environmental, spatial and temporal space (Section 2.3.2). As a next logical step in the investigation of ecosystems, spaces are compared in search for commonalities. Clarke (1993), introducing this strategy to analysis of benthic communities, puts it as follows: 'Having allowed the community data to "tell its own story", its relationship to matching environmental data is examined by superimposing the values of each abiotic variable separately onto the biotic ordination.' Hence, when talking about 'ecological patterns' in this chapter I address various methods of comparison, such as superimposing, correlating, variance partitioning and so on.

There is a specific class of methods, not yet addressed in this chapter, where one vector expressing performance of a species or a vegetation type is correlated with environmental variables, such as precipitation or temperature. Finding correlations of this kind is the ultimate goal in predictive modelling (cf. Guisan and Zimmermann 2000) and Chapter 9 is devoted to this due to the broad practical relevance, for example, when predicting environmental change.

In contrast to predictive modelling where maximizing model fit is the prevailing issue, striving for joint patterns is understood here as a step towards an understanding of casual mechanisms. For example, plant species interact and the change in their performance may be caused by competition or facilitation from another species rather than by a site factor directly. The same principle holds among environmental factors: water uptake of plants, for example, depends on temperature, and an ecologically sound interpretation requires a simultaneous view of these two factors. Hence, the focus is on the analysis of relationships within and between spaces – the biological, the ecological and the physical, whilst Chapter 10 is devoted specifically to temporal change. A typical group of methods doing this is constrained ordination, where variance is partitioned and that shared by two spaces is used for joint ordination, of vegetation and environmental data, for example (Section 7.4). Another example for the comparison of two multivariate spaces is the analysis of contingency, where data from different spaces are classified and the joint frequencies are evaluated in contingency tables (Section 7.2.4). These are analysed, the agreement quantified, tested and interpreted.

What is the implication of common properties in patterns? There are good reasons to hypothesize that response exists in either direction. Typically, vegetation responds to environmental conditions, which play the role of *independent* – and vegetation the *dependent* – variables, the view adopted in predictive modelling. In the ecological reality, the opposite can happen too. In a ruderal environment species composition can be a result of the seed bank and not so much of present site conditions. As time passes, it is expected that a correlation between species composition and site factors will emerge, a typical case of convergence.

A frequently observed phenomenon is *spatial autocorrelation*. This is the case when any two plots located in close neighbourhood are more similar than could be expected from the site conditions. Thus, autocorrelation is created by processes acting across small distances. In vegetation data spatial autocorrelation is often a result of plant dispersal and for this reason it is of ecological significance. As will be seen in the following, there are methods to identify autocorrelation, whilst many others are just hampered by this.

7.2 Evaluating groups

In this section I present the basic idea of analysis of variance as a tool for measuring strength of group pattern in an ecological context. Analysis of variance can serve two different purposes, the first being statistical where it is a means of hypothesis testing (by statistical inference) as described in Section 7.2.1. The same example also demonstrates how it is used for evaluating the performance of site factors in explaining group pattern of vegetation. Alternatively, as in Section 7.2.2, the analysis can be a purely descriptive tool for quantitatively measuring the resolving power of species in classified relevés. This is species ranking based on variance or indicator value. The result cannot be used for statistical testing, because classification is derived from the same body of data. But it addresses the idea of indicator species, a more practical point of view.

7.2.1 Variance testing

Forming groups of vegetation relevés is not trivial as many different solutions exist depending on the method chosen (Section 5.5). Analysis of variance offers a means to test whether a group structure found in vegetation can be confirmed by an external continuous variable, a site factor in this case. Data for the example are given in Table 7.1. In this the three groups result from clustering vegetation (the methods described in Chapter 5 and in Section 7.6). pH, altitude and slope are the site factors available for testing.

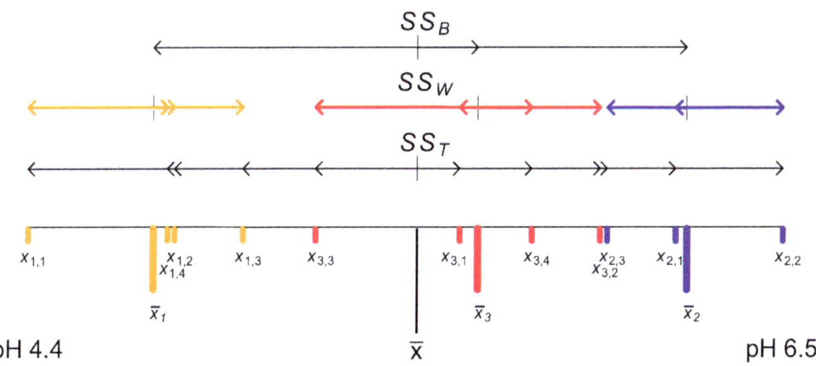

Figure 7.1: Distinctness of group structure along a one-dimensional gradient. SS_B, distances used for calculating between group sum of squares; SS_W, distances used for calculating within group sum of squares; and SS_T, distances used for calculating total sum of squares.

Analysis of variance is normally used to reject or accept the null-hypothesis stating lack of pattern: *there is no relation between pH and the group structure.* The underlying idea in this is in partitioning the total sum of squares, SS_T, according to:

$$SS_T = SS_W + SS_B \tag{7.1}$$

The principle of total (SS_T), between-group (SS_B) and within-group (SS_W) deviation from the mean is illustrated in Figure 7.1 using pH as site factor. This univariate case shows how the sum of the squared distances between the groups can be compared with the same within the groups, expressing the crispness of the classification. The following notation is used in the equations below:

n	sample size, total	
m	no. of groups	
i	current sampling unit	$i = 1, \ldots, n$
k	current group	$k = 1, \ldots, s$
l	current sampling unit in group k	
s	group size of current group k	

The total sum of squares is the summed squared deviations of all measurements from the overall mean, $\overline{x}$:

$$SS_T = \sum_{i=1}^{n} (x_i - \overline{x})^2 \tag{7.2}$$

Sum of squares within groups does literally the same, but it refers to the group means (the centroids) instead of the total. This is done for sample l in any current group and summed up for all m groups:

$$SS_W = \sum_{k=1}^{m} \sum_{l=1}^{s} (x_l - \overline{x}_k)^2 \tag{7.3}$$

In the example of pH values in Table 7.1 the mean of group 1 is $\overline{x}_k = (4.4 + 4.8 + 5.0 + 4.8)/4 = 4.75$. The squared differences from this are $SS_{W1} = 0.35^2 + 0.05^2 + 0.25^2 + 0.05^2 = 0.19$. Doing the same for the other two groups yields $SS_{W2} = 0.127$ and $SS_{W3} = 0.35$ and for all three groups it is $SS_W = 0.6667$. For sum of squares between groups, only the group means are compared with the total:

$$SS_B = \sum_{k=1}^{m} s_k(\overline{x}_k - \overline{x})^2 \tag{7.4}$$

The overall mean of the 11 pH values is $\overline{x} = 5.482$. Between groups sum of squares of group 1 is $SS_{B1} = 4 * ((4.75 - 5.482)^2) = 2.143$. For the other groups we get $SS_{B2} = 1.694$ and $SS_{B3} = 0.113$ with a total of $SS_B = 3.949$. Note that in order to consider all sampling units, the deviations of group means from the total mean have to be multiplied by group size, s_k. For the purpose of testing, the degrees of freedom are determined. As shown in Section 5.5, these are:

$$df1 = m - 1 \qquad \text{and} \qquad df2 = n - m \tag{7.5}$$

The test criterion, the F-value, is the proportion of mean squares between groups and the same within, that is:

$$F = \frac{SS_B/df1}{SS_W/df2} \tag{7.6}$$

For pH we then get $F = (3.949/2)/(0.6667/8) = 23.698$. This is a large F-value and if we were consulting an F-table in a statistical textbook we would find this being highly significant so that the null-hypothesis mentioned above can be rejected: pH apparently reflects the group structure with an error probability of $p < 0.01$.

Table 7.1: Selected variables from data frames nveg and snit presented in the form of a traditional synoptic table. Based on this classification, pH, slope and altitude are being subjected to analysis of variance (see main text).

	Relevé no.		49	25	2	6	39	18	50	9	4	10	27
	Group no.		1	1	1	1	3	3	3	3	2	2	2
1	pH		4.8	5	4.4	4.8	5.2	6	5.8	5.6	6.2	6.5	6
2	Altitude		550	480	450	420	500	400	520	580	500	560	450
3	Slope (deg.)		15	18	10	4.5	12.5	2.5	3	6	0.5	4	2.5
8	*Vaccinium myrtillus*	3	+		2	1							
6	*Sambucus racemosa*	3	+		2	1							
21	*Polytrichum formosum*	3	3	+	+	1							
13	*Veronica officinalis*	3	+	+	1	1							
12	*Luzula nemorosa*	3	2	1	2	+							
2	*Quercus petraea*	1	4	3	1	2	+	+		1			+
9	*Carex silvatica*	1	+		+	1	+	2		+	+		2
20	*Eurhynchium striatum*	1	+	+	1		+	+		1	+		+
11	*Viola silvestris*	1	+		+	+	+		+	1	+		+
1	*Fagus silvatica*	1	2	4	3	4	5	5	+	4	+	2	1
10	*Oxalis acetosella*	1		+	+	+	3	1		1	1	2	
14	*Galium odoratum*	1	+	+	+		+	2		1	+	+	1
15	*Lamium galeobdolon*	1			+	+	+			+	2	1	2
5	*Lonicera xylosteum*	1		+	+		+				1	+	1
3	*Acer pseudoplatanus*	2					+	+	1	1	1	2	4
4	*Fraxinus excelsior*	2						+	1	+	3	3	2
7	*Sambucus nigra*	4						+			+	2	1
18	*Arum maculatum*	4									1		+
19	*Ranunculus ficaria*	4									1	2	4
16	*Primula elatior*	4									+	2	2
17	*Allium ursinum*	4									4	2	+

For the three site factors we obtain:

pH	$F =$ 23.6980	$p =$	0.0004
Slope	$F =$ 3.7561	$p =$	0.0707
Altitude (m)	$F =$ 0.2235	$p =$	0.8046

The pH value qualifies as the best predictor for species composition as it reflects the group structure with an error probability of $p < 0.01$. For slope, significance is given only at $p = 0.0707$, while altitude does not mirror this pattern at all.

Doing analysis of variance in $\mathcal{R}$ merely requires typing a single line when data is ready. The relevé numbers, stored as row names, as well as the pH values are taken from object nsit:

```
names(nsit)
```

```
[1] "PH"        "ALTITUDE"  "SLOPE.deg" "X.AXIS"    "Y.AXIS"    "EXPOSURE"  "YEAR"
[8] "GROUP_NO"
```

```
nsit$PH
```

```
 [1] 4.4 6.2 4.8 5.6 6.5 6.0 5.0 6.0 5.2 4.8 5.8
```

```
row.names(nsit)
```

```
 [1] "2"  "4"  "6"  "9"  "10" "18" "25" "27" "39" "49" "50"
```

This is the raw order of relevés, not the one used in Table 7.1. Therefore we generate a vector **groups** of group membership of each of the relevés, starting with number 2 belonging to group number 1, followed by relevé 4 with membership 2, etc.:

```
groups<- as.factor(c(1,2,1,3,2,3,1,2,3,1,3))
```

It is important to declare **groups** a vector of what is called 'factor' in $\mathcal{R}$. Analysis of variance (function aov()) will recognize the elements as types rather than quantities. Starting with PH we get:

```
o.aov<- aov(PH~groups,data=nsit)
summary(o.aov)
```

```
            Df Sum Sq Mean Sq F value   Pr(>F)
groups       2  3.950  1.9748    23.7 0.000435 ***
Residuals    8  0.667  0.0833
---
Signif. codes:  0 *** 0.001 ** 0.01 * 0.05 . 0.1   1
```

The notation PH~**groups** is model syntax of $\mathcal{R}$ and it means 'analyse PH as a function of **groups**'. The summary also follows conventions. Because we know some of the results already, understanding the display is not too difficult: in the first column we see the degrees of freedom. The sum of squares of groups is our sum of squares between groups, the same of the residuals is the sum of squares within groups. Mean square is the same, but divided by the corresponding degrees of freedom. Then the result confirms the F-value seen above. The function conveniently computes the exact error probability, in this case $p = 0.000435$ confirming very high significance. For the two remaining site factors simply replace PH by SLOPE.deg and ALTITUDE in function aov(), respectively.

7.2.2 Variance ranking

The idea of ranking species based on a variance criterion has been proposed by Jancey (1979). Technically it is identical to the F-testing in analysis of variance shown in the previous section. When the variables tested are the species, then the same body of data is used as for the classification of the relevés. Therefore, the F-value cannot serve as a test criterion: it merely represents a measure for the relative resolving power of the species in distinguishing the relevé groups. Jancey (1979) suggests sorting the species list according to the F-values in decreasing order. Such a list is presented in Table 7.2, based on data and classification shown in Table 7.1. Species with sufficiently high F-values are candidates for building keys for vegetation mapping in the field, for example.

Table 7.2: Variance ranking of species (data frame nveg) based on the classification of relevés shown in Table 7.1.

Rank no.		Species	F-value	p
1	13	*Veronica.officinalis*	86.471	3.82e-06
2	19	*Ranunculus.ficaria*	81.781	4.73e-06
3	12	*Luzula.nemorosa*	60.853	1.45e-05
4	16	*Primula.elatior*	54.090	2.25e-05
5	3	*Acer.pseudoplatanus*	46.648	3.89e-05
6	17	*Allium.ursinum*	30.979	0.000171
7	21	*Polytrichum.formosum*	27.777	0.000251
8	4	*Fraxinus.excelsior*	22.093	0.000552
9	7	*Sambucus.nigra*	14.018	0.00243
10	2	*Quercus.petraea*	7.936	0.0126
11	6	*Sambucus.racemosa*	6.427	0.0217
12	8	*Vaccinium.myrtillus*	6.427	0.0217
13	15	*Lamium.galeobdolon*	5.353	0.0334
14	18	*Arum.maculatum*	5.346	0.0336
15	5	*Lonicera.xylosteum*	4.069	0.0604
16	1	*Fagus.silvatica*	2.007	0.197
17	14	*Galium.odoratum*	0.460	0.647
18	10	*Oxalis.acetosella*	0.371	0.701
19	20	*Eurhynchium.striatum*	0.107	0.899
20	11	*Viola.silvestris*	0.098	0.908
21	9	*Carex.silvatica*	0.013	0.987

Jancey (1979) also mentions that F-values can be computed for a restricted set of groups. In this case, the analysis yields the resolving power of species based on the selected relevé groups only. In fact, any new classification of relevés will alter the F-values and the list of ranks accordingly.

Computing the F-values of Table 7.2 in $\mathcal{R}$ requires the same function `aov()` as in Section 7.2.1. The 'independent' variable is now no longer pH or any other site factor but a species vector. Typing group membership (variable `groups`) as done there is feasible, but not practical when data sets are large. We would do better to access vegetation data (`nveg`), compute the distance matrix of relevés, `dd`, using function `vegdist()` from package `vegan`, followed by clustering (function `hclust()`). Function `cutree()` provides group membership automatically as explained in Section 5.5:

```
dd<- vegdist(nveg^0.5,method="euclid")
```

```
o.clust<- hclust(dd,method="ward.D2")
groups<- as.factor(cutree(o.clust,k=3))
```

To get the full list of ranked species, function `srank()` in the `dave` package is used. This prints the content of Table 7.2 into the $\mathcal{R}$ console:

```
o.srank<- srank(nveg,groups,method="jancey",y=0.5)
o.srank
```

Object `o.srank` is a list with the a content to be accessed for further processing, like restricting data frame `nveg` to the 10 best differentiating species, for example:

```
nveg.red<- nveg[,o.srank$species.no[1:10]]
```

The new, reduced data frame `nveg.red` contains the 10 species ranked highest in Table 7.2, starting with *Veronica.officinalis* all the way down to *Quercus.petraea*.

Table 7.3: Variance ranking of site factors based on the classification of relevés ('Schlaenggli' data sets `sveg` and `ssit` used).

Rank no.		Site factor	*F*-value	*p*
1	16	pH.water	71.154	3.08e-23
2	20	y.axis	62.249	7.74e-22
3	18	log.Ca.water	53.779	2.43e-20
4	9	Base.sat.perc	44.219	2.12e-18
5	1	pH.peat	39.755	2.22e-17
6	3	Ca_peat	27.579	4.5e-14
7	17	log.cond.water	22.908	1.58e-12
8	7	Acidity.peat	11.449	1.13e-07
9	13	Waterlev.min	9.472	1.3e-06
10	12	Waterlev.av	8.881	2.8e-06
11	14	log.peat.lev	8.587	4.13e-06
12	15	log.slope.deg	4.995	0.000731
13	4	Mg_peat	3.099	0.0153
14	10	P.peat	2.914	0.0207
15	11	Waterlev.max	2.820	0.0241
16	2	log.ash.perc	2.808	0.0246
17	8	CEC.peat	2.560	0.037
18	19	x.axis	2.518	0.0397
19	5	Na_peat	1.082	0.38
20	6	K_peat	0.964	0.448

A straightforward idea is to use the same procedure and even the same $\mathcal{R}$ function for ranking site factors. This is analysis of variance in the strict sense as explained in Section 7.2.1 and it yields a ranked list of site factors. It is beneficial only when a fairly large number of site factors is available and for purpose of demonstration the 'Schlaenggli' data sets, `sveg, ssit`, are taken with as many as 20 site factors. We first derive six groups based on vegetation, the same way as above when ranking species:

```
dd<- vegdist(sveg^0.5,method="euclid")
o.clust<- hclust(dd,method="ward.D2")
groups<- as.factor(cutree(o.clust,k=6))
```

But now access to site factors is needed. Function `srank()` can handle this as well:

```
o.srank<- srank(ssit,groups,method="jancey",y=1.0)
o.srank
```

Various conclusions can be drawn from the resulting Table 7.3. Surprisingly, 17 out of 20 site factors yield a significant F-value confirming a pronounced strength of the gradient captured by the sample. In first place there is pH measured in open water, followed by the y-axis. As will be shown in Section 7.3.3 this axis runs parallel to the main vegetation gradient. The third is calcium in water and the fourth base saturation. pH measured in dried and ground peat samples (pH.peat) has rank 5, a far more time consuming measurement than pH measured in open water. Another interesting fact is that dry conditions (Waterlev.min) affect vegetation more than wet conditions (Waterlev.max).

7.2.3 Ranking by indicator values

Identifying the best variables to assess classification has always been an issue in the context of data sets with excessive number of descriptors. Dufrêne and Legendre (1997) devised an index they called IndVal considering the occurrence of species in classifications of relevés. In more detail, Indval $d_{i,c}$ is:

$$f_{i,c} = \frac{\sum_{j \in c} p_{i,j}}{n_c} \tag{7.7}$$

$$a_{i,c} = \frac{(\sum_{j \in c} x_{i,j})/n_c}{\sum_{k=1}^{K}((\sum_{j \in k} x_{i,j})/n_k)} \tag{7.8}$$

$$d_{i,c} = f_{i,c} \times a_{i,c} \tag{7.9}$$

where $p_{i,j}$ is the presence–absence (1/0) of species i in sample j, $x_{i,j}$ is the abundance of species i in sample j and n_c is the number of sampling units in cluster c of set K.

In the equations, $f_{i,c}$ is relative presence–absence of species j in cluster c and $a_{i,c}$ is relative abundance of species j in cluster c compared with the mean abundance in all clusters. Indicator value $d_{i,c}$ is the product of these

Table 7.4: Ranking of species by indicator values based on the classification of relevés shown in Table 7.1.

Rank no.		Species	IndVal	Error p
1	12	*Luzula.nemorosa*	1.000	0.006
2	13	*Veronica.officinalis*	1.000	0.008
3	16	*Primula.elatior*	1.000	0.007
4	17	*Allium.ursinum*	1.000	0.003
5	19	*Ranunculus.ficaria*	1.000	0.006
6	21	*Polytrichum.formosum*	1.000	0.011
7	7	*Sambucus.nigra*	0.847	0.021
8	6	*Sambucus.racemosa*	0.750	0.060
9	8	*Vaccinium.myrtillus*	0.750	0.058
10	4	*Fraxinus.excelsior*	0.691	0.016
11	18	*Arum.maculatum*	0.667	0.056
12	5	*Lonicera.xylosteum*	0.630	0.075
13	15	*Lamium.galeobdolon*	0.619	0.029
14	2	*Quercus.petraea*	0.609	0.006
15	3	*Acer.pseudoplatanus*	0.598	0.057
16	14	*Galium.odoratum*	0.389	0.767
17	1	*Fagus.silvatica*	0.375	0.522
18	10	*Oxalis.acetosella*	0.301	0.825
19	11	*Viola.silvestris*	0.282	0.924
20	20	*Eurhynchium.striatum*	0.270	0.875
21	9	*Carex.silvatica*	0.259	0.985

two becoming large only if neither is small (Legendre and Legendre 2012, p. 398).

In $\mathcal{R}$ package `labdsv` there is function `indval()` doing these operations. $\mathcal{R}$ function `srank()` in the `dave` package offers a method `"indval"` for ranking species by indicator value and we can try this using the same example as in Section 7.2.2:

```
dd<- vegdist(nveg^0.5,method="euclid")
o.clust<- hclust(dd,method="ward.D2")
groups<- as.factor(cutree(o.clust,k=3))
o.srank<- srank(nveg,groups,method="indval",y=0.5)
o.srank
```

When comparing the ranking that results (Table 7.4) we can see that this is almost identical to the same in Table 7.2 despite the fact that it is based on totally different considerations. Note that computation of error probability requires random numbers and it will vary accordingly.

Just as in Jancey's ranking we can now reduce data frame `nveg` to the 10 species ranked highest by function `srank()`:

```
nveg.red<- nveg[,o.srank$species.no[1:10]]
dim(nveg.red)
```

```
[1] 11 10
```

Species names in `nveg.red` start with *Luzula.nemorosa* and end with *Fraxinus.excelsior*. Just as in Jancey's ranking IndVal depends on the number of groups involved in the computation. Wildi and Feldmeyer-Christe (2013) have shown that both behave similarly as long as the number of groups is low. But Jancey's ranking excels IndVal when increasing the group number within a given sample.

7.2.4 Analysis of concentration

Methods of classification and ordination are aimed at analysing data matrices from full samples. The basic idea presented below is to analyse data in summarized form; that is, after classification of relevés and species. From the point of view of pattern the aim is to reveal, for example, interaction of relevé classifications and species classifications. An element of such a contingency table contains counts of occurrence of a species group within a relevé group. The relevé group can originate from the same body of data. In this case the contingency table merely serves the analysis of vegetation data. However, in cases where the groups are derived from environmental data, then it is an assessment of the interaction between site types and species groups. In any case the onset are presence–absence scores yielding true frequency counts. These are the elements of the contingency table and the notation used for all elements of the table accords with:

$$\mathbf{F} = \begin{array}{ccccc|c} f_{11} & \cdots & f_{1j} & \cdots & f_{1n} & f_{1.} \\ \cdots & \cdots & \cdots & \cdots & \cdots & \cdots \\ f_{i1} & \cdots & f_{ij} & \cdots & f_{in} & f_{i.} \\ \cdots & \cdots & \cdots & \cdots & \cdots & \cdots \\ f_{p1} & \cdots & f_{pj} & \cdots & f_{pn} & f_{p.} \\ \hline f_{.1} & \cdots & f_{.j} & \cdots & f_{.n} & f_{..} \end{array} \tag{7.10}$$

where $f_{i.}$ is the sum of the n elements in row i, $f_{.j}$ is the sum of column j with p elements and $f_{..}$ is the grand total. There are n columns and p rows in this matrix.

The method was first proposed by Feoli and Orlóci (1979) under the name 'analysis of concentration' (AOC) and further developed in Orlóci (1991a). 'Concentration' refers to the uneven allocation of non-zero scores within the blocks of relevés and species groups. Concentration is high when the counts concentrate in a few blocks while other blocks are empty. It is low when the scores are dispersed all over the contingency table. The aim of the method is to measure concentration and to reveal interactions among and between the classifications of relevés and species.

The method is illustrated using the classified data set shown in Table 7.1. Counting the species scores within the blocks of the table yields the following contingency table with three relevé- and four species groups:

$$\mathbf{F} = \begin{array}{ccc|c} 18 & 0 & 0 & 18 \\ 27 & 25 & 21 & 73 \\ 0 & 7 & 6 & 13 \\ 0 & 1 & 14 & 15 \\ \hline 45 & 33 & 41 & 119 \end{array} \tag{7.11}$$

Typically, the sizes of the 12 blocks differ in Table 7.1, while in the final analysis each group is intended to have the same weight. Therefore, an appropriate adjustment is needed. In the first step, the number of relevés and species per group is counted to yield block size, Z:

$$\mathbf{Z} = \begin{array}{ccc|c} 20 & 20 & 15 & 55 \\ 36 & 36 & 27 & 99 \\ 8 & 8 & 6 & 22 \\ 20 & 20 & 15 & 55 \\ \hline 84 & 84 & 63 & 231 \end{array} \tag{7.12}$$

Block sizes range from 6 to 36. In the original publication, all frequencies were adjusted to the minimum block size. Orlóci and Kenkel (1985) and Orlóci (1991b) later proposed an adjustment to retain the grand total. This is the case when:

$$a_{ij} = \frac{\frac{f_{..}f_{ij}}{n_{ij}}}{\sum_{g=1}^{p}\sum_{h=1}^{q}\frac{f_{gh}}{n_{gh}}} \tag{7.13}$$

In the denominator of Equation 7.13, the sum of all frequencies weighted by the inverse of block size is used. In the above example, this value is 5.980. The adjustment of the first element thus yields:

$$a_{11} = \frac{\frac{119*18}{20}}{5.980} = 17.908$$

The matrix of the adjusted frequencies is:

$$\mathbf{A} = \begin{array}{ccc|c} 17.908 & 0.0 & 0.0 & 17.908 \\ 14.923 & 13.818 & 15.476 & 44.217 \\ 0.0 & 17.411 & 19.898 & 37.309 \\ 0.0 & 0.955 & 18.571 & 19.526 \\ \hline 32.831 & 32.184 & 53.945 & 119.0 \end{array} \tag{7.14}$$

It can be seen that this transformation changes the elements considerably in the case of both small (e.g. row 3) and large (e.g. row 2) groups. These adjusted frequencies are now further analysed by CA, as shown in Section 6.4.

This yields ordination coordinates for relevé and species groups. In the example above, there are two non-zero eigenvalues (according to the dimension of the data matrix minus 1):

$$\lambda_1 = 0.58834 \qquad \lambda_2 = 0.1221$$

The canonical correlations are the square roots of these:

$$R_1 = 0.76703 \qquad R_2 = 0.34947$$

CA is based on deviations from expectation. 'Expectation' here is the assumption that all frequencies are evenly dispersed across all blocks, which is the case when there is no structure in the table. The χ^2 by definition sums the squared deviations from expectation and is therefore a measure of concentration. It is obtained from the squared canonical correlations, multiplied by the grand total of F (or A, which is the same):

$$\chi^2 = \chi_1^2 + \ldots + \chi_q^2 = R_1^2 f_{..} + \ldots + R_q^2 f_{..} \tag{7.15}$$

This shows that the χ^2 is in fact the sum of q orthogonal components. In the example we get:

Component i	R_i	χ_i^2	$\lambda_i\%$
1	0.76703	70.01	82.8
2	0.34947	14.53	17.2
Total		84.54	100

Although Orlóci and Kenkel (1985) test χ^2 for significance, this may not be too helpful in practice. For comparison of classifications it is more convenient to consider the mean square contingency coefficient, C:

$$C = \frac{\chi^2}{A_{..}(m-1)} \tag{7.16}$$

where m is equal to the smaller dimension of the contingency table, in this example the number of relevé groups, that is $3 - 1$. C lies between 0 and 1; in the example above $C = 0.355$. This is a fairly high value, typical for very small sets of vegetation data.

Just like CA, this method yields coordinates. But the data points now refer to groups rather than to the individual relevés and species. This simplifies the interpretation considerably (Figure 7.2). In the present example, the ordination confirms that there is a correspondence between relevé group 1 and species group 4, between relevé group 2 and species group 3, and also between relevé group 3 and species group 2. Species group 1 is intermediate and not really indicative for any other, being located close to the origin of the coordinate system.

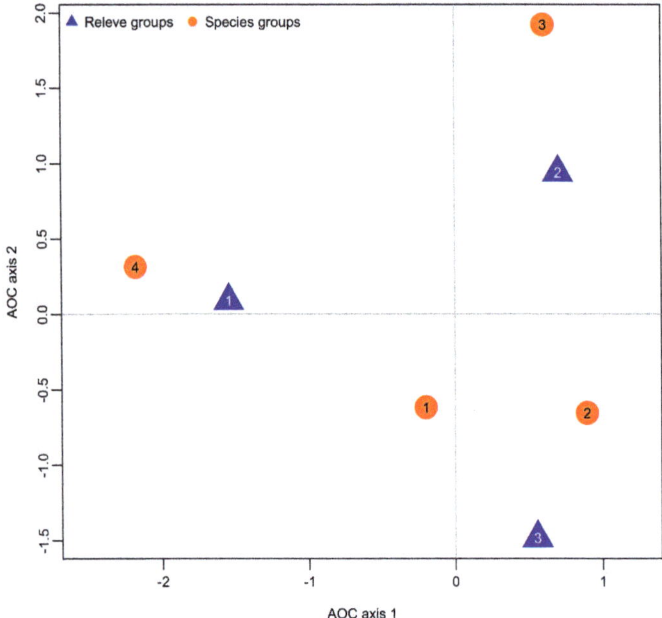

Figure 7.2: Ordination of group structure in the test data set nveg as derived from clustering relevés and species.

Analysis of concentration is available through $\mathcal{R}$ function aocc() implemented in the dave package. It operates on a vegetation matrix (rows as relevés and columns as species) and it applies signum transformation to the scores to derive frequency counts. Two vectors are needed to build the contingency table: group labels of relevés (o.rgr) and the same of species (o.sgr). Function aocc() assumes input from cluster analysis or manually typed labels as for instance taken from Table 7.1. Classifications for relevés are obtained as follows:

```
dr<- vegdist(nveg^0.5,method="bray")
o.clr<- hclust(dr,method="ward.D2")
o.rgr<- cutree(o.clr,k=3)
```

Forming the three groups is done by function cutree() as explained in Section 5.5. Then, classification of species is the analogue, but data frame nveg is used in transposed form (function t()):

```
ds<- vegdist(t(nveg^0.25),method="euclid")
o.cls<- hclust(ds,method="ward.D2")
o.sgr<- cutree(o.cls,k=4)
```

Function `aocc()` does all the computations and displays the main results shown above:

```
o.aocc<- aocc(nveg,o.rgr,o.sgr)
plot(o.aocc)
```

The plot method displays the content of Figure 7.2. With some additional effort function `aocc()` can also be used when vectors of the relevés and species memberships are available. For the relevés in Table 7.1 where the groups are labelled differently this would be:

```
o.rgr<- c(1,2,1,3,2,3,1,2,3,1,3)
o.sgr<- c(1,1,2,2,3,4,3,4,1,1,1,4,4,1,3,3,3,3,3,1,4)
```

This is group membership of the rows (`o.rgr`) and the columns (`o.sgr`) of `nveg`, respectively. A call of `aocc()` yields the arrangement of Table 7.1.

7.3 Correlating spaces

7.3.1 The Mantel test

We have seen so far how classifications help to relate vegetation and site conditions. A more general problem is to compare two (or more) unclassified multivariate spaces. One solution to this is canonical correlation, a method that relates n variables of one data set to m variables of another. I abstain from this, because in ecology one usually runs into problems with degrees of freedom; altogether, there are often too many variables involved compared with the number of sampling units: the system becomes over-determined. To compare the patterns of two spaces, a most elegant solution is the correlation of resemblance matrices. In this, all pairwise elements of any two similarity- or distance matrices are involved. It may concern the biological data space versus the environmental, but also the spatial (the arrangement of plots in space) and the temporal (the states in time).

The example presented here may appear trivial because one of the spaces is univariate only (the environmental). But a one-dimensional vector is sufficient to calculate a square distance matrix. Let us consider the following situation:

Relevé	1	2	3	4
pH	4.5	4.1	4.2	3.8
Species 1	0	1	1	2
Species 2	3	2	2	1

The pH vector represents the environmental space. The resemblance pattern of this is given by the distance matrix, D_e. Since this is one-dimensional, the distance between any pair of relevés is the difference of the respective pH values:

$$\mathbf{D_e} = \begin{matrix} 0 & & & \\ 0.4 & 0 & & \\ 0.3 & 0.1 & 0 & \\ 0.7 & 0.4 & 0.3 & 0 \end{matrix}$$

The two species vectors form the biological space. This pattern is inherent in the distance matrix, D_v. Using Equation (4.1) (Euclidean distance) we get:

$$\mathbf{D_v} = \begin{matrix} 0 & & & \\ 1.41 & 0 & & \\ 1.41 & 0 & 0 & \\ 2.5 & 1.41 & 1.41 & 0 \end{matrix}$$

For the purpose of comparison, the elements of the triangular matrices are now rearranged as vectors:

$$\mathbf{D_e}; \mathbf{D_v} = \begin{matrix} 0.4 & 1.41 \\ 0.3 & 1.41 \\ 0.1 & 0 \\ 0.7 & 2.5 \\ 0.4 & 1.41 \\ 0.3 & 1.41 \end{matrix}$$

The two vectors differ in scale, and for comparison they have to be adjusted. Mantel (1967) suggested using the correlation coefficient to measure the fit (see also Legendre and Fortin 1989), because, as shown in Table 4.3, the product moment correlation coefficient involves standardization. In the example above we get:

$$\begin{aligned} r(D_e; D_f) &= 0.965 \\ p &= 0.166 \end{aligned} \qquad (7.17)$$

Where does the error probability, p, come from? In statistical textbooks tables of significance levels for correlation coefficients can be found. In this situation, however, they are not valid because the elements in the two vectors are not normally distributed; nor do they represent a random sample. As in many other occasions in ecology a randomization test is more appropriate. This measures the probability that an equally high correlation could occur by chance. To this end the elements of one vector are reordered randomly and the calculation of the correlation coefficient is repeated. When $p = 0.166$ then this means that in 166 out of 1000 cases the random order yields an

$r \geq 0.965$. It may come as a surprise that the very high correlation coefficient of $r = 0.965$ does not yield a smaller p-value, but the sample size is definitely too low. More reliable results are obtained when $n \geq 10$, where the number of off-diagonal elements in the resemblance matrix is $m \geq 45$.

$\mathcal{R}$ package `vegan` offers function `mantel()` to compute and to test Mantel correlation. First, pH and vegetation are stored as matrices (in practice these are usually read from files):

```
pH <- matrix(c(4.5,4.1,4.2,3.8),nrow=1)
veg <- matrix(c(0,3,1,2,1,2,2,1),nrow=2)
```

Function `dist()` now calculates the distance matrices. Because this by default operates on rows the data matrices have to be transposed by function `t()`:

```
De <- dist(t(pH),method="euclidean")
Dv <- dist(t(veg),method="euclidean")
```

Function `mantel()`, among other things, yields Mantel correlation and the significance level based on 999 permutations:

```
mantel(De,Dv)
```

```
Mantel statistic r: 0.9649
      Significance: 0.177
```

Whenever repeating this example Mantel correlation should be exactly the same, but there is a good chance that the p-value differs. It is the result of a random process and it depends on the seed value used to generate the pseudo random numbers. To keep this fixed one types, say, `set.seed(57)` in front of any call of function `mantel()`.

7.3.2 Correlograms

Correlograms are mostly used for spatial or temporal analysis. They help to detect specific patterns, such as autocorrelation, temporal and spatial trend, periodicity and also nonlinearity. Basically, they can be applied to any type of ordinal or metric data. In geographical space various kinds of interactions of vegetation and site factors can be distinguished (Borcard and Legendre 2012; Legendre and Legendre 2012), for example:

Spatial (or temporal) dependence. This occurs when there is a gradient present in the data. Along this, plots in close proximity tend to be more similar than distant ones. The site conditions – and most likely also vegetation – change from one end of the gradient to the opposite. Because of the lack of an equilibrium such systems are called nonstationary.

Anisotropy. The above-mentioned case of spatial dependence often differs by direction. Along the main gradient, dependence is strong. Perpendicularly, it is weak or even lacking. If the same dependence exists in all directions, the system is called isotropic. Anisotropy is the rule in ecological systems, a fact plant ecologists strive for when trying to find gradients. In one-dimensional systems anisotropy does not exist.

Spatial autocorrelation. Even if the system is free of an overall gradient, correlation can occur at small distances. When analysing a system with inherent autocorrelation, the species composition in neighbouring plots is more similar than could be expected from the measured site conditions. Often, it is assumed that this is caused by species propagation: whatever the environmental factors are, it is easier for a species to reach a plot nearby than a remote one.

Correlograms respond to all these effects, the short ranging such as autocorrelation and the long ranging such as spatial trends. A correlogram is a plot of Mantel correlation $M(d)$ as a function of step length.

$$M(d) = \frac{\frac{1}{W}\sum_{h=1}^{n}\sum_{i=1}^{n} w_{hi}(y_h - \overline{y})(y_i - \overline{y})}{\frac{1}{n}\sum_{i=1}^{n}(y_i - \overline{y})^2} \quad \text{for} \quad h \neq i \qquad (7.18)$$

where W is a transformed distance matrix: it considers classes of distance, such as *short, medium, long.* When an element w_{hi} falls into the class to be analysed, it takes the value of one; otherwise it is zero. In our very small pH example below with $n = 10$ [Figure 7.3(a)], there are possible step lengths ranging from 1 to 9. These are taken as distance classes to compute the correlogram shown in Figure 7.3(b). There exist nine distances of length 1 in this data set and the corresponding Mantel correlation is 0.376 (which is highly significant according to $p = 0.007$ given, see below). Step length 2 occurs eight times only, step length 3 yields seven pairs and so on. In Figure 7.3(b), a clear trend of decreasing correlation is found up to step length 6, confirming the existence of spatial dependence. This not only levels off at longer distances, but positive correlations return at step lengths 8 and 9. This is how correlograms reflect periodicity, the one visible in Figure 7.3(a).

$\mathcal{R}$ package **vegan** not only offers a function for Mantel correlation but also a method for plotting a correlogram: `mantel.correlog()`. To generate the results seen in Figure 7.3 we first type pH and location of measurement sites in space:

```
pH <- c(4.7,4.5,4.4,4.6,4.9,5.1,5.1,5.3,5,4.8)
location <- c(1,2,3,4,5,6,7,8,9,10)
```

A plot of these two vectors can be seen in Figure 7.3(a), generated as follows:

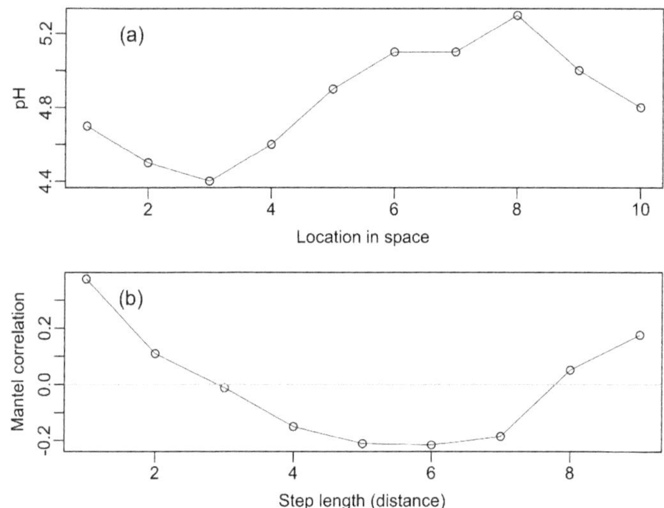

Figure 7.3: Plot of pH versus location in space (a) and correlogram of pH measurements according to Table 7.5 (b).

```
plot(location,pH,lwd=0.5)
lines(location,pH,lwd=0.5)
```

To precisely reproduce the results in Table 7.5 `mantel.correlog()` has to be forced to split the spatial space properly. Using the `seq()` function we generate a series of so-called break points ranging from 0.5 to 10.5 with an interval of 1.0, hence, 0.5-1.5, 1.5-2.5, etc.:

```
bp <- seq(0.5,10.5,1)
```

From vector pH the 'ecological' distance, De, is derived and from vector location the 'geographical', Dg, both processed subsequently in `mantel.correlog()`:

```
De <- dist(pH,method="euclidean")
Dg <- dist(location,method="euclidean")
o.corr <- mantel.correlog(De,Dg,cutoff=FALSE,break.pts=bp)
o.corr
```

This prints the content of Table 7.5. The object named `o.corr` is the list that `mantel.correlog()` generates. Function `mantel.correlog()` provides a plot function. To generate the correlogram in Figure 7.3(b) it is sufficient to type `plot(o.corr)`.

Table 7.5: Mantel correlogram resulting from function `mantel.correlog()`. The corresponding graphs are in Figure 7.3. Note that distance classes are counted twice.

	class.index	n.dist	Mantel.cor	Pr(Mantel)	Pr(corrected)
D.cl.1	1	18	0.37614	0.007	0.007
D.cl.2	2	16	0.11021	0.253	0.253
D.cl.3	3	14	-0.01200	0.476	0.506
D.cl.4	4	12	-0.14955	0.180	0.540
D.cl.5	5	10	-0.21132	0.126	0.504
D.cl.6	6	8	-0.21566	0.085	0.425
D.cl.7	7	6	-0.18453	0.126	0.630
D.cl.8	8	4	0.05111	0.411	0.822
D.cl.9	9	2	0.17555	0.152	0.882

7.3.3 More trends: 'Schlaenggli' data revisited

This section demonstrates various applications of Mantel correlation and correlograms to real world data. First we address correlation of vegetation with geographical space using the x- and y-coordinates of the locations of sampling units. Secondly, there is an evaluation of site factors as predictors for vegetation pattern based on Mantel correlation, an alternative to analysis of variance (Section 7.2.1) but without using a classification. And finally, correlograms are computed to achieve improved insight into correlation of vegetation and individual site factors, even in the case of nonlinearity.

The 'Schlaenggli' data set **sveg** originates from systematic sampling of a square grid with 1 m^2 plots spaced by 10 m (Figure 7.4). The investigation area is almost quadratic in shape and spatial trends can therefore be evaluated in different directions. From Table 7.3 we know already that there is a strong floristic gradient in the vertical direction represented by the y-axis (in the method shown below, $\alpha = 90\,°$). Hence, the vegetation pattern is space-dependent.

Since this dependence has its origin in a spatial gradient, direction is a major issue. To evaluate different directions, the spatial distances have to be projected onto one line. This reduces the space to one dimension and the new distances allow a new distance matrix to be derived. Finally, from any new direction, a Mantel correlation r is obtained. The way distances are projected at angles of $\alpha = 0\,°$, $45\,°$ and $90\,°$ is shown in Figure 7.4. For these directions, the Mantel test yields $r = 0.0727$, 0.4966 and 0.5975, respectively. At an angle of $\alpha = 0\,°$, where the vertical component of the space is suppressed and only the horizontal expansion considered, Mantel correlation is close to zero, indicating absence of vegetation trend. At $\alpha = 90\,°$ the vertical expansion is considered only and Mantel correlation suggests a strong vegetation trend.

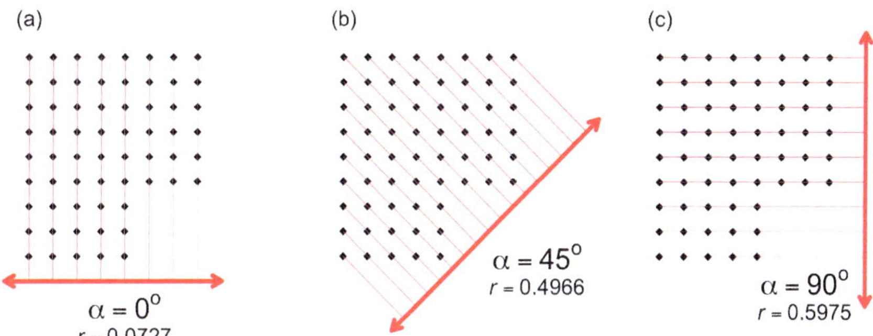

Figure 7.4: Projecting distances in three different directions. (a) $\alpha = 0°$; (b) $\alpha = 45°$; and (c) $\alpha = 90°$. Projection in the direction of $90°$ yields the highest Mantel correlation, r.

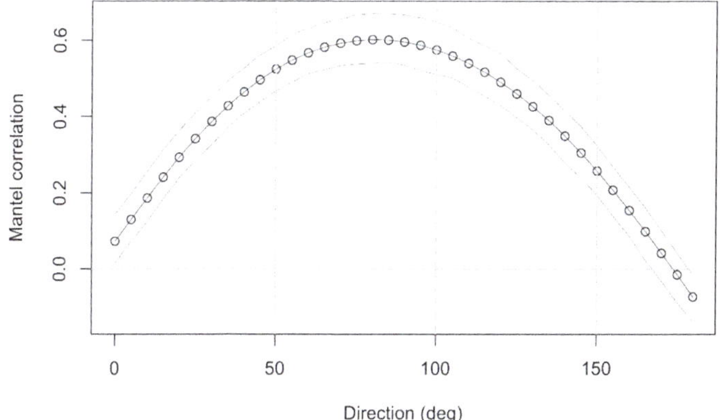

Figure 7.5: Evaluating the direction of the floristic gradient. The gradient is strongest at about $\alpha \approx 80°$. Light lines are the 5% confidence limits. This is the output of function `dircor()`.

A full evaluation of all directions in the range of $0° \leq \alpha \leq 180°$ allows identification of the direction in which the gradient is strongest. This is shown in Figure 7.5. The maximum is achieved around $\alpha \approx 80°$, indicating that the main gradient runs in almost vertical direction. Perpendicular to this, at $\alpha \approx 165°$, the floristic gradient vanishes ($r \approx 0.0$).

Figure 7.5 is generated by function `dircor()` in the `dave` package. As usual we use the data frame `sveg` holding vegetation data and `ssit` with site factors where the x- and y-axes reside:

```
o.dircor<- dircor(sveg,ssit$x.axis,ssit$y.axis,step=5)
plot(o.dircor)
```

Variable `step` defines the interval at which directions are computed, 5° in this case. The function requires some computation time to determine the confidence limit for each point. Object `o.dircor` is a list containing the raw data used for plotting.

Table 7.6: Mantel test of the site factors in data frame `ssit`. F-values from Section 7.2.2, Table 7.3 are added for comparison.

No.	Site factor	Mantel's r	Permutation p	F-value
16	pH.water	0.762	0.001	71.15
18	log.Ca.water	0.696	0.001	53.78
9	Base.sat.perc	0.677	0.001	44.22
1	pH.peat	0.602	0.001	39.75
3	Ca_peat	0.585	0.001	27.58
20	y.axis	0.584	0.001	62.25
17	log.cond.water	0.552	0.001	22.91
13	Waterlev.min	0.358	0.001	9.47
14	log.peat.lev	0.355	0.001	8.59
7	Acidity.peat	0.350	0.001	11.45
12	Waterlev.av	0.310	0.001	8.88
15	log.slope.deg	0.105	0.051	5.00
11	Waterlev.max	0.094	0.052	2.82
19	x.axis	0.085	0.035	2.52
4	Mg_peat	0.080	0.105	3.10
2	log.ash.perc	0.033	0.288	2.81
8	CEC.peat	0.032	0.282	2.56
10	P.peat	0.017	0.393	2.91
6	K_peat	-0.036	0.648	0.96
5	Na_peat	-0.062	0.837	1.08

Mantel correlation is equally operational with measured site factors. Unlike x- and y-axis in space these tend to be correlated and to avoid conflicts one single site factor should be taken at a time. Whereas in Section 7.2.2 site factors have been evaluated based on their potential to predict a specific classification of the relevés, Mantel correlation measures this independent of classification, based on the full similarity matrices of the relevés. When doing this with all site factors the Mantel correlations in Table 7.6 are obtained and also the corresponding error probabilities p resulting from the randomization test. For the purpose of comparison, the F-values from Section 7.2.2 are also added. It does not come as a surprise that Mantel correlation and analysis of variance find similar trends, but not quite the same. We have to recall that analysis of variance evaluates one specific classification out of many possible, whereas Mantel correlation will always be the same.

Computing Mantel correlation in $\mathcal{R}$ is done as shown in Section 7.3.1. Taking pH in water as an example and using data frames sveg and ssit from the above example we get:

```
Dv<- dist(sveg^0.5,method="euclidean")
De<- dist(ssit$pH.water,method="euclidean")
mantel(De,Dv)

Mantel statistic based on Pearson's product-moment correlation

Call:
mantel(xdis = De, ydis = Dv)

Mantel statistic r: 0.7621
      Significance: 0.001
...
```

Table 7.6 is a compilation of ordered Mantel correlations with all 20 site factors. A potential pitfall in this is to miss the proper orientation of the data matrix, in this example to get distance matrices of relevés. To switch from site factors to relevés (or vice versa) it is sufficient to transpose the data matrix, using t(sveg) for example.

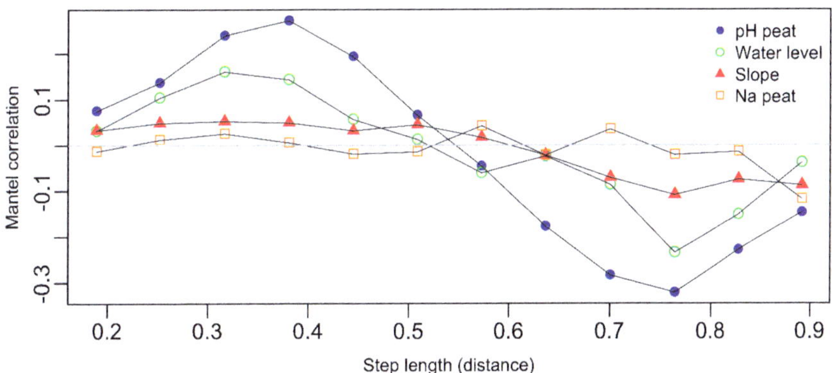

Figure 7.6: Correlograms of site factors with vegetation distance (step length). Small steps signify similar composition of vegetation, long steps different composition.

Mantel correlation is meaningful only when applied to metric measurements and it is distribution sensitive. It may be wise to apply some transformation of vectors prior to analysis, standardization for example. Correlograms reveal further details about correlation of site factors with vegetation. Figure 7.6 gives examples of four variables with distinct correlation properties. Unlike correlograms where geographical space is involved, vegetation usually exhibits low correlation with site factors at short step lengths, the maximum reached only at three to four steps in the present example: small distance here means small difference in vegetation composition and this is usually blurred by random variation. A typical example for strong correlation with vegetation is pH in peat: after reaching a maximum at step length 0.4 it decreases monotonically, levelling off towards the end where sample size gets scarce. A typical shape for low or lack of correlation is Na in peat, just showing faint (not significant) local autocorrelation. Also weak, and almost linear, is correlation with water level.

Provided vegetation data are available in data frame `sveg` and site factors in `ssit`, distance matrices of vegetation and pH peat, as an example, are derived to generate a correlogram by function `mantel.correlog()` as follows:

```
D.veg<- vegdist(sveg^0.5,method="bray")
D.pH <- vegdist(decostand(ssit$pH.peat,"stand"),method="euclid")
o.manpH<- mantel.correlog(D.pH,D.veg,cutoff=FALSE)
plot(o.manpH)
```

Bray–Curtis distance is taken for convenience. The square root of species scores is taken and pH in peat is standardized. Numerical results are available in object `o.manpH`. For the remaining curves in Figure 7.6 `pH.peat` is replaced by `Waterlev.av`, `log.slope.deg` and `Na_peat`, respectively, the proper names seen from output of `names(ssit)`.

7.4 Constrained ordination

Constrained ordination methods are multivariate linear models. They analyse two data spaces simultaneously; hence the frequently used umbrella term *canonical analysis* for all related methods. Typical results are ordinations with three types of data points: one for releveś, a second for environmental variables and a third for species. What distinguishes constrained ordinations from ordinary ones is the involvement of regression. Regression is partitioning dependent vectors, such as the species scores, into two components: the expected and the deviation from this, that is, the residuals. In linear regression the expected values are the projections of the scores onto the straight regression line. This is the input data used for ordination which is therefore named ‘constrained’. The main issues of constrained ordination are the following:

- The multiple regressions involved divide the total variance into an explained partition (the expected) and an unexplained partition (the residuals). The quotient of explained by total expresses how much variance is common to both data matrices, one holding vegetation description and the other the site factors.

- The explaining variables, such as the environmental vectors, can be subjected to permutation tests. When the elements of the vectors are randomly exchanged, correlation is expected to vanish. By repeating this process the significance of the environmental factors in contributing to the canonical ordination can be tested.

As always in ecology randomization tests have their limitations, as explained by Lepš and Šmilauer (2003) in the context of using the program package CANOCO. Significance just means that correlation probably exists, even though it may be spurious.

Many ordination methods can potentially be extended to a constrained version:

- Redundancy analysis (the constrained form of PCA) was first proposed by Rao (1964).
- Canonical correspondence analysis (CCA; constrained CA) was invented by ter Braak (1986).
- Constrained PCOA was proposed by Legendre and Anderson (1999) (although they called it 'distance-based redundancy analysis').
- Wagner (2004) devised a spatially constrained CA, revealing the effects of spatial dependence and spatial autocorrelation.

In all methods the coordinates generated by computer programs can be scaled differently and many offer options for this. Comparison with the output of different software is further complicated when randomization tests are involved, because random numbers differ between computer programs and also often in consecutive runs of the same program.

Below, the principle of constrained ordination is explained in detail by *redundancy analysis* (RDA), the constrained version of PCA. It differs from PCA in that expected (in some $\mathcal{R}$ functions also called 'fitted') species scores are used instead of the originals. Expectation $\hat{y}_j$ of any species j is derived through multiple regression, according to:

$$\hat{y}_j = a_j + b_{j,1}x_{j,1} + b_{j,2}x_{j,2} + \ldots + b_{j,k}x_{j,k} + \varepsilon_j \tag{7.19}$$

The $x_{j,k}$ values are the k measurements available for each plot. The $b_{j,k}$ values are the regression coefficients and a_j is the intercept. ε_j are random variables representing the residuals not explained by the linear regression. Usually the data matrices are centred by species such that the intercept vanishes. Denoting the species by relevé matrix Y, a new matrix of expected values, $\hat{Y}$, is obtained, where:

$$Y = \hat{Y} + Y_{res} \tag{7.20}$$

This relationship (Legendre and Legendre 2012) is of practical importance as it expresses the proportion of explained (canonical) variance, $\hat{Y}$, compared with the total, Y. Hence, the operations explained in Section 6.2 are performed on $\hat{Y}$ rather than Y.

Rao (1964) did not introduce RDA as a 'new method' but as an exercise to illustrate variance partitioning. In recent literature we now find step by step solutions to this using $\mathcal{R}$, aimed at demonstrating the principle. The computation of RDA applied below uses the small data frames `nveg` and `nsit` (Table 7.1) following a tutorial published by David Roberts (http://ecology.msu.montana.edu/labdsv/R/labs/lab7), although an alternative would be the one given in Borcard *et al.* (2011). Before starting the exercise we derive a reduced object of three site variables describing ecological conditions:

```
nsit3<- nsit[,c("PH","ALTITUDE","SLOPE.deg")]
```

The first comma within the square brackets means that all rows are taken. The new data frame `nsit3` now has 11 rows and 3 columns, that is, the site factors pH, altitude and slope. The aim of the following four steps is to obtain a matrix of 'canonical eigenvectors' rather than ordinary eigenvectors as in PCA (Section 6.2):

1. Vegetation data are standardized by species. Species will get zero mean and unit variance:

   ```
   c.nveg<- scale(nveg,center=TRUE,scale=TRUE)
   ```

   The first species vector, *Fagus silvatica*, before standardization is:

   ```
   nveg[,1]
   ```

   ```
    [1] 4 1 5 5 3 6 5 2 6 3 1
   ```

   And after standardization it is:

   ```
   round(c.nveg[,1],digits=1)
   ```

   ```
      2    4    6    9   10   18   25   27   39   49   50
    0.1 -1.5  0.7  0.7 -0.4  1.2  0.7 -0.9  1.2 -0.4 -1.5
   ```

2. Site factors, data frame `nsit3`, are standardized the same way:

   ```
   c.nsit3<- scale(nsit3,center=TRUE,scale=TRUE)
   ```

The three site factors, in the raw data measured on different scales, have now zero mean and unit variance.

3. Multiple regression uses the species data as dependent, the environmental as explaining variables. Function `lm()` recognizes `c.nveg` as a matrix and does multiple regressions for all the species. Function `fitted()` calculates the expected (fitted) species scores:

```
lm.veg<- lm(c.nveg~c.nsit3)
fit.veg<- fitted(lm.veg)
round(fit.veg[,1],digits=1)
```

```
   2    4    6    9   10   18   25   27   39   49   50
 0.5 -0.6  0.2 -0.7 -0.7  0.2  1.0 -0.1  0.4  0.3 -0.5
```

The first column of the fitted values, `fit.veg[,1]` contains the expected species scores of the first species, *Fagus silvatica.*

4. To conclude the basic operations we perform PCA on the fitted values:

```
o.rda<- pca(fit.veg,cor=TRUE)
plot(o.rda)
```

Plotting the scores displays an ordination of the relevés in canonical space. This space has three dimensions only because it relies on the three site factors used. To check this we can have a look at the first seven eigenvalues, for instance:

```
round(o.rda$sdev[1:7],digits=2)
```

```
[1] 3.95 1.75 1.52 0.00 0.00 0.00 0.00
```

As expected there are only three non-zero eigenvalues.

The example of the species vector of *Fagus silvatica* demonstrates how a constrained ordination can differ from an unconstrained. The most influential change in pattern usually comes from the use of fitted values. This reduces the total variance in the data and we can determine this by taking the sum (function `sum()`) of the diagonal values (function `diag()`) of the covariance matrix (function `cov()`) of data frames `c.veg` and `fit.veg`, respectively:

```
tot.var.1 <- sum(diag(cov(c.nveg)))
tot.var.2 <- sum(diag(cov(fit.veg)))
```

We get $tot.var.1 = 21.0$ and $tot.var.2 = 10.197$. This means that constrained variance in this example is 48.6% of the unconstrained.

PCA as used above derives the ordination scores Z in the space of the ecological variables from the data matrix of expectations, $\hat{Y}$, and it multiplies these with the canonical eigenvectors U by matrix multiplication (see Section 6.2):

$$Z = \hat{Y}U \tag{7.21}$$

The same eigenvectors can serve to get unconstrained ordination scores F when taking the standardized vegetation scores Y instead of the expectations:

$$F = YU \tag{7.22}$$

For plotting the environmental variables the correlations of the site factors with matrix F are taken. Relevé, species and site factor scores fit into one single ordination only after proper scaling, as summarized in Legendre and Legendre (2012, p. 666). All this suggests that constrained ordinations deserve careful interpretation taking into account that scatter plots are further restricted by reduced dimensionality.

Probably the most frequently used method of constrained ordination is CCA, the constrained version of CA. It differs from RDA in that the transformation used in CA [Equation (6.5)] is applied prior to multiple regression. As in the unconstrained version, CCA is considered appropriate when species response is unimodal.

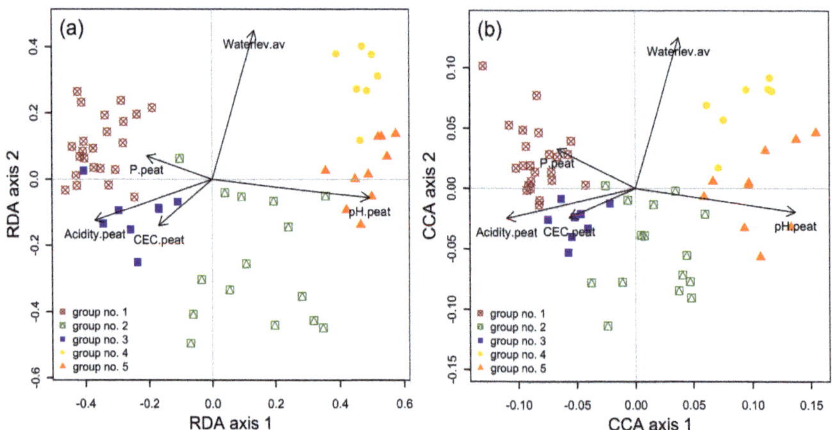

Figure 7.7: Comparison of RDA (a) and CCA (b), data set 'Schlaenggli' (Appendix 14). Groups are superimposed to facilitate comparison.

To demonstrate the interpretation of constrained ordination I compare RDA and CCA using data set 'Schlaenggli', data frame `ssit`. From the 20 environmental factors I use 5 for simplicity:

- pH of peat
- Acidity (mval/100 g peat)
- Cation-exchange capacity (CEC; mval/100 g peat)
- Phosphorus (mg/100 g peat)
- Water level (average; cm below surface).

The results are shown in Figure 7.7. Overall performance of the ordinations is as follows:

Method	Constrained variance	Unconstrained variance	Total	Percentage constrained
RDA	12.65	25.66	38.31	33
CCA	0.670	1.70	2.37	28

Performance of both ordinations is rather high, with RDA being slightly better (33% explained variance) than CCA (28%). In other words, the linear model succeeds in revealing correlation between vegetation and environmental factors despite the strong horseshoe pattern (Section 6.5). Two almost independent factors dominate the system: pH in peat and the depth of the average water table. CEC and acidity are highly correlated, while phosphorus has its maximum in peat bog vegetation on the left of the ordinations, a fact already recognized in the original investigation (Wildi 1977). The overall pattern of RDA is rather similar to the unconstrained version of PCA, as can be seen in comparison with Figure 6.7(b), whereas CCA is similar to CA [Figure 6.7(a)]. In conclusion, both methods in this example perform very well and choosing either of these is probably just a matter of preference.

Table 7.7: Storage location of selected parameters from functions `rda()` and `cca()`, package `vegan`.

Parameter	RDA	CCA
Eigenvalues	o.rdaCCAeig	o.ccaCCAeig
Site factor scores	o.rdaCCAbiplot	o.ccaCCAbiplot
Unconstrained variance	o.rda$tot.chi	o.cca$tot.chi
Constrained variance	o.rdaCCAtot.chi	o.ccaCCAtot.chi

The user interface of $\mathcal{R}$ functions `rda()` and `cca()` in package `vegan` is intuitive, but retrieving specific results is not. To reproduce Figure 7.7 some supplementary hints are useful. The 'Schlaenggli' data used for this is in `sveg` and `ssit`. First I combine the five site factors specified in `s5`:

```
s5<- c("pH.peat","P.peat","Waterlev.av","CEC.peat","Acidity.peat")
```

```
ssit5<- ssit[s5]
o.rda<- rda(sveg~.,data=ssit5)
plot(o.rda)
```

Function rda() uses the formula interface of $\mathcal{R}$. This reads as 'analyse sveg as a function of pH.peat and P.peat and Waterlev.av and CEC.peat and Acidity.peat'. More precisely, the dot after the tilde means 'take all site factors located in ssit5'. For CCA we alter the lines as follows:

```
o.cca<- cca(sveg~.,data=ssit5)
plot(o.cca)
```

The two plots obtained from this are almost identical to Figure 7.7(a) and (b), but there is a difference in scaling of the relevé scores. I personally prefer axes that are scaled proportional to the eigenvalues. Self-made plotting, however, requires access to various parameters with locations listed in Table 7.7. To get the scaling of relevés as in Figure 7.7(a) and (b) the following adjustments are done:

```
sc <- scores(o.rda)
s1<- (o.rda$CCA$eig[1]^0.5)*sc$sites[,1]/(sum(sc$sites[,1]^2))^0.5
s2<- (o.rda$CCA$eig[2]^0.5)*sc$sites[,2]/(sum(sc$sites[,2]^2))^0.5
plot(s1,s2,asp=1)
```

Function scores() is a method provided to retrieve the scores. Vectors s1 and s2 are relevé scores adjusted to eigenvalues. Then, one may add species with coordinates in sc$species:

```
points(sc$species[,1],sc$species[,2],pch=17)
```

Table 7.7 also gives the addresses where total unconstrained and constrained variances are found. We finally add the site factors:

```
sf.scores<- o.rda$CCA$biplot*0.5
points(sf.scores,pch=3)
text(sf.scores[,1],sf.scores[,2],names(ssit5),pos=1)
```

Scaling by 0.5 is done by trial and error and the resulting data points, the locations of the arrow tips in Figure 7.7, are annotated by site factor names.

7.5 Nonparametric multiple analysis of variance

7.5.1 Method and example

Just as constrained ordination, nonparametric multiple analysis of variance (NP-MANOVA) too is a multivariate linear method. It answers a fundamental question in plant ecology: how much of the variance of a full vegetation sample is explained by one or several factors simultaneously? This method was devised by Anderson (2001). The principle is most easily understood when thinking of an extension of analysis of variance (Section 7.2.1) as sketched in Figure 7.1 for the one-dimensional case, and also in Figure 5.4 for two dimensions. Anderson (2001) mentions commonalities and differences to analysis of variance forming key issues of the method:

- Analysis of variance and NP-MANOVA decompose variation of vegetation into two partitions, the first explained by factors that either originate from sampling design (management type, for example) or from measured site variables, the second summarizing unexplained variance.
- In NP-MANOVA vegetation space is represented in the form of a distance matrix of relevés. (In cases where a relevé by species matrix is provided, deriving the distance matrix constitutes the first step in the analysis.)
- As a consequence the number of descriptors (species) contributing to distance may be excessively large without inflating calculation time or hampering statistical inference.
- In NP-MANOVA the test for significance is entirely based on permutation. Therefore, not only metric distance measures (Euclidean distance, for instance) can be used but also semimetric (Bray–Curtis distance, for instance).

The use of a distance matrix as the response variable resides in the fact that for computing variation within and between groups, centroids need not be supplied, but variance can be computed directly from inter-point distances (Figure 7.8; Anderson 2001). This even applies to semimetric distance measures where it happens that points cannot be displayed graphically because the distances violate the triangle inequality. In such a case an unequivocal centroid simply does not exist but the method shown in Figure 7.8(b), comes into play: within-group sum of squares is equal to the sum of squares of all connecting distances divided by group size. Freedom in the choice of distance measure means, for example, that all kinds of distances shown in Chapter 4 are allowed.

NP-MANOVA belongs to the family of linear models (LMs). Technically speaking LMs are basically multiple regressions of the form shown in Equation (7.19). In NP-MANOVA the explaining variables can either be continuous (pH, for example) or nominal (management type, exposure) and also combinations of these. NP-MANOVA is implemented in $\mathcal{R}$ function `adonis()` located in the `vegan` package. It uses the model interface of $\mathcal{R}$ which is just a short-cut formulation of normal mathematical notation. Linear regression $y = a + bx$, for example, is written as `y~x` and multiple

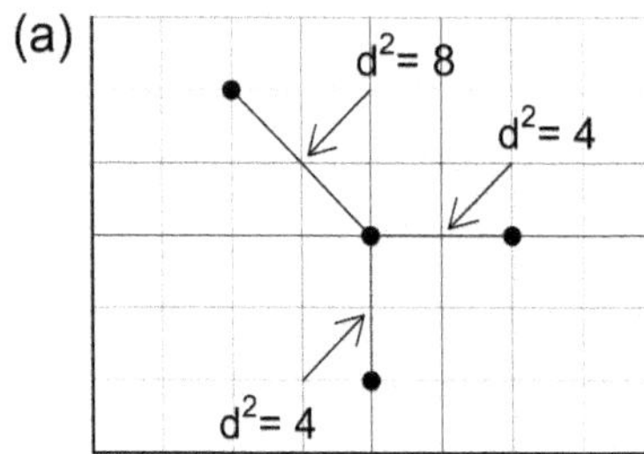

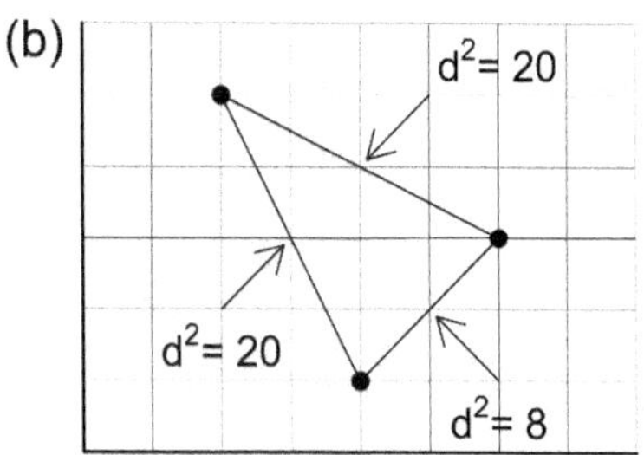

Figure 7.8: Using distance matrices in NP-MANOVA. (a) The conventional way of computing within-group variation, taking the sum of squared distances from all members to the centroid, $SS_{[T]} = 8 + 4 + 4 = 16$. (b) In NP-MANOVA the same is obtained by dividing the sum of all squared distances among group members by group size, $SS_{[T]} = (8 + 20 + 20)/3 = 16$. Adapted from Anderson (2001).

regression $y = a + bx + cz$ becomes `y~x+z`. The plus sign does not mean 'add', but 'include' the subsequent variable. Another important convention can be seen in model `y~x*z`. This is a three-fold regression with terms x and z and including an interaction term $x : z$. A more comprehensive overview of model formula in $\mathcal{R}$ is presented in Crawley (2005).

For applications in ecology, sample size should be sufficiently large, even when models are built with one or two independent variables only. The following example with just 11 relevés and 21 species serves technical explications concerning $\mathcal{R}$ function `adonis()` only. The first data set needed is vegetation (`nveg`), the second is site factors. Typing `names(nsit)` reveals that there is a variable called `PH` which we shall use in the small example:

```
o.adonis<- adonis(nveg^0.5~nsit$PH,method="bray")
o.adonis
```

```
Call:
adonis(formula = nveg^0.5 ~ PH, method = "bray")

          Df SumsOfSqs MeanSqs F.Model      R2 Pr(>F)
PH         1   0.72356 0.72356  8.4851 0.48528  0.001 ***
Residuals  9   0.76747 0.08527         0.51472
Total     10   1.49103                 1.00000
```

The call of `adonis()` means 'model square root transformed `nveg` as a function of `PH`'. We do not provide `adonis()` a distance matrix, but the original data in `nveg`. It therefore computes distances itself using Bray–Curtis function (Equation 4.5). The results report that there are 10 degrees of freedom in total, that is sample size minus one. According to the F-statistic the contribution of `PH` in explaining total variation in vegetation is highly significant. The most interesting column is `R2`, decomposing total variation of vegetation: `PH` accounts for 48.5%, the residuals for the remaining 51.5%.

The analysis provides further information about the sources of variability. In `o.adonis`, we find `o.adonis$coefficients` where the rows are the sources of variation (`PH` in the present case) and the columns are the species. Each column expresses the fit of the corresponding species to the linear model. For *Fagus silvatica*, the first species, we type:

```
o.adonis$coefficients[,1]
```

```
(Intercept)      nsit$PH
  3.7273564  -0.3402226
```

We get an intercept of 3.727 and a slope of −0.340. This is the same as when computing ordinary linear regression:

```
o.lm<- lm(nveg[,1]^0.5~PH,data=nsit)
o.lm
```

These coefficients are not available if `adonis()` is only supplied with a distance matrix.

There are also coefficients available of sites (relevés). These are located in `o.adonis$coef.sites`. Each column expresses the fit of a relevé to the linear model. For the first relevé, number 2, we get:

```
o.adonis$coef.sites[,1]
```

```
(Intercept)      nsit$PH
 -1.3537855    0.3240976
```

This is what we get when we compute regression with the first column of the distance matrix `dd[1,]` as the dependent variable and `PH` as the predictor. Because `adonis()` computed the distance matrix internally it is not yet available and must be reproduced prior to regression:

```
dd<- as.matrix(vegdist(nveg^0.5,method="bray"))
lm(dd[1,]~nsit$PH)
```

```
...
Coefficients:
(Intercept)      nsit$PH
    -1.3538       0.3241
```

The first row of the distance matrix expresses the distance of the first relevé to all others and the linear model relates this to the measured pH value.

For the sake of clarity it is also worth looking at `o.adonis$model.matrix` where all independent variables used in the model are separately listed. In the above example there are only two, intercept and PH. When including a *nominal variable*, for example `EXPOSURE` with states N, S and E, one can observe that this adds two so-called dummy variables to the model. In fact each nominal variable accounts for $m-1$ degrees of freedom where m is the number of states, suggesting this kind of predictor should be used with care.

7.5.2 Data transformation revisited

In this case study we probe the effect of transforming species scores on the performance of NP-MANOVA. To this end not only transformation of scores is varied but also the distance measure used as it may account for intrinsic transformations. The result is of course valid only for this very example. It is shown in Table 7.8 using the `sveg` and `ssit` data frames for a simple model with the variable correlating best with vegetation: `pH.water`. The numbers in the table are proportions of variation explained by `pH.water`, taken from the output of `adonis()`. The result is striking: Euclidean distance (Equation 4.1) offers the worst fit while Manhattan distance (Equation 4.2) performs much better. As frequently stated in the literature (Clarke 1993; Anderson 2001), Bray–Curtis (Equation 4.5) is a good choice, in the case of presence–absence even the best, outperforming the rather similar Canberra distance (Equation 4.4). In view of this it comes as a surprise that all are excelled by correlation used as distance [Equation (4.6)]. As transformation of cover values is concerned the models tend to yield improved fit when approaching presence-absence. The original rank scale in the data (range 0-6), used untransformed ($x' = x^{1.0}$), offers low model fit, generally improving when taking the square root ($x' = x^{0.5}$, see also Table 3.4). In conclusion the result suggests that high investments in field work for measuring cover precisely hardly pay off.

Table 7.8: Evaluating data transformation and the choice of distance function based on model performance of NP-MANOVA, using pH measured in water in ssit and vegetation data in sveg. Results are proportions of explained variance.

Distance type	$x' = x^{2.0}$	$x' = x^{1.0}$	$x' = x^{0.5}$	$x' = x^{0.25}$	x=sign(x)
Euclidean	0.136	0.230	0.249	0.245	0.237
Manhattan	0.264	0.358	0.391	0.404	0.415
Bray–Curtis	0.281	0.389	0.430	0.446	0.460
Canberra	0.244	0.274	0.295	0.306	0.319
Correlation as distance	0.242	0.414	0.459	0.454	0.438
Range of code	(0) 1-36	(0) 1-6	(0) 1-2.5	(0) 1-1.57	0/1

The model interface of $\mathcal{R}$ is used to set up models yielding the content of Table 7.8. The model for Euclidean distance and quadratic transformation, the first row in Table 7.8, is the following:

```
o.adonis<- adonis(sveg^2.0~ssit$pH.water,method="euclidean")
o.adonis
```

```
adonis(formula = sveg^2 ~ ssit$pH.water, method = "euclidean")

Terms added sequentially (first to last)

              Df SumsOfSqs MeanSqs F.Model      R2 Pr(>F)
ssit$pH.water  1    3664.8  3664.8  9.5763 0.13569  0.001 ***
```

```
Residuals     61   23344.1   382.7          0.86431
Total         62   27008.8                  1.00000
---
Signif. codes:  0 *** 0.001 ** 0.01 * 0.05 . 0.1   1
```

The proportion of variance explained by pH is found under `R2` or directly under `o.adonis$aov.tab$R2[1]`.

To obtain signum-transformed species scores function `sign()` is available, that is `sign(sveg)`, for example. Unlike all other distance measures, correlation as distance is not directly available in `vegan`. This distance matrix has to be calculated first and then used in the model:

```
dc<- as.dist((1-cor(t(sveg^2.0)))/2)
o.adonis<- adonis(dc~ssit$pH.water)
o.adonis
```

Because the species by relevé matrix is not involved in the calculation this version does not yield coefficients for species.

7.5.3 Clustering revisited

NP-MANOVA not only handles continuous, but also nominal variables, such as group memberships of sampling units. This class of data is called 'factors' in the realm of $\mathcal{R}$. Whenever a vector of factors is supplied as the independent variable, function `adonis()` splits these up into $g - 1$ 'dummy variables' where g is the number of factors, that is, the number of states. How this is done is illustrated in Table 9.1, for example.

This now offers the opportunity to measure the performance of classifications in explaining the variance of an entire vegetation sample. The example given below is the same as presented in Section 5.8, but performance is evaluated quantitatively rather than just qualitatively. Hence, in Figure 7.9 data transformation, the choice of resemblance measure and the clustering methods are varied simultaneously. The interpretation of results is straightforward: the larger the proportion of variance explained by the classification is, the more successful is clustering. The bar charts confirm the visual interpretations of the silhouette plots (Figure 5.8): Ward's method based on correlation as distance is an example where grouping is most successful, single linkage analysis is the worst choice.

To get the results in Figure 7.9 vegetation data `sveg` is always clustered first and the resulting classification is subjected to function `adonis()`. For the very first bar in the first barplot we get:

```
ddr<- as.dist((1-cor(t(sveg^3)))/2)
o.hclr<- hclust(ddr,method="single")
o.relgr<- cutree(o.hclr,k=5)
o.adonis<- adonis(ddr~as.factor(o.relgr))
o.adonis$aov.tab$SumsOfSqs[c(1,3)]
```

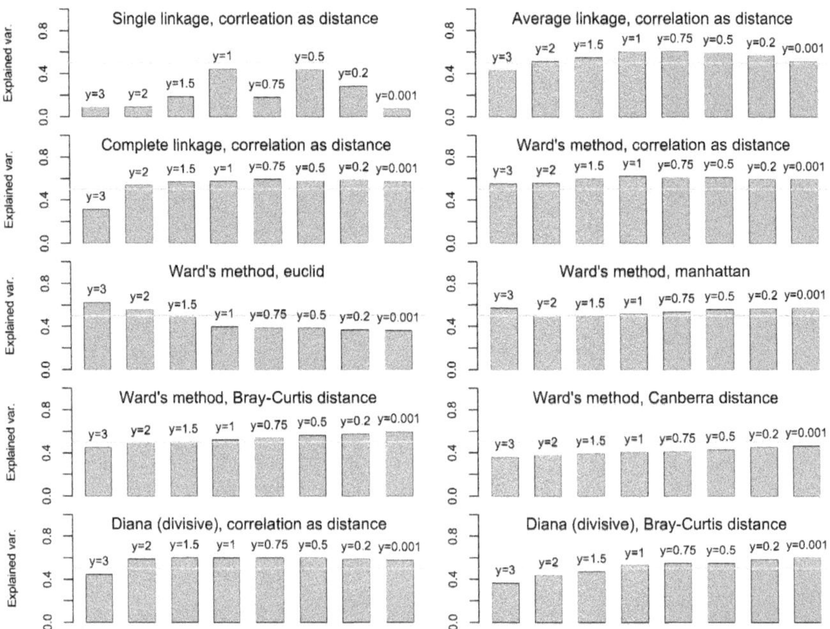

Figure 7.9: Performance of clustering methods in explaining total variance of vegetation data. The examples are the same as in Section 5.8, Figure 5.8. Transformation of species scores accords with $x^{'} = x^{y}$.

```
[1] 0.5540857
[1] 6.341194
```

The quantity 0.554 is explained variance, whereas 6.341 is the total. This ends up in a really modest 8.7% of explained variance, whereas in good classifications this regularly exceeds 50%.

This example demonstrates a useful way to evaluate clustering methods. But it is essential not to confound this with statistical inference: the source of classification is the same body of data as the dependent part of the model. What we do here is nothing else but partitioning variance.

7.6 Synoptic vegetation tables

7.6.1 The aim of ordering tables

In early times of plant ecology the predominant method of vegetation data analysis consisted in the rearrangement of synoptic tables: rows and columns were moved to achieve an order for relevés and species reflecting the similarity pattern of the sample. A description of the method was published by Ellenberg (1956) (English version in Mueller-Dombois and Ellenberg 1974).

It implemented some rules for simultaneous ordering of rows and columns. When multivariate clustering became operational for large data sets, several approaches were developed for substituting the manual process. Examples are the computer programs TABORD (van der Maarel *et al.* 1978) and TWINSPAN (Hill 1979b). While TABORD is heuristic and finds a solution through iteration, TWINSPAN includes in its first steps divisive clustering applied to the result of CA. This regularly leads to misclassifications as the similarity space is not considered in all dimensions simultaneously.

A second series of approaches appeared some 10 years later. They include the method described below (Wildi 1989), a similar but extended strategy by Podani and Feoli (1991), the program ESPRESSO by Bruelheide and Flintrop (1994) and probably others. Whereas ESPRESSO still intends to optimize the 'two way arrangement' of vegetation tables (Borcard *et al.* 2011) the method devised by Wildi (1989), implemented in MULVA (Wildi and Orlóci 1996), is entirely based on numerical analysis.

Ordered vegetation tables are more than just clustered two-dimensional arrangements. They are aimed at visualizing complex patterns, that is, groups, gradients or a combination of these. Patterns may occur on different scales, for example, when a series of small-scale groups exists forming a gradient at large scale. Or, groups may prevail which internally have gradient structure. Because all methods attempt to concentrate high presence scores close to the diagonal of the table, for each relevé or relevé group (traditionally in the columns) the characteristic species occurrences (in the rows) can immediately be found. As a further convention, species with low predictive power are either excluded or moved to the bottom part of the table to facilitate interpretation. This explains why more than one method is needed to fulfil all these conditions. The method I describe also serves as an example for combining different multivariate methods, rather than just representing a good solution for generating 'nice' tables.

7.6.2 Steps involved in sorting tables

The steps described below follow the suggestions published in Wildi (1989), with some minor adaptations. They are summarized in Table 7.9.

Step 1. The procedure starts with clustering of relevés. If the data are percentages, like the data set of Ellenberg (1956) shown in Section 7.6.3, then the scores should be transformed to closer reflect presence–absence. I suggest the use of $x' = x^{0.2}$ or similar. In Figure 7.10(b-f) this is $x' = x^{0.5}$ because the scores are ranks. Using correlation as distance implicitly standardizes the relevés to compensate for differences in species richness. Minimum-variance clustering allows for groups largely differing in size.

Step 2. Before proceeding to the analysis of species, the relevés are subjected to ordination to find the dominating gradient in the sample. This will be used later to rearrange relevés and species within the groups. While the scalar transformation remains the same, CA is usually a good choice for identifying a single dominating trend of relevés and species. Such a solution is shown in Figure 7.10(c).

Step 3. For clustering species, the methods used differ due to the fact that correlation of species is generally rather low and nonlinear (see Figure 3.3). To enhance joint occurrence, the scores are transformed close to presence–absence, using $x' = x^{0.2}$. I suggest normalizing (but not standardizing!) the species vectors. Since groups consisting of frequent but also rare species are generally not welcome, a resemblance measure has to be used that does not centre the vectors. In Table 7.9 Manhattan distance is proposed. Choosing complete linkage clustering, evenly sized groups can be expected. The outcome of clustering relevés and species is shown in Figure 7.10(b).

Step 4. Because the relevé and species groups are generated from dendrograms, their order is arbitrary as illustrated in Figure 5.2. Blocks of high score density will be dispersed all across the table. AOC, a method including an ordination (Section 7.2.4), now counts the frequency of non-zero scores in each block, thereby forming a contingency table of groups. According to the first ordination axis of AOC, the procedure arranges relevé groups and species groups along a main gradient. It results in what is shown in Figure 7.10(d).

Table 7.9: Steps involved in sorting synoptic tables by multivariate methods according to Wildi (1989).

Step	Trans-formation	Method	Effect
1 Clustering relevés	$x' = x^{0.5}$	Minimum variance clustering (Ward's method) based on correlation as distance	Forming relevé groups
2 Ordinating relevés and species	$x' = x^{0.5}$	Correspondence analysis	Determining major trend for order within groups
3 Clustering species	$x' = x^{0.2}$	Complete linkage (Manhattan distance), normalizing species	Forming species groups
4 Analysis of concentration	presence–absence	Analysis of concentration	Determining major trend among relevé/species groups
5 Rearrange relevés and species within groups	see step 2	Order taken from step 2	Reveal gradients within groups
6 Species ranking	$x' = x^{0.2}$	Analysis of variance (F-values)	Separating species with high *vs* low resolving power
7 Printing	none	Apply ordering criteria from steps 1-5	Sorting, as suggested by Ellenberg (1956)

Step 5. Frequently gradients exist within groups. To account for these, relevés and also species are ordered (within groups only) according to their sequence revealed in CA, step 2. This is the final order when considering all species irrespective of resolving power.

Step 6. The species are subjected to Jancey's ranking (Jancey 1979, Section 7.2.2). For formal reasons transformation should be the same as when clustering species. The species with the highest F-values are taken for the upper, discriminating part of the table. It is a matter of preference how many are to be included. In any case, species with low F-values (approaching $F = 1.0$) should of course be suppressed and moved to the bottom part of the table. In Figure 7.10(e) (12 differentiating species) and Figure 7.10(f) (18 differentiating species) these carry no group label.

Step 7. Plot the table choosing the appropriate method (`"normal"` or `"condensed"`).

When all the row and column vectors are arranged as explained in the previous steps, the table can be displayed. Since the sign of ordination axes is arbitrary, it can happen that the diagonal of high scores points from the upper right to the lower left.

All graphs of Figure 7.10 result from using function `Mtabs()` doing the complex task of applying all the steps mentioned above. It offers six options, namely `"raw"` [Figure 7.10(a)], `"clust"` [Figure 7.10(b)], `"ca"` [Figure 7.10(c)], `"aoc"` [Figure 7.10(d)], `"mulva"` [Figure 7.10(e) and (f)] and, not shown in Figure 7.10, `"sort"`, ordering rows and columns by decreasing frequency. The number of differentiating species, `ndiffs`, is used by method `"mulva"`. The exponents `y.r` and `y.s` are applied when structuring relevés and species, respectively, and `k.r`, `k.s` are corresponding group numbers:

```
y.r<- 0.5 ; y.s<- 0.2
k.r <- 3 ; k.s <- 4
ndiffs <- 18
o.Mt<-Mtabs(nveg,method="mulva" ,y.r,y.s,k.r,k.s,ndiffs)
plot(o.Mt,method="normal")
```

```
Call:
Mtabs.default(veg = nveg, method = "mulva", y.r = y.r, y.s = y.s,
    k.r = k.r, k.s = k.s, ndiffs = ndiffs)

CA eigenvalues, %:        52.82 16.61 9.78
AOC eigenvalues, %:       87.76 12.24 NA

Mean square contingency coefficient: 0.20763
```

This does all the computations and plots the graph requested by `method`. Data sets of size up to 80 rows and columns, respectively, will fit on a single page. When exceeding this size it is split into as many pages as needed.

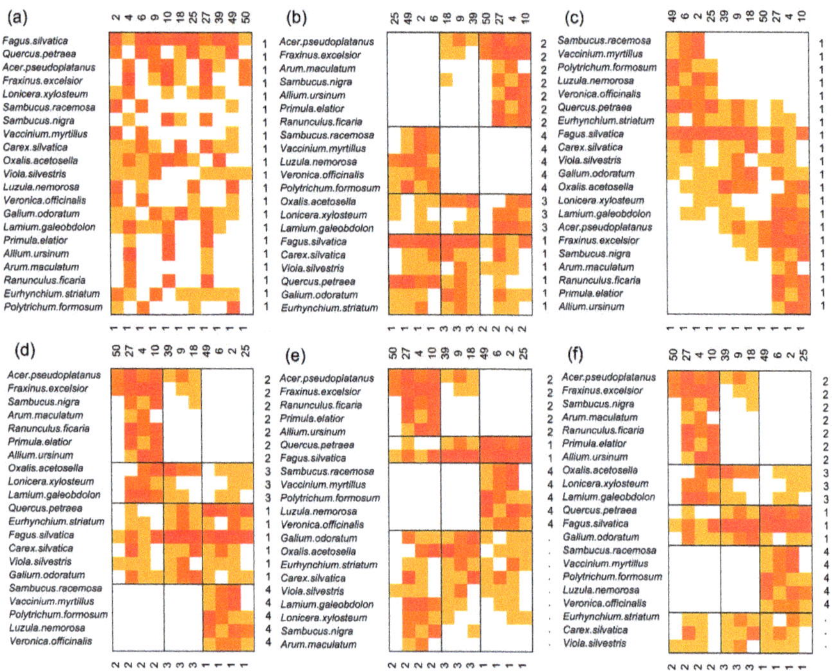

Figure 7.10: Graphical display of synoptic, structured vegetation tables. (a) Raw data table, method "raw". (b) Rows and columns arranged by cluster analysis, method "clust". (c) Rows and columns ordered according to the first axis of CA, method "ca". (d) Same as (b), but entire groups rearranged according to AOC, method "aoc". (e) The "mulva" solution with 12 differentiating species, method "mulva". (f) The "mulva" solution with 18 differentiating species, method "mulva".

Beyond several hundreds of relevés it may be better to set plot parameter `method` to `"compressed"` squeezing the entire table to fit on one page. Details like species names and relevé numbers are no longer printed, only success of simultaneous classification remains visible.

One may now wish to inspect the role of individual species. The traditional way to do that is by deriving frequency tables where group centroids rather than relevés are printed. Such a table, when displayed in conventional form, has as many columns as there are relevé groups and the number of rows accords with the number of species as shown in Table 7.10 for the data displayed in Figure 7.10(f). Frequency counts of the kind are implemented in the summary method of function `Mtabs()`:

```
summary(o.Mt,range=c(1,18))
```

By using the mandatory parameter `range` we limit the list to the first 18 species. In this presentation zeros in the matrix are replaced by dots to

Table 7.10: Frequency table derived from structured synoptic vegetation table, Figure 7.10(f), restricted to the first 18 species.

	2	3	1
Acer.pseudoplatanus	1	1	.
Fraxinus.excelsior	1	0.67	.
Sambucus.nigra	0.75	0.33	.
Arum.maculatum	0.5	.	.
Ranunculus.ficaria	0.75	.	.
Primula.elatior	0.75	.	.
Allium.ursinum	0.75	.	.
Oxalis.acetosella	0.5	1	0.75
Lonicera.xylosteum	0.75	0.33	0.5
Lamium.galeobdolon	0.75	0.67	0.5
Quercus.petraea	0.25	1	1
Fagus.silvatica	1	1	1
Galium.odoratum	0.75	1	0.75
Sambucus.racemosa	.	.	0.75
Vaccinium.myrtillus	.	.	0.75
Polytrichum.formosum	.	.	1
Luzula.nemorosa	.	.	1
Veronica.officinalis	.	.	1

improve overview. It can easily be seen that the order of relevé groups and species is the same as in Figure 7.10(f). In cases where the relevés are not classified, as for instance when choosing the `"raw"`, the `"sort"` or the `"ca"` method, then it is assumed that all belong to the same group and the result is a single centroid vector of species probability.

In cases where users wish to get their ordered vegetation data back for import into a spreadsheet program, function `Mtabs()` provides access to the new arrangements of relevés and species in the list `o.Mt`. In a first step the vegetation data frame, `nveg` in this case, is reordered in the following way:

```
ord.r <- order(o.Mt$order.rel)
ord.s<- order(o.Mt$order.sp)
nvegt<- t(nveg[ord.r,ord.s])
```

The third line furthermore transposes data frame `nveg` to appear in the traditional orientation with species as rows and relevés as columns (see Figure 7.1). Function `write.csv` will generate a file in the working directory of $\mathcal{R}$ (Section 2.4):

```
write.csv(nvegt,"nvegt.csv")
```

File `nvegt.csv` is now ready for import by any computer program accepting spreadsheets.

7.6.3 Example: ordering Ellenberg's data

For the purpose of illustration the data set of Ellenberg (1956) (also published in Mueller-Dombois and Ellenberg 1974) is used in Figure 7.11 to demonstrate the effect of ordering real world data. The number of relevé groups chosen is three, the same as Ellenberg used. In many cases, taking the square root of the number of relevés in a given sample turned out to yield pleasing results (i.e. about 15 for a table of size 200, 30 for a table of size 1000, etc.). There are 10 species groups in this example. Because the correlation among species is generally low, the groups should be small in size (i.e. three to six species). Under many circumstances, dividing the number of species by about four may be a good choice for the number of species groups.

Ellenberg's data (data frame `mveg`) include 25 relevés and 93 species. In order to save some space this is reduced first by suppressing species that occur just once. There exists an elegant but at the same time enigmatic way to do this in $\mathcal{R}$. The key issue is the use of function `apply()`. In this the first parameter is the data frame processed, considering presence–absence using function `sign()`. The second parameter, 2, directs `apply()` to operate on columns (species, in this case). Finally, parameter `sum` invokes function `sum()` forming the vector `f.s` with frequency counts. The new data frame `mveg1` holds species with frequency two or more only:

```
f.s<-apply(sign(mveg),2,sum)
mveg1<- mveg[,f.s > 1]
```

In Figure 7.11(a), raw data are displayed. Obviously frequencies differ greatly among species vectors, but the relevés are rather balanced. Hence, applying some vector transformation to species (Chapter 3) would alter the result considerably, whereas doing the same with relevés would hardly affect the outcome. In Figure 7.11(b) the CA solution is shown. This reveals the end-points of a main gradient, but in between there is no striking pattern emerging. This suggests trying to find a group structure within the relevés as done in Figure 7.11(c). It is the `"mulva"` solution explained in Section 7.6.2 with three relevé and 10 species groups and 25 differentiating species on top of the list. From visual inspection it seems that the group structure is fairly strong and the three groups are almost identical to those found in Mueller-Dombois and Ellenberg (1974).

All parts of Figure 7.11 are generated by function `Mtabs()`. For Figure 7.11(c) this requires the following instructions:

```
y.r<- 0.25 ;  y.s<- 0.10
k.r <- 3 ; k.s <- 10
ndiffs <- 25
o.Mt<- Mtabs(mveg1,method="mulva",y.r,y.s,k.r,k.s,ndiffs)
plot(o.Mt,method="normal")

Call:
Mtabs.default(veg = mveg1, method = "mulva", y.r = y.r, y.s = y.s,
    k.r = k.r, k.s = k.s, ndiffs = ndiffs)
```

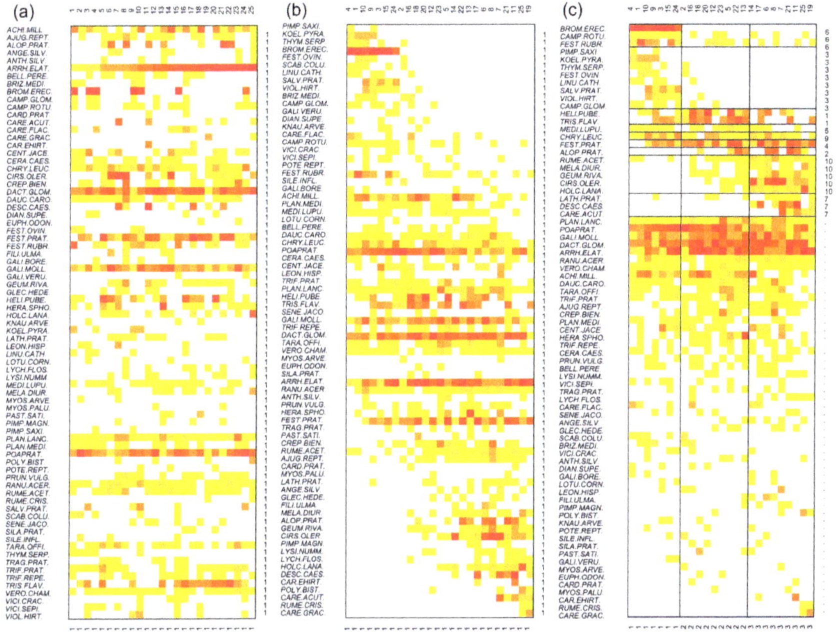

Figure 7.11: Structuring the meadow data set of Ellenberg (Mueller-Dombois and Ellenberg 1974). (a) Raw data. (b) Order according to CA (rows and columns). (c) The "mulva" solution.

```
CA eigenvalues, %:        21.35 9.91 7.27
AOC eigenvalues, %:       83.99 16.01 NA

Mean square contingency coefficient: 0.07341
```

The meaning of these parameters is explained in Section 7.6.2. Note that `Mtabs()` is now using `mveg1` rather than `mveg`. To obtain Figure 7.11(a) and (b) the method has to be changed, to `"raw"` and `"ca"`, respectively.

8 Traits and indicators

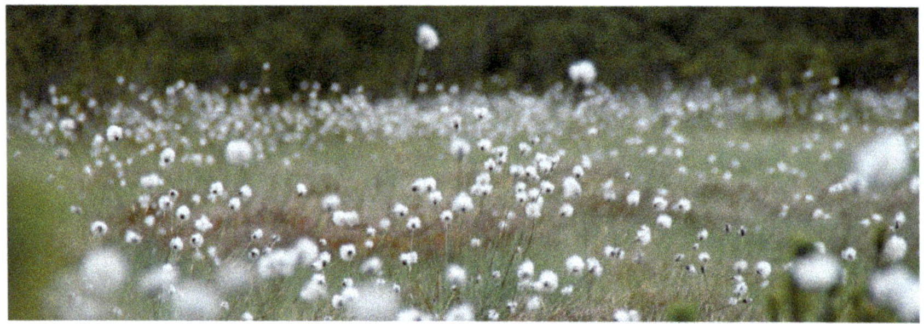

In large-scale and particularly global assessments of vegetation, species pools change in space and time. Descriptors other than species have to be found. Broad validity of results can be achieved when using indicator values, functional traits, life strategies or combinations of these as descriptors.

8.1 Vegetation beyond the species concept

Species populations are excellent descriptors of vegetation composition within geographically constrained areas. But as soon as these are extended the comparability of results deteriorates. The limited areas of species generate phytogeographical boundaries (Pillar and Sosinski 2003) and remote sites eventually have no more species in common. When striving for global patterns vegetation scientists soon became aware that descriptors other than species might be needed. An example are the life forms of Raunkiaer (1937) in which the location of buds (above, on or below ground) plays a crucial role in the survival of plants during unfavourable seasons, or the scheme of plant strategies introduced by Grime (2001) distinguishing 'competitors', 'ruderals' and 'stress-tolerants'. Box (1981) presented a worldwide vegetation model in which he used abstract 'model plants' that later were called 'plant functional types', the term emerging in the papers of Díaz (1995) and Box (1995, 1996) and being widely used thereafter.

Plant functional types are an alternative to taxonomic types (Pillar and Sosinski 2003). If any two plants are identical with regard to specific properties, that is, the traits, then they are members of the same plant functional type. Two cases can be distinguished. If any two plants belong to the same taxonomic unit (species, for instance), then the traits are considered species specific. If they belong to different taxonomic units, the traits concern plant individuals. In the examples below, I use species-specific traits only, because they can be retrieved from databases and applied to the many relevé by species data sets available these days. In ongoing investigations traits of individuals measured either in the field or in the laboratory offer some more flexibility as they can take within-species variation into account. Cornelissen *et al.* (2003) in their review present a panoply of possible measurements suited as plant functional traits. They distinguish vegetative traits (whole-plant, leaf, stem and belowground-traits) and regenerative traits (concerning dispersal and resprouting capacity), a structure also found in the list of Landolt *et al.* (2010). The most comprehensive source of traits by species presently available is the 'Plant Trait Database' (TRY database) (www.try-db.org), a network supported by vegetation scientists all around the world.

In this chapter I not only address the issue of traits as descriptors, but also indicator values, considered an alternative choice for vegetation analysis. Species indicator values, intended to be a practical aid in the interpretation of plot-based vegetation data (Ellenberg 1974; Landolt 1977) unexpectedly found their way into the scientific literature with an increasing number of papers published since the year 2000. Zelený and Schaffers (2012) in their critical review-paper count 95 respective articles appearing in the *Journal of Vegetation Science* and *Applied Vegetation Science* within the first decennium of the 21st century. Unfortunately indicator values, like species, have limited geographical validity too. But adaptations of lists of indicator values to different regions of the world (Borhidi 1995; Hill *et al.* 1999; Lawesson 2003; Pignatti 2005; Landolt *et al.* 2010) confirm their appeal to practitioners as well as scientists. Indicator values reflect the opinion of (few) experts (Diekmann 2003) and therefore it is widely agreed that they are a means of ecological interpretation, although not substituting measured environmental factors (Wildi 2016).

Traits- and indicator-based data are subjects of analysis and interpretation like species-based data. As shown in Chapter 3.1 data type is a central issue in the choice of the analytical method. Measurements of plant functional traits can be of any type, metric, ordinal, nominal or binary. Furthermore, when retrieving traits or indicator values from databases, missing values may become abundant and these have to be circumvented with care. In this Chapter I first present a general framework showing how relevé by trait or indicator matrices respectively are derived, how they are analysed and interpreted. Then I explain the role of further matrix operations involved in the derivation of alternative data spaces and their interactions.

By revisiting the 'Schlaenggli' data set as an example I present the transformations needed in practice, the results of data analysis and subsequent interpretation.

In a concluding remark I argue that vegetation ecology is still underway in finding a universal tool of description with world-wide validity. As a matter of fact a global consistent classification of vegetation types is as yet missing (Peet and Roberts 2013).

8.2 Analytical framework

The aim of the analysis of traits and indicator data is to answer questions about the structure and functioning of the plant-environment system. As Figure 8.1 illustrates this involves the investigation of various interactions among elements assessed by different data sets. Comparing panels (a) and (b) also reveals that the basic operations are the same for traits and indicator values, differing solely in a few technical details. Steps involved when moving from species to traits or indicator values as descriptors are the following:

Step 1. Deriving new matrices from existing: the relevé by traits matrix T and the relevé by indicators matrix J [Figure 8.1, panels (a) and (b), upper part]. Whereas in most cases T and J are considered an alternative to a traditional vegetation data matrix, V, it is essential to recognize traits and indicator values as new sources of information. These too represent vegetation, but in different variable spaces.

Step 2. We now concentrate on matrices T and J with the aim to explore the resulting patterns, for instance by methods of classification and ordination as described in Chapters 6 and 7. This may reveal similarity patterns in the data, like continuous trends, group structures or others, either also present in relevé by species data or possibly new. In cases where the patterns differ, this will not necessarily mean that one set of descriptors 'outperforms' the other. As the sets of descriptors differ in content they reflect different properties of the same system.

Step 3. The result of pattern recognition notwithstanding, most investigators wish to measure the difference between alternative vegetation descriptions in quantitative terms. Pattern, as explained in Section 4.5, is comprehensively captured by resemblance matrices, the triangles in Figure 8.1. The distance matrix of the relevé by traits matrix is denoted by D_t and this is compared to vegetation data matrix V with distance matrix D_v. As the descriptors involved in V and T differ in type, the distances in D_v and D_t will differ in scale. But we are interested in the relative positions of data points in resemblance space only and this is assessed by Mantel correlation, for instance. As explained

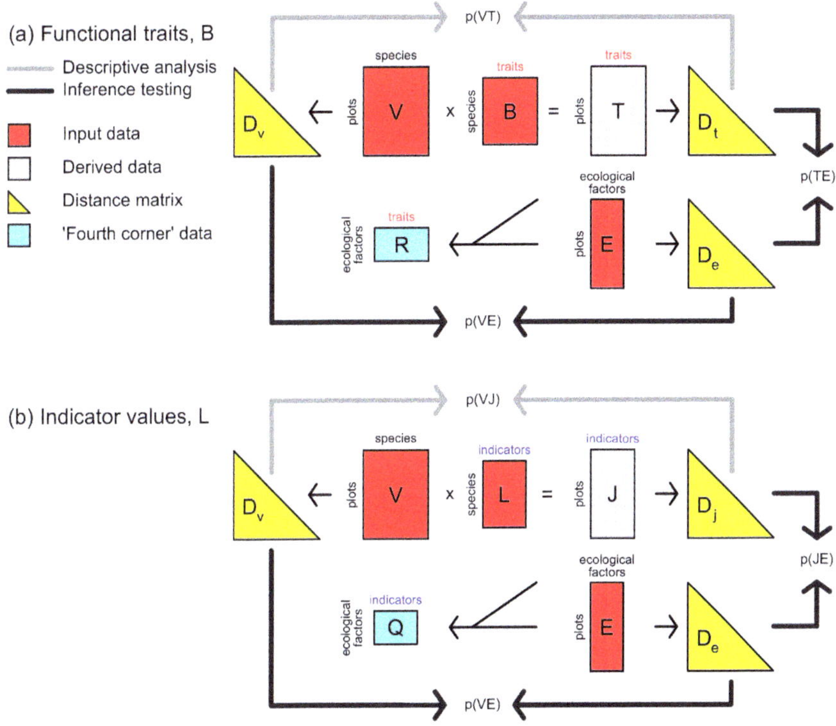

Figure 8.1: Analytical framework for processing species-, traits- and indicator-based vegetation data as well as environmental measurements (Wildi 2016). Rectangles are data matrices, triangles distance matrices.

in Section 7.3.1 this yields a squared correlation coefficient and its significance (error probability) is derived by a permutation test. Other methods involving the same matrices offer a more detailed view of patterns. These are, for example, redundancy analysis and CCA (Section 7.4). Both try to explain pattern in vegetation data, V, by traits or indicators found in matrices T and J respectively.

Step 4. This brings plot-based environmental measurements, matrix E in Figure 8.1, into play. Matrix E is formally independent from vegetation data, V, and statistical inference can reveal relationships between vegetation and environment. Again, the patterns in matrix E are best assessed by a distance matrix, D_e. This is compared to matrices V, T and J respectively. The methods to do this are the same as in step 3: Mantel correlation, redundancy analysis or CCA, and so on. However, these methods now are means of statistical inference.

Step 5. There are two more derived matrices in Figure 8.1, namely R and Q. Their computation is related to the so-called 'fourth corner problem' introduced by Legendre *et al.* (1997) and discussed in detail in Legendre and Legendre (2012). These are the ecological factors by traits matrix, R, in panel (a) and the ecological factors by indicator value matrix, Q, in panel (b). The fourth corner problem is explained in more detail in Section 8.3.

It is sometimes believed that the Mantel test is a method to generally assess statistical inference of traits or indicator values versus species as descriptors (Zéleny and Schaffers 2012). This, however, is a misinterpretation of the analytical concept. The relevé by indicator matrix, T, as well as the relevé by indicator matrix, J, are derived from the relevé by species matrix, V, and they lack independence as would be required for a proper statistical test. Therefore, in the scheme of Figure 8.1 this step is termed 'descriptive analysis' (Wildi 2016). This not only concerns Mantel correlation but all methods involving the three matrices (see step 3).

The framework in Figure 8.1 does not include all the possible questions concerning the pattern and functioning of the system. Others are analyses across elements of panels (a) and (b). Subjects could be matrices T and J to compare traits space and indicator value space, or matrices B and L with the aim to determine correlation between traits and indicator values.

8.3 Matrix operations in a nutshell

Deriving all the data matrices shown in Figure 8.1 is an exercise of matrix operations combined with proper adjustment of row- and column vectors. This is shown below in a little, artificial example suited for longhand calculation. All this starts by generating a conventional vegetation data set, V, where plots are the sampling units and species the descriptors. This has two rows (the relevés) and three columns (the species):

$$\mathbf{V} = \begin{array}{cccc} & sp1 & sp2 & sp3 \\ r1 & 0.1 & 0.5 & 0.4 \\ r2 & 0.5 & 0.5 & 0.0 \end{array}$$

Input in $\mathcal{R}$ is done through the following code:

```
V<- matrix(c(0.1,0.5,0.5,0.5,0.4,0.0),nrow=2)
rownames(V)<- c("r1","r2")
colnames(V)<- c("sp1","sp2","sp3")
```

In this matrix the rows must have unit sum to avoid changes of scale in subsequent operations (see Section 8.4 for a more detailed example). This condition is fulfilled here already. If this was not yet the case, the following $\mathcal{R}$ code would do it:

```
rs <- apply(V,1,sum)
V <- V/rs
```

Then, we enter a species by traits matrix, B:

$$\mathbf{B} = \begin{array}{cccc} & pf1 & pf2 & pf3 \\ sp1 & 1 & 1 & 0 \\ sp2 & 0 & 1 & 0 \\ sp3 & 1 & 0 & 1 \end{array}$$

Input into $\mathcal{R}$ is as follows:

```
B<- matrix(c(1,0,1,1,1,0,0,0,1),nrow=3)
colnames(B)<- c("pf1","pf2","pf3")
rownames(B)<- c("sp1","sp2","sp3")
```

These are presence–absence data, that is, a trait is either present or it is non-existent. We usually need not be concerned about missing values as we are interested in presence only, but not in the cause of eventual absence. In the course of application a different issue may arise. When taking traits from a database, which is the case considered here, some will have zero or very low frequency, such as one, two or alike. We may therefore want to implement a mechanism omitting low frequency traits. In Section 8.4 it is explained how this is done in $\mathcal{R}$.

We are now ready to derive the plot by traits matrix, T [Figure 8.1(a)]:

$$\mathbf{T} = \mathbf{VB} = \begin{array}{cccc} & pf1 & pf2 & pf3 \\ r1 & 0.5 & 0.6 & 0.4 \\ r2 & 0.5 & 1.0 & 0.0 \end{array}$$

An element of T is the relative frequency of a trait in any one relevé. For example, the first element in T is the sum of the first row elements in V multiplied by the first column elements of B, $(0.1*1)+(0.5*0)+(0.4*1)=0.5$. In $\mathcal{R}$ we write:

```
T <- V %*% B
```

The next matrix to be considered, L, is a description of species by indicator values (Figure 8.1(b)):

$$\mathbf{L} = \begin{array}{ccccc} & ind1 & ind2 & ind3 & ind4 \\ sp1 & 2 & 2 & 5 & 1 \\ sp2 & 1 & 2 & 5 & 0 \\ sp3 & 5 & 2 & 4 & 1 \end{array}$$

Matrix L is entered as follows:

```
L<- matrix(c(2,1,5,2,2,2,5,5,4,1,0,1),nrow=3)
colnames(L)<- c("ind1","ind2","ind3","ind4")
rownames(L)<- c("sp1","sp2","sp3")
```

We interpret indicator values as continuous metric variables suitable for many kinds of operations. As long as there are no missing values, a relevé by indicator matrix is readily derived. But this will rarely be the case, as the database of Landolt *et al.* (2010) includes scores (letter 'x') where the indicator value remains undefined. The meaning of this is that either the experts do not know the proper value or the species does not respond to the factor in mind. As an intermediate step we globally replace 'x' by 0 (zero). The new matrix J', to which undefined values still contribute, is the scalar product of V and L:

$$\mathbf{J}' = \mathbf{VL} = \begin{array}{ccccc} & ind1 & ind2 & ind3 & ind4 \\ r1 & 2.7 & 2 & 4.6 & 0.5 \\ r2 & 1.5 & 2 & 5.0 & 0.5 \end{array}$$

In $\mathcal{R}$ code this is:

```
J.dot<- V %*% L
```

The elements of J' are mean indicator values. In cases where some of these have zero (undefined) contributions we have to correct the computation of means using the frequencies of valid indicators in matrix L only. Obviously, ind4 is incorrect (0.5), a value not compatible with the definitions in Landolt *et al.* (2010). We therefore derive a matrix of relative indicator frequency, Ci, and divide J' in an element by element operation by this:

```
Ci <- V %*% sign(L)
J <- J.dot / Ci
J

   ind1 ind2 ind3 ind4
r1  2.7    2  4.6    1
r2  1.5    2  5.0    1
```

In the adjusted matrix, J, ind4 is now correct with elements equal to 1.0.

All operations presented so far concerned new sets of variables substituting species as vegetation descriptors. A data source independent of vegetation constitutes the site factors measured within the same plots, matrix E in Figure 8.1. Correlating species vectors with site factors reveals the niches the species occupy in the environment. When species are substituted by their traits then an analogue issue is finding the niches of the traits – if these exist. In Figure 8.1(a) the matrix holding this information, R, has as many rows as there are site factors in matrix E, and as many columns as there are columns in matrix B. Deriving and testing matrix R is called the 'fourth corner problem', devised by Legendre *et al.* (1997) and extended by Dray and Legendre (2008). A detailed presentation is found in Legendre and Legendre (2012), p. 613, where matrices V, B, E and R in Figure 8.1(b) are called A, B, C and D respectively. As in previous steps of analysis the solution is again found by matrix algebra:

$$R = (VB)'E = T'E \tag{8.1}$$

Equation 8.1 is also applicable to the small example in this section (which, for testing significance of relationship, is far too small). But first we need a matrix of environmental factors, E, which in this case are metric:

$$\mathbf{E} = \begin{array}{cc} & pH \\ r1 & 4.5 \\ r2 & 5.9 \end{array}$$

Input into $\mathcal{R}$ proceeds as usual:

```
E<- matrix(c(4.5,5.9),ncol=1)
rownames(E)<- c("r1","r2")
colnames(E)<- c("pH")
```

Matrices R and Q should be equally scaled as the original measurements, pH in this example. pH is the sum of contributions by the traits in matrix T and therefore traits vectors are adjusted to unit sum, the new matrix now denoted by Tn:

$$\mathbf{T} = \begin{array}{cccc} & pf1 & pf2 & pf3 \\ r1 & 0.5 & 0.6 & 0.4 \\ r2 & 0.5 & 1.0 & 0.0 \end{array} \quad \mathbf{Tn} = \begin{array}{cccc} & pf1 & pf2 & pf3 \\ r1 & 0.5 & 0.375 & 1.0 \\ r2 & 0.5 & 0.625 & 0.0 \end{array}$$

This adaptation is achieved by the following $\mathcal{R}$ code:

```
rs <- apply(T,2,sum)
Tn <- t(t(T)/rs)
```

To ensure that the division by rs is applied to the columns, matrix T has to be transposed, and, after the operation, back-transposed. It is now sufficient to form the scalar product of Tn' and E to obtain matrix R:

$$R = Tn'E \tag{8.2}$$

Or, in terms of the $\mathcal{R}$ code:

```
R <- t(Tn) %*% E
```

Matrix R has the same number of columns as matrix E, holding the pH values:

$$\mathbf{R} = \begin{array}{cc} & pH \\ pf1 & 5.200 \\ pf2 & 5.375 \\ pf3 & 4.500 \end{array}$$

This reads as follows: the mean pH values at sites where functional type pf1 occurs, is 5.2. When pf2 occurs, it is 5.375, and so on. As this calculation depends on matrix T derived above, it intrinsically takes species scores (cover-abundance) into account. The elements in R are again linear combinations and the first one, for example, is obtained as (0.5*4.5)+(0.5*5.9)=5.2.

Figure 8.1 suggests that the fourth-corner problem also exists in the context of species indicator values. All data needed for our example is ready from the preceding steps: matrices V, J (the relevé by indicator matrix, compensated for missing values) and E. Just as in the case of traits the columns of matrix J are adjusted to unit sum:

$$\mathbf{J} = \begin{array}{ccccc} & ind1 & ind2 & ind3 & ind4 \\ r1 & 2.7 & 2.5 & 4.6 & 1.0 \\ r2 & 1.5 & 2.5 & 5.0 & 1.0 \end{array} \quad \mathbf{Jn} = \begin{array}{ccccc} & ind1 & ind2 & ind3 & ind4 \\ r1 & 0.64 & 0.5 & 0.48 & 0.5 \\ r2 & 0.36 & 0.5 & 0.52 & 0.5 \end{array}$$

The respective $\mathcal{R}$ code is:

```
rs <- apply(J,2,sum)
Jn <- t(t(J)/rs)
```

Finally, we get Q as the scalar product of transposed matrix Jn and E:

$$\mathbf{Q} = \mathbf{Jn'E} = \begin{array}{cc} & pH \\ ind1 & 5.00 \\ ind2 & 5.20 \\ ind3 & 5.23 \\ ind4 & 5.20 \end{array}$$

The notation in $\mathcal{R}$ is:

```
Q <- t(Jn) %*% E
```

The meaning of these scores is straightforward: high values of `ind1` result in a mean pH value of 5.00, etc. Finding a really useful application, however, may be difficult as the elements of Q are composed of contributions of high indicator scores. In the context of indicator values by Ellenberg (1974) or Landolt *et al.* (2010) matrix Q is meaningful only as long as an indicator value is interpreted as a continuous expression of one environmental factor (nutrient availability, for instance).

8.4 Schlaenggli data example

The 'Schlaenggli' example is the same as used in various sections of Chapters 6 and 7 (63 plots, 119 species). All data sources needed are implemented in $\mathcal{R}$ package `dave`, version 2.0 and later. The objective of this section is to explore data analysis in view of the extended possibilities when projecting data into the new spaces of indicator values and species specific traits respectively. As shown in Section 8.3 the analytical operations are rather simple, but properly preparing these requires additional effort. Therefore, in the first subsection I outline some of the preparatory tasks preceding analysis and interpretation.

8.4.1 Preparing data matrices

In order to keep the notation consistent we first assign the data frames to the variable names used in Figure 8.1 and we eventually combine this step with adaptations devised in Section 8.3:

```
V <- as.matrix(sveg^0.5)
rs <- apply(V,1,sum)
V <- V/rs
```

This first assigns the plot by species data frame `sveg` to matrix V while simultaneously taking the square root of species scores (Section 3.2). Then the relevé vectors are adjusted to unit sum.

In the next step data frame `sspft`, the plots by traits matrix is prepared. It has as many as 23 columns (the functional traits). Among these we are interested in joint occurrences only, that is, in the more frequent. We therefore restrict the matrix to columns with minimum frequency 5, for example:

```
col.freq <- apply(sign(sspft),2,sum)
B <- as.matrix(sspft[,col.freq >= 5])
```

Typing `dim(B)` reveals that 12 traits vectors are now left.

Then the indicator values excerpted from Landolt *et al.* (2010) are copied to become matrix L:

```
L<- as.matrix(ssind)
```

L may still include missing values, but these will be compensated in subsequent operations.

The plot-based environmental factors constitute the last matrix to be prepared. Not all variables may be suited for further analysis, for instance, if they show a seriously skewed distribution despite some transformation previously applied (Section 3.2). Or, there may be measurements below the detection threshold causing abundant zero elements. It is good practice to routinely inspect the distributions of all site factors by displaying the respective histograms. For data frame `ssit` this is done as follows:

```
par(mfrow=c(5,5))
for(i in 1:NCOL(ssit)) hist(ssit[,i],main=names(ssit)[i])
```

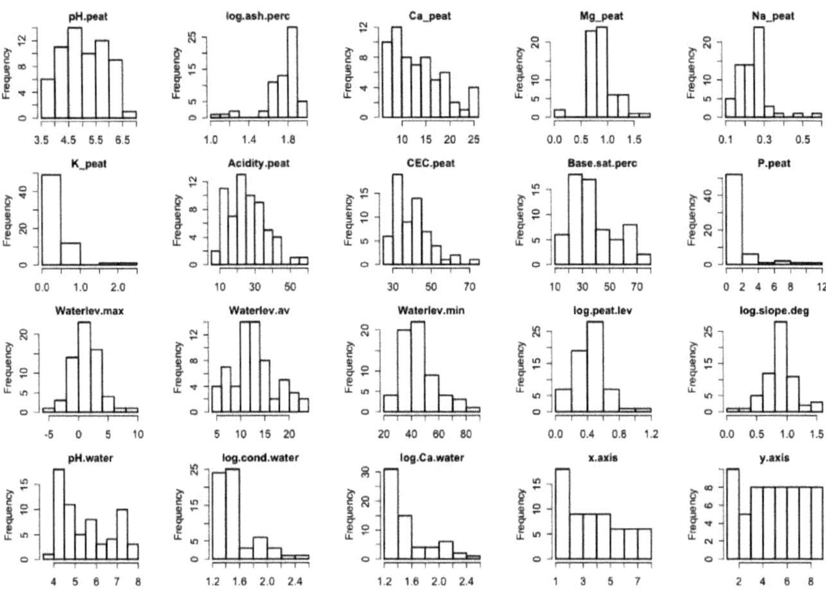

Figure 8.2: Histograms of all 20 site factors in data frame `ssit`.

The first line reserves space for 25 graphs on one page, the second plots all histograms with the variable names as titles, using function `hist()`. This delivers Figure 8.2. Some site factors show almost symmetric distributions whereas others are still left- or right-skewed. Removing these now, however, would be premature as we would run the risk of losing useful information. Because the number of site factors is far too large in view of the 63 plots, we try to get a restrictive selection optimally reflecting the resemblance pattern in the data. As a first step we omit variables 19 and 20, the spatial coordinates. A method to efficiently reduce the number of variables, not yet considered in Section 6.7, is orthogonal ranking of site factors. The result of ranking is subsequently used to build the final matrix E:

```
E.red <- as.matrix((ssit[,c(-19,-20)]))
o.orank <- orank(as.data.frame(E.red),use="columns",rlimit=10)
E <- E.red[,o.orank$var.names]
```

The result (not shown below) reveals that 10 factors are sufficient to account for roughly 95% of total variation.

Table 8.1: Definitions of variables of matrices B (data frame `sspft`), L (`ssind`) and E (`ssit`) in the 'Schlaenggli' example according to Figure 8.1. Plant traits and indicator values are taken from Landolt *et al.* (2010), where abbreviation LF is used for life form, FS for reproduction and c, r and s are life strategies according to Grime (2001). Site factors are taken from Wildi (1977).

(a) Plant traits

Name	Symbol used	Definition
Geophyte	LF.g	With resting buds below ground
Hemicryptophyte (long-lived)	LF.h	Buds on or directly below ground (rosettes, tussocks)
Therophyte	LF.t	Dying back, surviving as seed or annual hemicryptophyte
Moss	LF.moss	Mosses list (Landolt *et al.* 2010)
Sphagnum	LF.sph	Moss, Sphagnum-like
Hermaphrodictic, normal sexual	FS.zw	Pollination necessary to reproduce
Unisexual and dioecious	FS.di	Only male or female organs
Unisexual and monoecious	FS.mo	Male, female and bisexual flowers on one individual
Polysexual	FS.ve	Male and/or female and bisexual flowers on one individual
Competitive strategists	c	Competitive, long-lived
Ruderal strategists	r	Pioneer species, short-lived
Stress-tolerant strategists	s	Adapted to harsh conditions

(b) Indicator values

Topic	Name	Lowest score	Highest score	Ranks
Climate	**T** Temperature	1 alpine and nival	5 very warm, colline	9
Climate	**K** Continentality	1 oceanic	5 continental	5
Climate	**L** Light	1 deep shade	5 full light	5
Soil	**M** Moisture	1 very dry	5 submerged	9
Soil	**W** Moisture availab.	1 low variation	3 high variation	3
Soil	**R** Reaction	1 extremely acid	5 alkaline	5
Soil	**N** Nutrients	1 very infertile	5 very fertile	9
Soil	**H** Humus	1 little/no humus	5 high humus cont.	3
Soil	**D** Aeration	1 bad aeration	5 good aeration	3

(c) Site factors

Name	Abbreviation	Description
Acidity.peat	Ac.p	Total hydrolytic acidity (mval/100g peat)
Base.sat.perc	B.sat	Percentage saturation with basic cations
Waterlev.av	w.av	Mean water level below surface, cm
K_peat	K.p	K content of peat (mval/100g peat)
log.slope.deg	slp	Slope in degrees, logarithm
log.peat.lev	peat	Logarithm of peat layer in cm
Mg_peat	Mg.p	Mg content of peat (mval/100g peat)
Waterlev.min	w.min	Minimum water level below surface, cm
log.cond.water	cond	Logarithm of conductivity of water
pH.peat	pH.p	pH measured in peat (dissolved in water)

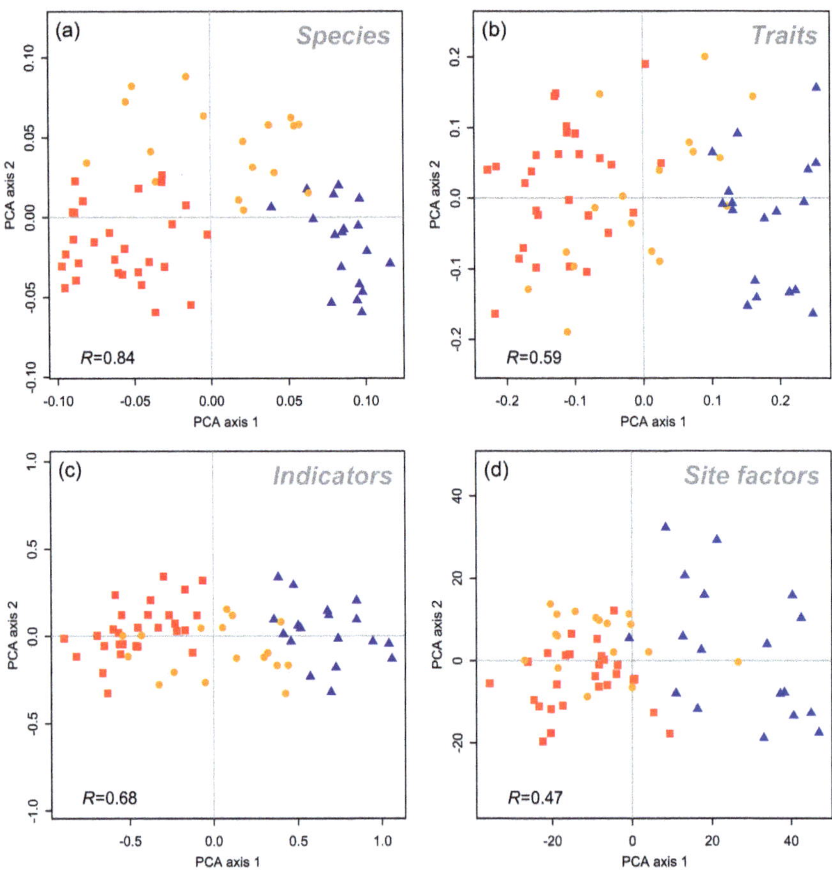

Figure 8.3: Ordinations of plot by species data (a), plot by traits (b), plot by indicator values (c) and plots by environmental factors (d) in two-dimensional Euclidean space (PCA solutions). Groups are distinguished for ease of interpretation only.

After preparing the data sets for species traits, indicator values and site factors, there are now 12, 9 and 10 variables respectively remaining for further processing, analysis and interpretation. The variable names and definitions of all these are listed in Table 8.1.

8.4.2 Deriving and projecting new data spaces

All data matrices serving as sources of information are now ready for further processing. The next step involves deriving matrices T (relevés by traits) and J (relevés by indicators) through matrix multiplication. For matrix T of species traits no further correction is needed and it is sufficient to form the matrix product of V and B:

```
T <- V %*% B
```

In matrix L there are some zero entries which are missing values. We compensate for these as explained in Section 8.3:

```
J.dot <- V %*% L
Ci <- V %*% sign(L)
J <- J.dot / Ci
```

As outlined in Section 8.1 the same relevés are now described in four different descriptor spaces by matrices V, T, J and E respectively. In order to get a visual representation of the resemblance patterns involved we generate the respective ordinations in Euclidean space using PCA (see Section 6.2). The result of this is shown in Figure 8.3. All data points in these graphs are similarly classified for ease of comparison. The three groups are generated by clustering the relevé by species matrix, V. Accordingly, they are best resolved in panel (a), whereas in (b), (c) and (d) they partly overlap. Although the classification in species space is not perfectly carried over to the indicator, traits and site factor space, the pattern is still apparent. The distinction of the groups is further facilitated by the R-values originating from analysis of similarities (ANOSIM, see Chapter 6.4). As expected the plots by species data used for clustering achieves the best resolution in two-dimensional ordination with R=0.84, followed by the indicators [(c), R=0.68], the traits [(b), R=0.59] and the environmental factors [(d), R=0.47]. Interesting differences emerge in the shapes of the point clouds. In panel (a), for instance, this is horseshoe-shaped (see Section 6.5), indicating strong nonlinearity occurring in species space. The same holds for the traits space, panel (b). Most remarkably, the indicator space [panel (c)] is almost linear. This is probably one of the reasons why mean indicator values are frequently experienced as being surprisingly effective (Diekmann 2003).

Deriving the graphs in Figure 8.3 starts with classifying the vegetation by species matrix, V, as explained in more detail in Chapter 5:

```
mde <- vegdist(V,method="euclidean")
o.hcl <-hclust(mde,method="ward.D")
o.grel<- cutree(o.hcl,k=3)
```

The three groups serve as a means for ease of interpretation only. Panel (a) is a plot of PCA scores in which group membership determines the kind of symbols (`pch=o.grel`) and also their colour (`col=o.grel`):

```
o.pca<- pca(V)
plot(o.pca$scores[,1],o.pca$scores[,2],asp=1,pch=o.grel,col=o.grel)
```

For the remaining graphs it is sufficient to replace `V` by `T`, `J` and `E` respectively. Figure 8.3 confirms some agreement of patterns in the four spaces, underpinning what similarity theory would predict (Section 1.2).

Table 8.2: Mean values of site factors [columns, Table 8.1, (c)], for plant traits [rows, Table 8.1, (a)] in the 'Schlaenggli' data set. The upper part is matrix R according to the terminology given in Figure 8.1. The last three rows are parts of matrix Q, where K, F, and R are indicator values [Table 8.1, (b)].

	Ac.p	B.sat	w.av	K.p	slope	peat	Mg.p	w.min	cond	pH.p
LF.g	25.89	38.81	13.17	0.43	0.92	0.46	0.87	45.36	1.53	5.16
LF.h	24.50	41.02	13.40	0.42	0.93	0.44	0.88	46.85	1.55	5.26
LF.t	17.11	56.02	15.43	0.34	1.07	0.30	0.97	52.96	1.76	5.96
LF.moss	24.90	40.45	13.29	0.40	0.94	0.43	0.88	45.90	1.54	5.25
LF.sph	30.43	29.91	12.04	0.50	0.84	0.54	0.85	41.29	1.41	4.75
FS.zw	25.10	39.77	13.30	0.42	0.92	0.45	0.88	46.28	1.53	5.21
FS.di	23.33	43.09	13.46	0.40	0.94	0.42	0.87	48.00	1.57	5.35
FS.mo	24.74	40.80	13.21	0.42	0.94	0.43	0.87	45.81	1.55	5.25
FS.ve	29.68	33.11	13.55	0.51	0.85	0.52	0.84	46.78	1.49	4.79
c	26.13	38.07	13.16	0.43	0.91	0.46	0.87	45.57	1.52	5.12
r	19.28	50.88	14.47	0.37	1.03	0.36	0.92	50.82	1.69	5.71
s	26.05	38.24	13.16	0.44	0.91	0.46	0.87	45.57	1.52	5.12
K	25.66	38.98	13.20	0.43	0.92	0.45	0.87	45.74	1.53	5.17
F	26.04	38.32	13.05	0.43	0.91	0.46	0.87	45.25	1.52	5.14
R	24.44	41.27	13.32	0.41	0.94	0.43	0.89	46.43	1.55	5.28

Table 8.3: Mean values of indicator values [columns, Table 8.1, (b)] for plant traits [rows, Table 8.1, (a)] in the 'Schlaenggli' data set. This is matrix S, not shown in Figure 8.1.

	T	K	L	F	W	R	N	H	D
LF.g	2.92	2.88	3.56	3.74	2.60	2.55	2.14	4.54	1.20
LF.h	2.92	2.89	3.57	3.73	2.60	2.62	2.16	4.50	1.20
LF.t	2.92	2.94	3.58	3.62	2.63	3.06	2.26	4.30	1.20
LF.moss	2.92	2.89	3.56	3.72	2.60	2.62	2.16	4.49	1.20
LF.sph	2.92	2.84	3.56	3.82	2.57	2.28	2.08	4.65	1.19
FS.zw	2.92	2.88	3.56	3.73	2.60	2.58	2.16	4.51	1.20
FS.di	2.93	2.90	3.56	3.71	2.61	2.68	2.18	4.48	1.20
FS.mo	2.93	2.88	3.56	3.73	2.60	2.60	2.16	4.52	1.20
FS.ve	2.93	2.86	3.54	3.75	2.57	2.35	2.11	4.62	1.20
c	2.92	2.87	3.56	3.74	2.59	2.53	2.14	4.54	1.20
r	2.92	2.92	3.58	3.66	2.62	2.91	2.22	4.38	1.20
s	2.92	2.87	3.56	3.74	2.59	2.53	2.14	4.54	1.20

Further matrix operations allow the derivation of matrices R and Q (Figure 8.1), for which we use the same instructions and symbols as in Section 8.3. We first adjust the vectors to unit sum, perform the matrix operations and round the elements in R and Q for convenience:

```
rs <- apply(T,2,sum) ; Tn <- t(t(T)/rs)
R <- t(Tn) %*% E ; round(R,digits=2)

rs <- apply(J,2,sum) ; Jn <- t(t(J)/rs)
Q <- t(Jn) %*% E ; round(Q,digits=2)
```

The result is shown in Table 8.2. The elements in the table are mean environmental factors for each species trait in matrix T. In the last column, for example, we see that when the life form is 'moss' (LF.moss) then the expected mean pH value is 5.25, whereas in the case of life form 'sphagnum' (LF.sph) it is 4.75.

As mentioned in Section 8.3 the Q matrix, relating indicator values and site factors, is probably not really useful. First, because this investigation is a local one, many indicator values are almost identical, such as temperature (T), continentality (K) and light (L). Second, our matrix operations assume indicator values to measure the expression of one specific property, whereas in reality they often address different issues (for example, poor aeration in wet, but lack of water in dry conditions).

An interesting extension, although beyond the general framework in Figure 8.1 is when relating species traits and indicator values. For this purpose we implicitly assume the latter to be site factors (which they are not) and the corresponding matrix, S, is calculated as follows:

```
rs <- apply(T,2,sum) ; Tn <- t(t(T)/rs)
S<- t(Tn) %*% J ; round(S,digits=2)
```

This code is the same as used for deriving matrix R, but indicators (J) are used instead of site factors (E). The content of S is shown in Table 8.3 and it provides mean indicator values for all species traits considered in the analysis. Just as in the case of matrix Q some indicator values vary with species traits, such as the F-value and the R-value, whereas others just exhibit minor random fluctuation (L, K and D, for example). Notably, these results are valid for the current example only and from other data sets we would get different results.

8.4.3 Measuring convergence

For measuring agreement of spaces (convergence according to Section 1.2) a good choice offers Mantel correlation (see Section 7.3.1). When computing all combinations of Mantel correlations the following similarity matrix is obtained:

$$
\mathbf{r}_{\text{Mantel}} = \begin{array}{c|cccc} & \mathbf{V} & \mathbf{J} & \mathbf{T} & \mathbf{E} \\ \mathbf{V} & 1.0 & & & \\ \mathbf{J} & 0.82 & 1.0 & & \\ \mathbf{T} & 0.79 & 0.78 & 1.0 & \\ \mathbf{E} & 0.56 & 0.62 & 0.59 & 1.0 \end{array}
$$

In this, the first element in the second row, for example, is obtained by function `mantel()`:

```
mdv <- vegdist(V,method="euclidean")
mdj <- vegdist(J,method="euclidean")
mantel(mdv,mdj)
```

The most interesting issue is correlation of vegetation with environmental factors, because the latter are supposed to be statistically independent. Mantel correlation is 0.56, 0.62 and 0.59 for V, J and T respectively. These are all high values and at the same time almost identical, signifying that all three spaces equally predict environmental factors. We also learn that indicator values, J, fit best with plots by species data, V (R=0.82), but species traits are almost as successful, with a Mantel correlation of 0.79. Again, this is not just due to the suitability of alternative vegetation descriptors, but also a consequence of the strong gradient involved in the data.

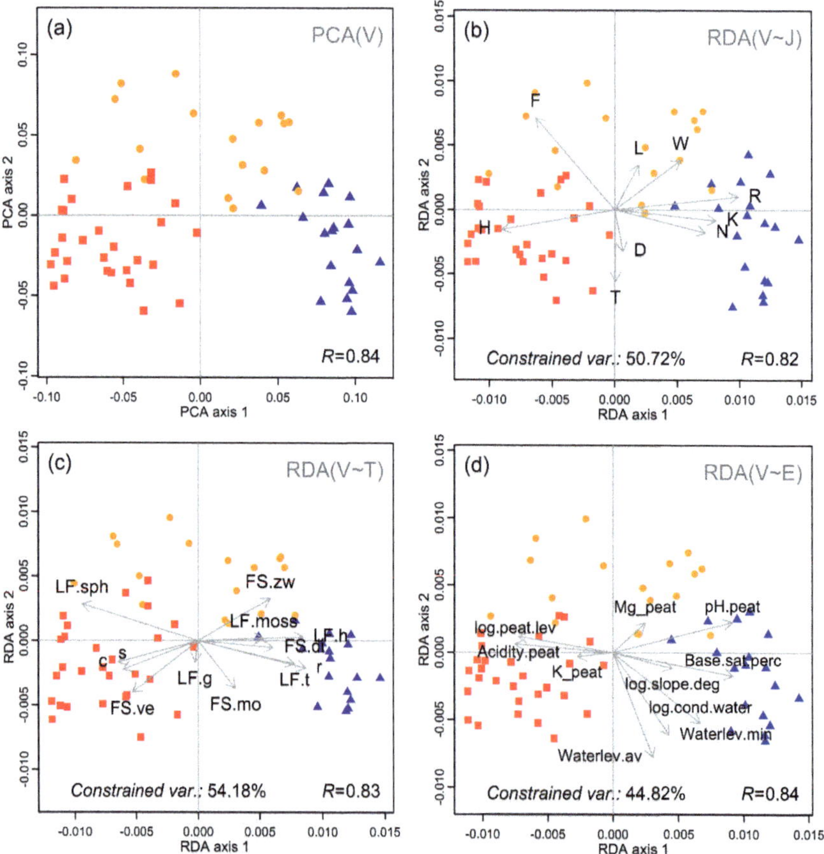

Figure 8.4: Convergence of species- (V), indicators- (J), traits- (T) and site factor (E) spaces using PCA and RDA solutions. (a) PCA of matrix V. (b) RDA of V constrained by indicator values. (c) RDA of V constrained by traits. (d) RDA of V constrained by site factors.

A more detailed interpretation is obtained when considering the role of the new descriptor variables individually, the traits and the indicator values and, of course, all these compared to real site factors. A short description of the traits as excerpted from Landolt *et al.* (2010) is given in Table 8.1, part (a) and the same for the indicator values in (b). The abbreviations of names and definitions of the site factors are shown in Table 8.1, part (c).

There are many different ways of showing correlations of spaces, and one of these is used in Figure 8.4, panels (a) through (d). As in Figure 8.3 all are based on principal component analysis (PCA). Panel (a) is PCA of plot by species data, matrix V, and it is identical to Figure 8.3 (a). The remaining graphs are ordinations of constrained variance in redundancy analysis (RDA, Section 7.4). All three sets of variables roughly constrain half of the variance, with traits having the highest proportion (54.18%) and environmental factors the lowest (44.82%). Not surprisingly, the resulting shape of point clouds show but minor variation. The R-values from analysis of similarities (ANOSIM, see Chapter 6.4) are all very high confirming the efficiency of the two-dimensional display.

In all three panels, (b) through (d), the majority of variables is correlated with the first ordination axis. In the graph of indicator values, (b), temperature, indicator value T, is orthogonal to this and humidity, F, also partly. In the panel of traits, (c), orthogonal components are practically missing. In the environmental factors, (d), various measures of water level contribute to the second axis.

This example supports widespread findings that all three spaces (species, traits and indicator values) capture the vegetation patterns surprisingly well. Obviously the strong trend in the data reflects a property of the system the descriptive variables notwithstanding. The indicator values are the only ones devised by experts – with the obvious advantage that some suggest conditions not even measured: temperature (T), for instance.

The ordinations in Figure 8.4 are generated as explained in Chapters 6 and 7. Just as in Figure 8.3 we first establish three groups, based on matrix V, to facilitate the recognition of types:

```
mde <- vegdist(V,method="euclidean")
o.hcl <-hclust(mde,method="ward.D")
o.grel<- cutree(o.hcl,k=3)
```

Panel (a) is plotted as explained in Section 8.4.2. RDA in panels (b) through (d) is done as explained in Section 7.4 where I suggest adjusting the coordinates to the magnitude of the corresponding eigenvalues. For computation of RDA in the **vegan** package the data matrices have to be converted to data frames (function `as.data.frame()`). Then, the scores are retrieved by method `scores()` to obtain x and y:

```
Vd<- as.data.frame(V) ; Jd<- as.data.frame(J)
```

```
Td<- as.data.frame(T) ; Ed<- as.data.frame(E)
o.rda<- rda(Vd~.,data=Jd)
sc <- scores(o.rda)
x <- sc$sites[,1] ; y <- sc$sites[,2]
```

We can adjust the scores to the corresponding eigenvalues, which are then plotted and the indicator values added in the form of arrows:

```
s1<- (o.rda$CCA$eig[1]^0.5)*x/(sum(x^2))^0.5
s2<- (o.rda$CCA$eig[2]^0.5)*y/(sum(y^2))^0.5
plot(s1,s2,asp=1,col=o.grel,pch=o.grel)
sf.scores<- o.rda$CCA$biplot*0.01
arrows(0,0,sf.scores[,1],sf.scores[,2],length=0.08)
text(sf.scores[,1],sf.scores[,2],colnames(Jd),pos=1)
```

The `sf.scores` are scaled arbitrarily to fit the graph. For panels (c) and (d) we replace Jd by Td and Ed twice, first for redundancy analysis, then for the labels of the arrows.

Arrows are not always convenient for the interpretation of ordinations. In Figure 6.3 surface plots were introduced for use with site factors. The frequency of traits (matrix T) or the mean indicator values (matrix J) are equally suited, as long as they show a clear trend within the ordination. Two examples are given in Figure 8.5. Panel (a) shows the trend surface of the frequency of life-form 'sphagnum' superimposed to a PCA ordination. There is a clear trend from zero (lower right) towards frequency 0.2 (upper right). In panel (b) the R value exhibits range 2.0 through 3.2. To reproduce panel (a), PCA is computed and drawn with plots by species data, V, followed by a call of function `ordisurf()`:

```
o.pca<- pca(V)
plot(o.pca$scores[,1],o.pca$scores[,2],asp=1)
ordisurf(o.pca$scores[,1:2],Td[,"LF.sph"],add=TRUE)
```

For panel (b) we replace `Td[,"LF.sph"]` by `Jd[,"R"]`.

A final question remains: which of these interactions are significant? This is far from trivial as a proper solution depends on the data types used. A convenient solution is offered in $\mathcal{R}$ package 'ade4', function `fourthcorner()`. This identifies data types and calculates significance through permutation. Function `fourthcorner()` is not accessed by package `dave` and it has to be loaded separately.

8.5 Rebuilding community ecology?

Annoyed by the slow progress and the lack of generality vegetation ecology recently achieved, McGill *et al.* (2006) suggested rebuilding community ecology from functional traits. In their view the culprits for the unpleasant

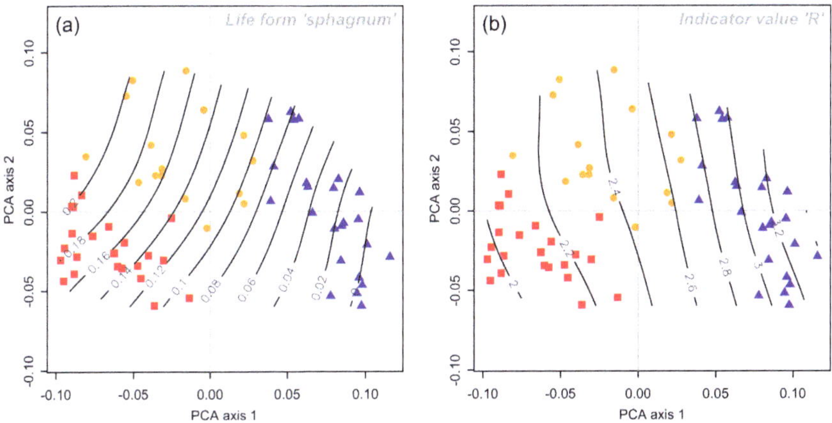

Figure 8.5: PCA ordinations with trend surfaces. (a) frequency of life form 'sphagnum', (b) indicator value 'R' used.

situation are the species. And in fact whenever an investigation addresses a specific species then the results are valid only within its area, that is, the geographical range of occurrence. This is why recent efforts have shifted from the species view towards plant traits (Kleyer *et al.* 2012).

In this chapter I have shown that substituting species by traits or indicator values is a logical step in a desired direction and it is easily done with the floristic databases available. Furthermore it can even be applied to the millions of historic relevés stored worldwide (2.4 million according to Dengler *et al.* 2011). Yet, the use of floristic databases has one serious implication: the similarity pattern of species is partly carried over to traits and indicator values (Wildi 2016), and as pointed out in Section 8.2 it limits the possibility of independent statistical testing. This raises the question of whether the step ahead desperately asked for by McGill *et al.* (2006) is realistic at all.

I first consider the example of plant traits as presented in Section 8.4. This suggests that even a crude selection [the traits used here are merely a preferential choice of what Landolt *et al.* (2010) offer in their book] delivers surprising agreements of patterns among species, traits and environmental factors. The strong floristic gradient inherent in the data automatically translates into the gradient of traits, whether the latter have ecological significance or not. On the upside the chance for successfully describing vegetation patterns worldwide seems to be granted and many of the traits (life forms, for example) may be of general validity.

The performance of indicator values once more proved to be stunning, a fact frequently causing scepticism (see, for example, Zelený and Schaffers 2012). Recalling just some of the favourite properties of indicator values reveals that this does not come as a surprise:

- Scaling. Indicator values of Landolt *et al.* (2010) all have the same range (1 to 5). Transformation to adapt scale is hardly ever needed.
- Dimensionality. Multidimensionality is limited to a maximum of nine variables, resulting in high explaining power of two-dimensional ordinations.
- Comprehensiveness. Unlike environmental factors indicator values are unlikely to miss the most important site conditions (light, nutrients, water, warmth, for instance).
- Species dependence. Because mean indicator values depend on plot by species data, any pattern (groups, gradients) translates directly into the indicator space.
- Expert knowledge. Indicator values result from a consensus of expert opinion in vegetation ecology.

However, indicator values as they are used these days offer no means of measuring or testing. Accordingly, they are probably restricted to local studies as is the case for the species. Making simultaneous use of a variety of options, plant traits, indicator values and site factors is probably still the most flexible strategy in local investigations. When it comes to the comparison of geographically separated areas we can expect that traits gain importance in measuring convergence of vegetation types while similarity based on species lists tends to deteriorate (Orlóci 1991b).

9 Static predictive modelling

What is the most likely outcome of vegetation composition under a real or hypothetical set of environmental parameters? This is the question static modelling is attempting to answer by means of statistical inference.

9.1 Predictive or explanatory?

The term modelling is used in a broad context and it deserves closer specification. Loehle (1983) proposed a classification of models, finding them to be either logical, theoretical or predictive. Logical and theoretical models have their strength in the universal validity: the systems described are assumed to be governed by natural rules and laws, although the parameters may still come from measurements. Functions involved attempt to mimic what happens in the reality or what is believed to happen. In vegetation ecology one would address, for instance, the flow of water between the soil, the plant and the atmosphere, or the same for nutrients where processes in the realm of bacteria and fungi play a crucial role. Yet, finding proper parameters to describe all this may be elusive.

The models shown in this chapter are modest in their claim as they operate in the context of the data world from which probabilities of occurrence of states are derived and Loehle (1983) would therefore classify these as 'predictive'. Because the relationships addressed are correlative by nature, De' ath (2002) in his introduction to the use of regression trees for modelling argues that the focus is on prediction rather than explanation. This has consequences already in the phase of acquisition of environmental variables where efficiency of measurement is often more important than ease of interpretation. Because the ultimate goal is maximizing predictive power at high resolution in space and time, variables integrating various properties of the environment are preferred, for example altitude (frequently integrating a temperature, precipitation and radiation gradient) or mean annual temperature (partly correlated with warmth in the growing season and the risk of frost damage). Consequently, static modelling is a rather technical discipline with focus on digital terrain models, spatial databases and advanced mathematical tools for data processing. Unlike applications shown in Chapter 7, understanding the underlying ecological process is welcome but not a priority.

Static models operate with probabilities, for which they are sometimes termed probability models (Guisan and Zimmermann, 2000). As is often the case in ecology they inherit uncertainty: that found in the underlying investigations. In regression, for example, uncertainty arises from the variance not explained by the straight regression line. More generally speaking, within the framework of modelling plant or animal occurrence, it causes the models to fail under certain circumstances.

In the case of vegetation a decision is needed on what should be predicted, either individual plant species, species traits, or entire vegetation types, in analogy to the historic debate in plant ecology with well-known publications by Clements (1916) voting for types and Gleason (1926, 1939) for individual species. Chapman and Purse (2011) argue that this is a largely technical question irrespective of objective. Once various species distributions are modelled then these can be subjected to classification, resulting in vegetation types and ultimately in probabilities for the occurrence of vegetation types. In the reverse case, if vegetation types are modelled directly, relative species abundance within types in the initial state deliver an estimate for occurrence probability after modelling.

9.2 Evaluating environmental predictors

As shown in Chapter 4, a resemblance matrix of relevés is a comprehensive representation of vegetation pattern. In probability modelling biotic data are related to environmental data. But there exists no general agreement for an assessment of the environment, forcing the modeller to evaluate the potential of his or her data in more detail. Commonly, a large number of site variables is available, some measured in field plots, others taken from external sources like high-resolution terrain models, climate databases or remote sensing data. The reasons why careful evaluation is needed are manifold (Guisan and Zimmermann 2000):

- *Insufficient resolution.* Models can be applied only to systems with sufficiently resolved site variables. It is easy to interpolate mean temperature, for instance, based on digital terrain models, but this does not usually work when trying to get detailed soil data or reliable information on land use.

- *Redundancy in the data.* Many variables reflect the dominating trends in the investigation area. In mountainous areas, for example, elevation is usually related to temperature and precipitation, both jointly affecting soil formation. When several variables are highly correlated, only one should be used for modelling: including all will maximize model fit, but at the expense of increased degrees of freedom, thereby hampering predicting power. The presence of correlated predictive variables in regression models is also known as *collinearity*.

- *Distribution problems.* Skewed distribution and most of all outliers (Section 12.3.1) mimic high correlation when this is low or absent. Data transformation or elimination of outliers is needed to correct for this.

- *Nonlinearity of relationships among variables* causing the opposite: it mimics absence of correlation when in reality it exists. Again, specific transformation may serve as the remedy.

Similarly behaving site factors are revealed by inspecting a correlation matrix or performing PCA (Section 6.2). An opportunity to even detect nonlinear correlations offers function `pairs()` in $\mathcal{R}$. To illustrate this the 'Schlaenggli' data, available as data frame `ssit` in package `dave`, serves as a real world example. Function `names()` displays all 20 variables:

```
names(ssit)
```

```
 [1] "pH.peat"       "log.ash.perc"    "Ca_peat"        "Mg_peat"        "Na_peat"
 [6] "K_peat"        "Acidity.peat"    "CEC.peat"       "Base.sat.perc"  "P.peat"
[11] "Waterlev.max"  "Waterlev.av"     "Waterlev.min"   "log.peat.lev"   "log.slope.deg"
[16] "pH.water"      "log.cond.water" "log.Ca.water"   "x.axis"         "y.axis"
```

These are far too many for modelling in view of the sample size of 63. Apparently there are measurements assessing similar properties of the environment, for example `pH.peat` and `pH.water`, the first measured in the laboratory in peat samples, the second in open water in the field. Then there are three measurements of water level, all resulting from the same hydrological process. The `pairs()` function would generate 400 graphs, too many for visual inspection. We had better select the variables we are interested in, for example, numbers 1 (`pH.peat`), 16 (`pH.water`), 13 (`Waterlev.min`), 12 (`Waterlev.av`) and 11 (`Waterlev.max`) to form a reduced object `site1`, followed by checking the correlation matrix [function `cor()`], rounded by function `round()`:

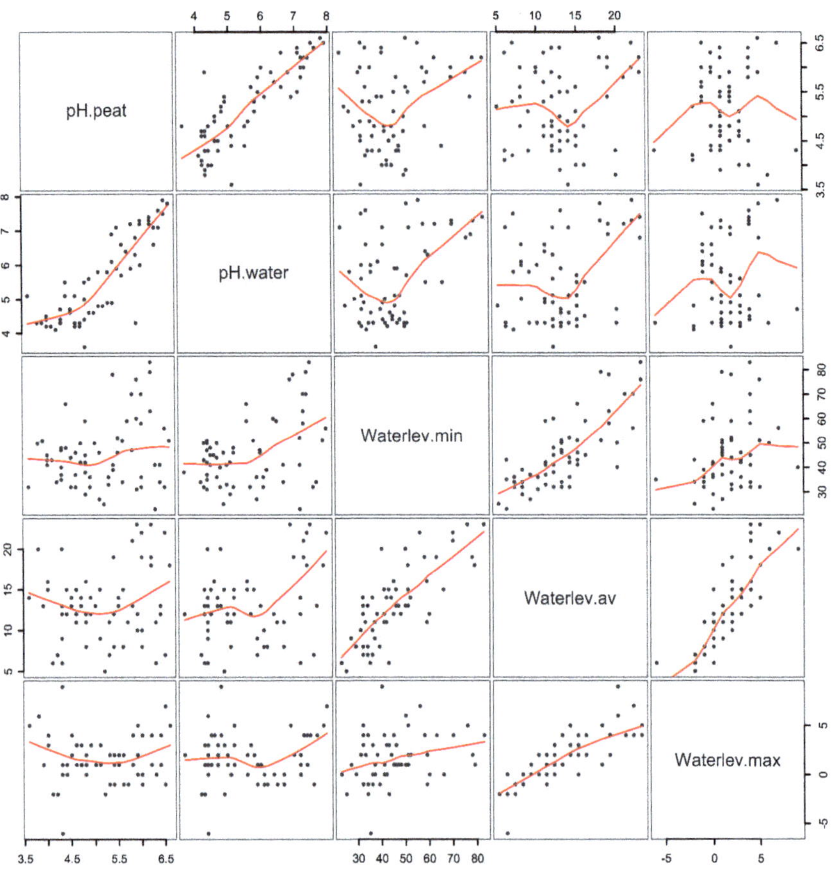

Figure 9.1: Pairwise plot of selected site variables. Regression lines result from a smoothing function.

```
site1<- ssit[,c(1,16,13,12,11)]
round(cor(site1),digits=3)
```

```
             pH.peat pH.water Waterlev.min Waterlev.av Waterlev.max
pH.peat        1.000    0.851        0.240       0.146        0.020
pH.water       0.851    1.000        0.423       0.345        0.170
Waterlev.min   0.240    0.423        1.000       0.768        0.296
Waterlev.av    0.146    0.345        0.768       1.000        0.811
Waterlev.max   0.020    0.170        0.296       0.811        1.000
```

This confirms our suspicion of pH values being highly correlated ($r = 0.851$), but average water level and, for example, pH peat just slightly ($r = 0.146$). The same data are displayed by function `pairs()`:

```
pairs(site1,panel=panel.smooth,gap=0)
```

Figure 9.1 is the result, a graphical correlation matrix. The regression lines stem from a smoothing function attempting to fit a nonlinear function to the variables involved. This should be interpreted with care as it may exaggerate trends. The plot in row two and column one, for example, confirms the perfect linear correlation of pH peat and pH water, carrying much redundant variance. The same holds for maximum and average water level, as can be seen in row five and column four. The lack of strong correlation found among pH peat and average water level, row one and column four, is in line with dispersion found in the point cloud whereas only faint nonlinearity and minor skewness can be detected. Another example of obvious independence is found in row five and column one for pH peat and maximum water level where the smooth function also identifies some local trends.

As discussed for instance by Mellert *et al.* (2011) there are two distinct strategies to select predictors among a large number of variables: either hypothesis-driven or data-driven. In the *hypothesis-driven* approach model building is predominantly based on expert knowledge which can be verified in the course of modelling. In the *data-driven* approach the selection considers model performance only, for instance based on trial and error. As an advantage one can expect that the best possible model performance is finally achieved, but the variables involved may be more difficult to interpret than those devised by experts. Mellert *et al.* (2011), as an example, decided to use hypothesis-driven models for simulating tree distribution in the Bavarian Alps.

9.3 Generalized linear models

The goal when using generalized linear models (GLMs) is to fit the biotic component of ecosystems (species occurrence, for example) to one or several site factors. Linear models (LMs) fail to handle this type of correlation which is so important in plant ecology. A simple example may reveal the pitfall. From the 'Schlaenggli' example (`ssit`) we take variable `pH.peat`, and from `sveg` species *Sphagnum recurvum* (variable name `Sphagnum.recurvum`, one of the column names of `sveg`). Only presence–absence shall be considered, using function `sign()` when plotting the scores as a function of `pH.peat`:

```
plot(ssit$pH.peat,sign(sveg$Sphagnum.recurvum))
```

Sphagnum recurvum mainly occurs on the left-hand side of Figure 9.2, that is, at low pH values whereas at high values it is regularly missing. The straight line from the upper left to the lower right (drawn by function `abline()` while the plot window is kept open) comes from linear regression, function `lm()`:

```
o.lmod<- lm(sign(sveg$Sphagnum.recurvum)~ssit$pH.peat)
abline(reg=o.lmod)
```

The LM also tests for significance of fit, the result displayed when typing `summary(o.lmod)`, yielding an R^2 of 0.5504 which is highly significant. Unfortunately, the line passes through the one probability line, allowing probabilities higher than one, and the zero line, suggesting negative probabilities. Neither of these exist in reality. Furthermore, linear regression assumes constant variance over the entire range of pH. When probability is zero and also when it is one, variance must be zero too. Otherwise it would still allow probabilities exceeding the range of zero to one.

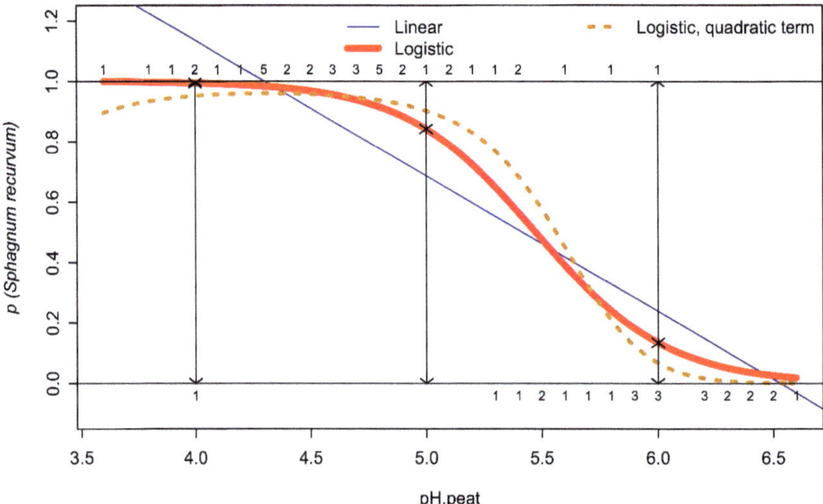

Figure 9.2: Linear and logistic regression of pH measurements in peat and *Sphagnum recurvum* (presence–absence scores) as the dependent variable. The dotted line is model response when including a quadratic term in the logistic model. Arrows illustrate deviance.

The solution to this is found in GLMs. The first to deal with is logistic regression, fitting an S-shaped or a unimodal response curve bound to the range of zero to one (Figure 9.2). The function to be fitted (Jongman *et al.* 1995) is:

$$\hat{y} = p = [\exp(b_0 + b_1 x)]/[1 + \exp(b_0 + b_1 x)] \tag{9.1}$$

Because the expected value for the simulated species, $\hat{y}$, has a range of zero to one it can be regarded as a probability for which symbol p is used. For the interpretation of results it is essential to recognize the following constraints:

- As pointed out by Jongman *et al.* (1995) there exists no ecological reason for choosing this equation, but 'mathematical convenience' only.
- Unlike in linear regression there is no explicit solution for solving the equation, but various algorithms approximating the parameters do

this. As a consequence results may vary among different software solutions.

The quantity to be minimized is called deviance (rather than variance as in linear regression). In Figure 9.2 deviance is illustrated by pairs of vertical arrows, the first pointing from the zero line to the expected species score (the probability of occurrence), the second from the latter up to the probability 1 level. Deviance is calculated according to:

$$dev = -2 * (p * \log(p) + (1 - p) * \log(1 - p)) \tag{9.2}$$

If probability p approaches 0 then deviance also becomes 0. The same happens when p approaches 1, because then $1-p$ tends towards 0. Deviance is maximum at $p = 0.5$, that is:

$$dev = -2 * (0.5 * \log(0.5) + (1 - 0.5) * \log(1 - 0.5)) = 1.386294 \tag{9.3}$$

Deviance values above 1 are considered high, meaning there is much uncertainty about the occurrence of the species modelled.

In vegetation ecology the most relevant application of this type of model is simulating species occurrence in space. We compare the spatial (geographic) distribution of species as observed in the field versus the expected as obtained by logistic regression:

```
spec1<- sign(sveg$Sphagnum.recurvum)
o.glm1<- glm(spec1~pH.peat,data=ssit,family=binomial)
summary(o.glm1)
```

```
Call:
glm(formula = spec1 ~ pH.peat, family = binomial, data = ssit)

Deviance Residuals:
    Min       1Q   Median       3Q      Max
-3.2314  -0.3836   0.1766   0.4200   2.0043

Coefficients:
            Estimate Std. Error z value Pr(>|z|)
(Intercept)   19.376      4.591   4.220 2.44e-05 ***
pH.peat       -3.540      0.841  -4.209 2.56e-05 ***
---
Signif. codes:  0 *** 0.001 ** 0.01 * 0.05 . 0.1   1

(Dispersion parameter for binomial family taken to be 1)

    Null deviance: 83.731  on 62  degrees of freedom
Residual deviance: 38.372  on 61  degrees of freedom
AIC: 42.372

Number of Fisher Scoring iterations: 6
```

With parameter `data=ssit` we can tell the function where `pH.peat` is taken from. Everything else looks like in linear regression, except for the term `family=binomial`. It forces the model to yield the deviation pattern explained above, with 0 at $p = 0$ and $p = 1$ and a maximum at $p = 0.5$, a property it shares with binomial distribution. The intercept of 19.376 is parameter b_0 in Equation (9.1) and -3.540 is b_1. The summary also reports deviance reduction achieved by the model and the Akaike information criterion (AIC), a measure of model quality explained in more detail in Section 13.3.3. We can now visualize the fit of the function as in Figure 9.2. We plot presence–absence of `Sphagnum.recurvum`, followed by adding the fitted values:

```
plot(ssit$pH.peat,sign(sveg$Sphagnum.recurvum))
points(ssit$pH.peat,fitted(o.glm1),pch=3)
```

With `pch=3` we choose crosses as symbols for the fitted values.

For a plot of expected occurrence in space (a map) we use x- and y-axes from `ssit` and draw circles (symbol type `pch=1`) with size `cex` proportional to the square root of the expectations, obtained from function `fitted()` for all locations:

```
x<- ssit$x.axis ; y<- ssit$y.axis
plot(x,y,pch=16,cex=2*fitted(o.glm1)^0.5,col="gray",asp=1)
```

The map of Figure 9.3(b) results. Setting parameter `asp=1` scales the axes equally thereby avoiding distortion of the map. To display species occurrence in the raw data, Figure 9.3(a), symbol size `pch` is set proportional to the signum-transformed scores of `Sphagnum.recurvum` and a triangle is chosen as the symbol (`pch=17`):

```
spec1<- sign(sveg$Sphagnum.recurvum)
plot(x,y,pch=17,cex=spec1)
```

Does this model fit the data? Obviously, the map of occurrence probability in Figure 9.3(b) nicely reproduces the pattern found in the field data [Figure 9.3(a)], all this confirmed by a hit rate of 90.5%. A hit is considered a plot where either occurrence probability is $p \geq 0.5$ and the species is present or where $p < 0.5$ and the species is absent. In $\mathcal{R}$ agreements and disagreements are put into a contingency table of hits (function `table()`):

```
hits<- table(spec1,fitted(o.glm1)>0.5)
sum(diag(hits))/sum(hits)*100

[1] 90.47619
```

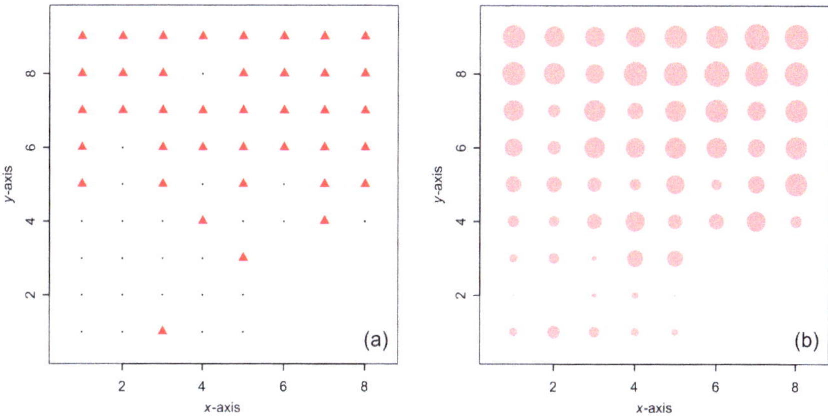

Figure 9.3: Occurrence of *Sphagnum recurvum* in the field (a) and expectation predicted by GLM with pH peat as the predictor variable (b).

In the second line the agreements, that is the total of the diagonal values accessed by function `diag()`, are compared with the total number of plots, multiplied by 100. In this example we get 90.47%, a very good result indeed!

There is one more, as yet unmentioned curve in Figure 9.2, the dotted one. This is the result of a model attempting to match a unimodal response. It results when a linear and a quadratic term enter into logistic regression, in the notation of $\mathcal{R}$:

```
sp1<- sign(sveg$Sphagnum.recurvum)
o.glm2<- glm(sp1~pH.peat+I(pH.peat^2),data=ssit,family=binomial)
summary(o.glm2)
```

```
Call:
glm(formula = sp1 ~ pH.peat + I(pH.peat^2), family = binomial,
    data = ssit)

Deviance Residuals:
    Min       1Q   Median       3Q      Max
-2.4663  -0.1812   0.2871   0.3588   2.3260

Coefficients:
             Estimate Std. Error z value Pr(>|z|)
(Intercept)   -34.994     26.140  -1.339   0.1807
pH.peat        17.704     10.607   1.669   0.0951 .
I(pH.peat^2)   -2.052      1.064  -1.928   0.0538 .
---
Signif. codes:  0 *** 0.001 ** 0.01 * 0.05 . 0.1   1

(Dispersion parameter for binomial family taken to be 1)

    Null deviance: 83.731  on 62  degrees of freedom
Residual deviance: 34.998  on 60  degrees of freedom
AIC: 40.998
```

```
Number of Fisher Scoring iterations: 6
```

This is very similar to the formulation of multiple regression, just using glm() instead of lm() and adding family=binomial. Function I() tells $\mathcal{R}$ that pH.peat shall be squared, the model now including the two independent variables pH.peat and pH.peat^2. Neither of the two regression parameters is significant at $p \leq 0.05$ level meaning that this model is clearly inferior to the univariate one (a finding confirmed when calculating the hit rate to be 87.30% in this case).

Just as in linear regression models, two or more independent variables can be included, as for example Waterlev.av. In $\mathcal{R}$ such a model would be written as:

```
o.glm3<- glm(sp1~pH.peat+Waterlev.av,data=ssit,family=binomial)
```

Again, the plus sign does not mean arithmetic addition, but it is part of the $\mathcal{R}$ syntax used for model formulation (Crawley 2005). It says that two independent variables are to be included: pH.peat and Waterlevel.av. Due to the limiting sample size of $n = 63$ this model does not perform better than the univariate one, as can be seen when typing summary(o.glm3).

9.4 Generalized additive models

Generalized additive models (GAMs) are an alternative to GLMs, having both strengths and drawbacks in comparison (Jongman *et al.* 1995). Whereas in all models presented so far the first step in the analysis is choosing a mathematical function to be fitted, this is not needed in GAMs. Instead, an algorithm tries to assess a smooth curve fitting the point cloud. This curve can be of any type: linear, S-shaped, unimodal, bimodal or other. Clearly, flexibility is the strength of the method compared with GLMs. On the downside, it cannot handle interactions. Multivariate models are fitted by averaging the various univariate solutions, hence the term 'additive' in the name. Developing fitting methods and algorithms is a mathematical field in progress from which ecology has profited in recent years.

From the point of view of even the experienced user GAMs are black boxes, although sometimes highly performing ones (Mellert *et al.* 2011). The main reason for difficulty in interpretation resides in the use of smoothing. Figure 9.4(a) illustrates this with the 'Schlaenggli' data. The fitting process is far more flexible than, for instance, a straight line or a bell-shaped response function. GAMs offer no mathematical functions of these lines which therefore are descriptive by nature.

As pointed out by Venables and Ripley (2010), additivity is analogous to multiple regression:

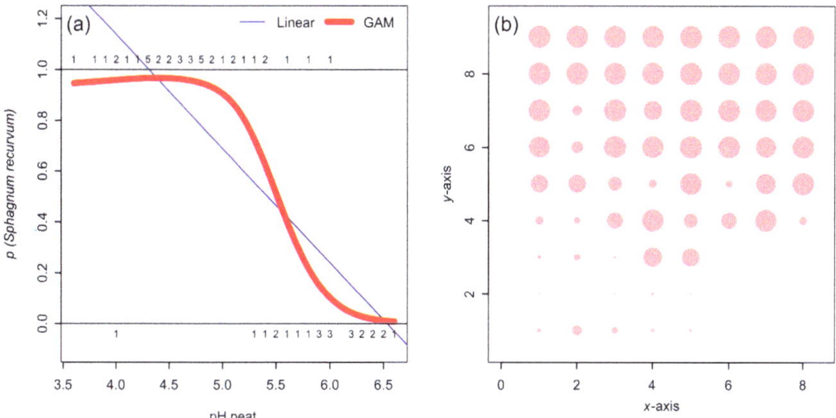

Figure 9.4: Fitting of *Sphagnum recurvum* with pH in peat using linear regression and GAM (a). Prediction of occurrence in space (b).

$$\hat{y}_j = a_j + b_{j,1}x_{j,1} + b_{j,2}x_{j,2} + \ldots + b_{j,k}x_{j,k} \tag{9.4}$$

In this parameters b have to be determined and the linear terms are added. In a GAM this becomes:

$$\hat{y}_j = a_j + f_1(x_{j,1}) + f_2(x_{j,2}) + \ldots + f_k(x_{j,k}) \tag{9.5}$$

Unlike in multiple regression the smooth functions $\mathtt{f}_1, \ldots, \mathtt{f}_k$ are nonlinear. In $\mathcal{R}$ function `gam()` parameters are the same as in `glm()` shown in Section 9.3. Making use of the 'Schlaenggli' data frames `sveg` and `ssit` a GAM is fitted and plotted as follows:

```
spec1<- sign(sveg$Sphagnum.recurvum)
o.gam1<- gam(spec1~s(pH.peat),data=ssit,family=binomial)
plot(ssit$pH.peat,spec1)
points(ssit$pH.peat,fitted(o.gam1),pch=3)
```

This is identical to the use of the `glm()` function except that the predicted variables are arguments of function `s()` telling `gam()` to use a smooth function. The solution is somewhat between an S-shaped response and a unimodal one [Figure 9.4(a)], with a hit rate slightly inferior (87.3%) compared with the same of logistic regression (90.5%). However, as explained in more detail in Section 9.7, there are better ways to measure accuracy of simulated maps. Printing the map in Figure 9.4(b) relies on the predictions:

```
x<- ssit$x.axis ; y<- ssit$y.axis
plot(x,y,pch=16,cex=3*fitted(o.gam1)^0.5,col=gray(0.7))
```

The fitted values are multiplied by three to obtain larger symbols. When compared with Figure 9.3 one can see that the GAM solution is a bit more restrictive than the same of the GLM, most likely a consequence of enhanced flexibility.

9.5 Classification and regression trees

This is a third family of methods for predicting species distribution based on environmental factors, introduced and explained by De' ath (2002). Classification and regression trees (CART) are reminders of early times of computer simulation when heuristic models were built based on binary trees built by experts (Binz and Wildi 1988). The principle is also familiar to taxonomists using species identification keys based on a series of binary (yes/no) decisions. Nowadays, although the trees are the outcome of a fitting process, interpretation remains straightforward and appealing. Having available data frames `sveg` and `ssit` holding the 'Schlaenggli' sample, the implementation in $\mathcal{R}$ is similar to that of GLMs and GAMs when using function `tree()`:

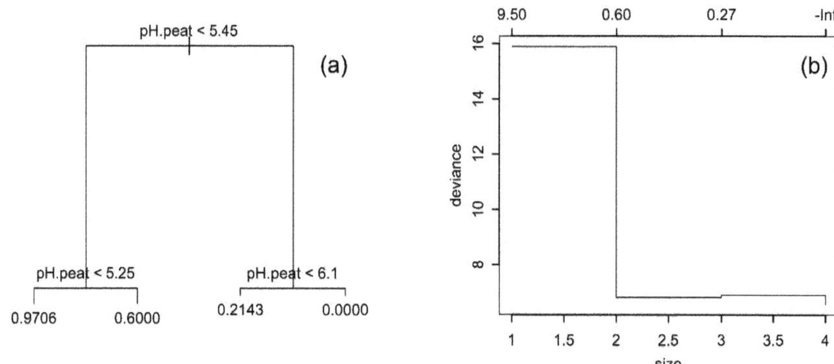

Figure 9.5: (a) Regression tree of pH measurements in peat predicting *Sphagnum recurvum* (presence–absence scores) as the dependent variable. (b) Deviance reduction of the four nodes of the tree ('size' of the tree, x-axis).

```
spec1<- sign(sveg$Sphagnum.recurvum)
o.tree<- tree(spec1~pH.peat,data=ssit,minsize=2)
summary(o.tree) ; o.tree
plot(o.tree); text(o.tree)
```

Functions `plot()` and `text()` generate Figure 9.5(a). Function `tree()` tries to find a threshold value for pH in peat where the distinction of presence and absence of *Sphagnum recurvum* is best. Function `summary(o.tree)` reports success of the model with the number of terminal nodes of the tree,

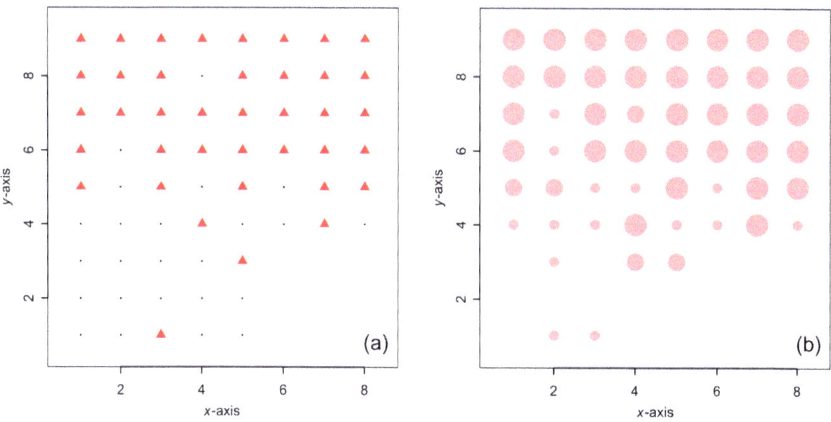

Figure 9.6: *Sphagnum recurvum* occurrence in space. (a) Field observation. (b) Predicted by pH in peat using a regression tree.

4 in this case, the residual mean deviance and distribution parameters of the residuals:

```
Classification tree:
Regression tree:
tree(formula = spec ~ pH.peat, data = ssit, minsize = 2)
Number of terminal nodes:  4
Residual mean deviance:  0.07674 = 4.528 / 59
Distribution of residuals:
    Min.  1st Qu.   Median     Mean  3rd Qu.     Max.
-0.97060  0.00000  0.02941  0.00000  0.02941  0.78570
```

Typing `o.tree` shows the entire structure of the tree:

```
node), split, n, deviance, yval
      * denotes terminal node

1) root 63 14.8600 0.6190
  2) pH.peat < 5.45 39  2.7690 0.9231
    4) pH.peat < 5.25 34  0.9706 0.9706 *
    5) pH.peat > 5.25 5  1.2000 0.6000 *
  3) pH.peat > 5.45 24  2.6250 0.1250
    6) pH.peat < 6.1 14  2.3570 0.2143 *
    7) pH.peat > 6.1 10  0.0000 0.0000 *
```

The first two lines are annotations of the elements of the following lines describing the seven nodes, exactly as seen in Figure 9.5(a). Interpretation of the tree is intuitive as the figure suggests: it starts on top by suggesting a pH threshold of 5.45. At higher levels, consider a threshold of 6.1 above which *Sphagnum recurvum* ultimately is assumed as being absent ($p = 0.000$). Below pH = 6.1 the probability of occurrence is $p = 0.2143$, etc. Unfortunately the resulting models tend to be tremendously over-fitted and just as in GLMs and GAMs a look at deviance reduction is needed. The `tree` package provides function `cv.tree()` for cross-validation of results. This performs an evaluation of results through a randomization test done as follows:

```
set.seed(57)
o.cv<- cv.tree(o.tree)
plot(o.cv)
```

The result is the graph of Figure 9.5(b) expressing deviance reduction when proceeding stepwise from the top to bottom of the tree. The best fit is obtained when considering all four nodes, but only two contribute much to the result. When striving for a parsimonious solution the `tree` package offers a 'pruning' function, `prune.tree()`:

```
o.ptree<- prune.tree(o.tree,best=2)
plot(o.ptree) ; text(o.ptree)
```

This further simplifies the tree shown in Figure 9.5(a). The full description of the new, reduced tree is now in `o.ptree`.

The predicted spatial distribution of *Sphagnum recurvum* is plotted using function `predict()` as follows:

```
x<- ssit$x.axis ; y<- ssit$y.axis
plot(x,y,pch=16,cex=(predict(o.tree)^0.5)*2,asp=1)
```

Taking the square root of the predicted values for symbol size, `cex`, adjusts symbol surface to the probability of occurrence with the result seen in Figure 9.6(b). I multiply these by two to enlarge symbols.

There are other functions available in $\mathcal{R}$ doing almost the same as `tree()`, for instance `rpart()` in package `rpart` (Venables and Ripley 2010). This controls over-fitting directly. In any case it is good practice to set a seed for cross-validation as done above: `set.seed(57)`. This forces the algorithm to always use the same random numbers. Without a seed one observes that the result in Figure 9.5(b) changes slightly with each run of the analysis.

Why should we avoid over-fitting in our models? Over-fitting means that the solution is adapted to the present data set and even a minor change leads to a different result. Boosted regression trees is a recent development promising to overcome this (Elith *et al.* 2008). In these, various models are combined with the aim to increase robustness of results. Boosting means combining models (Ridgeway 1999). Functions doing this are available in the `dismo` and `gbm` packages.

9.6 Testing and building scenarios

In the age of environmental change, prediction is the ultimate task of vegetation analysis. In the previous sections we 'fitted' models using the predicting variables. This tells us the model view of the data we analyse, but often we may want to investigate scenarios answering questions such as 'what would happen if temperature increases', or 'how would the system change if pH increased due to environmental pollution'. Under these circumstances our model should be tested using independent data, not coming from model building. But where should these be found?

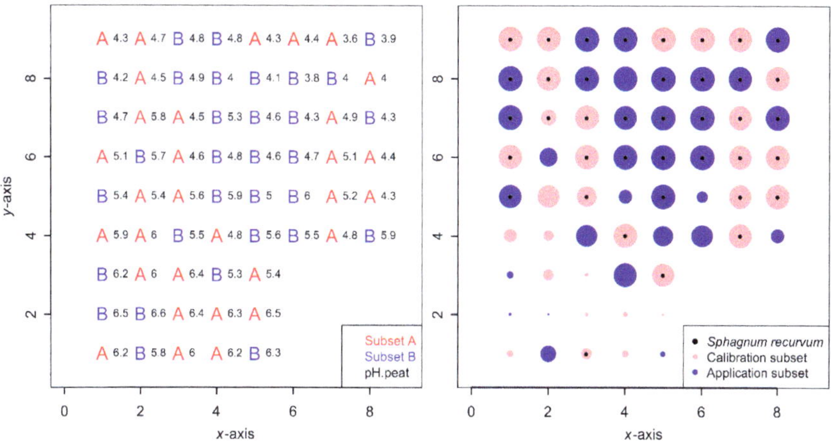

Figure 9.7: Testing a model using independent variables. Panel (a): Dividing the sample into sub-sample A for model building and sub-sample B for testing. Panel (b): predicted values based on data used for modelling (sub-sample A) and independent data (sub-sample B).

In cases where a sample of sufficient size is available a well-established practice is subdividing this into a random sub-sample for model calibration, A, and the complement of this for model testing, B (Albert *et al.* 2010). This is demonstrated in Figure 9.7 using the *Sphagnum recurvum* example from the previous sections. $\mathcal{R}$ provides function `sample()` for this working with random numbers. These cause the result to change with every new run. In the script below this is avoided by first setting a seed with the aim to get results that can be reproduced. Then from the 63 row indices a subset of 31 is formed which is used to generate sub-samples of data frame `ssit`, `ssit.A` and `ssit.B` and the same for vegetation data `sveg`. Analogous divisions are made for `pH.peat` and *Sphagnum.recurvum*:

```
set.seed(57) ; sub<- sample(1:63,31)
ssit.A<- ssit[sub,] ; ssit.B<- ssit[-sub,]
sveg.A<- sveg[sub,] ; sveg.B<- sveg[-sub,]
```

```
pH.A<- ssit.A[,"pH.peat"] ; pH.B<- ssit.B[,"pH.peat"]
Sp.A<- sign(sveg.A[,"Sphagnum.recurvum"])
```

The negative index, `-sub`, signifies 'take all rows except those occurring in `sub`'. The outcome of this division of sample is shown in Figure 9.7(a), obtained when first plotting letter 'A' with coordinates in subsample A, then adding letter 'B' accordingly without closing the plot window:

```
x.A<- ssit.A[,"x.axis"] ; y.A<- ssit.A[,"y.axis"]
x.B<- ssit.B[,"x.axis"] ; y.B<- ssit.B[,"y.axis"]
plot(x.A,y.A,pch="A",col="red",asp=1)
points(x.B,y.B,pch="B",col="blue")
```

In this graph the pH-values are also added by function `text()` but taking the full sample of coordinates and pH-values. Sub-sample A is now used for model building as explained in Section 9.3 by function `glm()`. From this the predicted (fitted) probabilities are plotted into panel (b). In the subsequent step function `predict()` is again calculating probabilities, but this time using the pH-values from sub-sample B:

```
o.glm.A<- glm(Sp.A~pH.A,family=binomial)
o.p1<- predict(o.glm.A,list(pH.A),type="response")
o.p2<- predict(o.glm.A,list(pH.A=pH.B),type="response")
```

What remains to do is to plot the two different predictions, the first solely showing model fit, the second illustrating the model performance when using data not included in model building:

```
plot(x.A,y.A,cex=o.p1^0.5*3,col=gray(0.7),pch=16)
points(x.B,y.B,cex=o.p2^0.5*3,col=gray(0.4),pch=16)
```

Occurrence of *Sphagnum recurvum* in the field is added in the same way as in Figure 9.6(a). From Figure 9.7(a) we conclude that the model in most cases properly predicts the occurrence of that species, even in subsample B with pH being an external predictor.

We now build a scenario with the methods from Sections 9.3, 9.4 and 9.5 with the aim to simulate environmental change. Let us assume that pH in peat increases by 1.5 units due to environmental pollution. This lets us expect that *Sphagnum recurvum* will retreat towards the upper, more acid part of the slope. First, the models are developed from the original data:

```
o.glm<- glm(spec1~pH.peat,data=ssit,family=binomial)
o.gam<- gam(spec1~s(pH.peat),data=ssit,family=binomial)
o.tree<- tree(spec1~ pH.peat,data=ssit)
```

Figure 9.8(a) uses the predictions from the original data, that is from `ssit`. To derive the remaining maps of Figure 9.8 we have to use pH values that are increased by 1.5 in the course of calculating predictions:

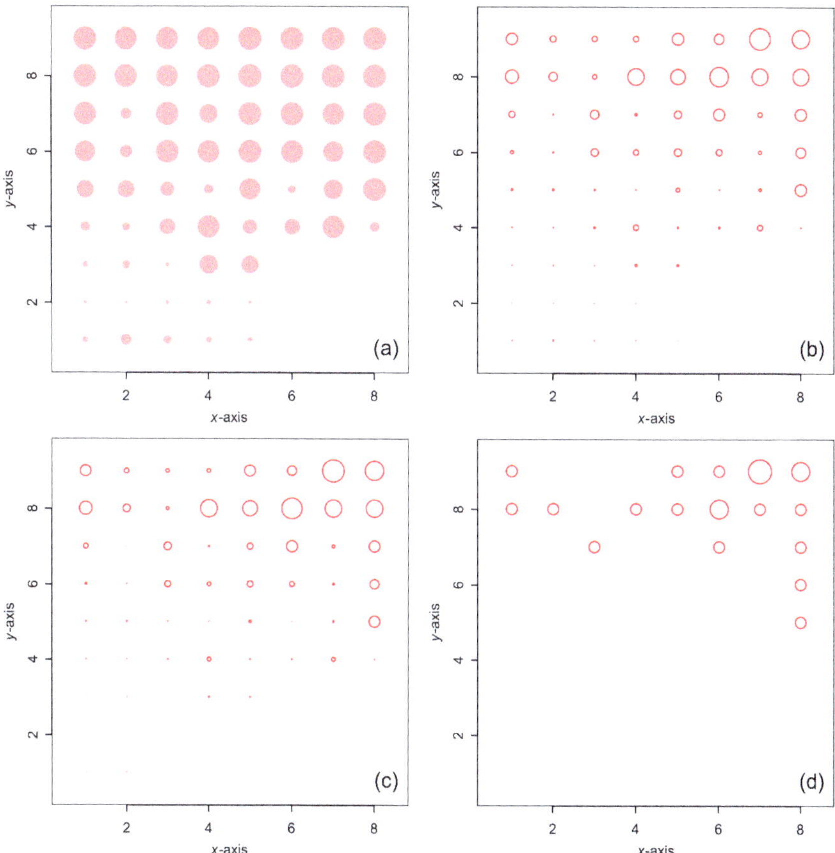

Figure 9.8: Scenario of *Sphagnum recurvum* occurrence in space when increasing pH in peat by 1.5 units. (a) GLM, no change. (b) GLM, pH increased. (c) GAM, pH increased. (d) CART, pH increased.

```
pH.peat<- ssit$pH.peat
o.p1<- predict(o.glm,list(pH.peat=pH.peat+1.5),type="response")
o.p2<- predict(o.gam,list(pH.peat=pH.peat+1.5),type="response")
o.p3<- predict(o.tree,list(pH.peat=pH.peat+1.5))
```

To visualize the effect of the scenario, plots in geographic space are needed (Figure 9.8). Figure 9.8(b) is obtained as follows:

```
x<- ssit$x.axis ; y<- ssit$y.axis
plot(x,y,pch=16,cex=o.p1^0.5*3,col=gray(0.7),asp=1)
```

As usual, symbol size `cex` is set to the square root of the predicted values multiplied by 3 to enlarge symbols. Using `o.p2` yields Figure 9.8(c) and `o.p3` Figure 9.8(d). Interestingly, results of GLM and GAM are very similar whereas CART presents a more coarse view of the situation.

Scenarios are especially appealing these days as we can manipulate variables related to climate, precipitation, mean temperature or radiation, for example, and also any combination of site factors. But the more variables are involved in a scenario the more complex becomes the aspect of model reliability and there is an increased risk of simulating situations that occur nowhere in the real world. Fitzpatrick and Hargrove (2009) address this as 'the problem of non-analog climate' and they present an example where they compare the Caspian Sea with the Great Lakes. They conclude that most parts are non-analogues and similarity only exists along some shore areas. Under these conditions studying the risk of species invasion, for example, may become unreliable.

9.7 Modelling vegetation types

Probability functions not only allow the occurrence of a species population or a life form to be described (McGill *et al.* 2006), but also the spatial extent of vegetation types. Early examples of this were the simulation of alpine vegetation by Fischer (1990) or assessment of potential natural forest vegetation (PNV) of Switzerland (Brzeziecki *et al.* 1993) as well as a similar study in Southern Germany by Lindacher (1996). Plant ecologists often hesitate to leave the firm ground of the species approach (see, for example, Guisan and Zimmermann 2000), because the observed presence of a species is considered reliable (wrong determinations of species are infrequent), although absence is far more uncertain as it is easy to miss a species. When modelling vegetation types, uncertainty is omnipresent because vegetation is continuous and the assessment of discrete types (Chapter 5) is an often coarse abstraction of reality.

The spatial extent of a vegetation type can be modelled like a species population. Hence, we could compute a logistic regression for each vegetation type, m, and then combine the resulting m probability layers to one multivariate. But multinomial log-linear models do this in one step (Venables and Ripley 2010). They derive response vectors containing the occurrence probabilities of all vegetation types, and these add up to $p = 1.0$.

In Table 9.1 a very small example (data frames `nveg` and `nsit`) illustrates this. In the classification before modelling [panel (c)] each relevé belongs to just one group (left-hand side in Table 9.1), that is, with a probability of $p = 1.0$ respectively. After fitting the model to one or several site factors, probabilities are obtained for the relevés to belong to all of the vegetation types involved. The rows of the probabilities (fitted values) again sum up to one, as for prior to modelling.

Table 9.1: Input and output of multivariate logistic regression with data from Table 7.1. Group membership is the dependent variable interpreted by the model as presence–absence vectors ($g1$–$g3$) of vegetation types. Fitted values are from multivariate logistic regression when using pH as an independent variable.

Model input						Fitted values			
Relevés	Group	$g1$	$g2$	$g3$	pH	$g1$	$g2$	$g3$	$\sum$
2	1	1	0	0	4.4	1.000	0.000	0.000	1.000
4	2	0	1	0	6.2	0.000	0.971	0.029	1.000
6	1	1	0	0	4.8	0.999	0.000	0.001	1.000
9	3	0	0	1	5.6	0.000	0.001	0.999	1.000
10	2	0	1	0	6.5	0.000	1.000	0.000	1.000
18	3	0	0	1	6.0	0.000	0.484	0.516	1.000
25	1	1	0	0	5.0	0.926	0.000	0.074	1.000
27	2	0	1	0	6.0	0.000	0.484	0.516	1.000
39	3	0	0	1	5.2	0.082	0.000	0.918	1.000
49	1	1	0	0	4.8	0.999	0.000	0.001	1.000
50	3	0	0	1	5.8	0.000	0.026	0.974	1.000
$\sum$	–	4	3	4	–	4.006	2.967	4.028	11.000

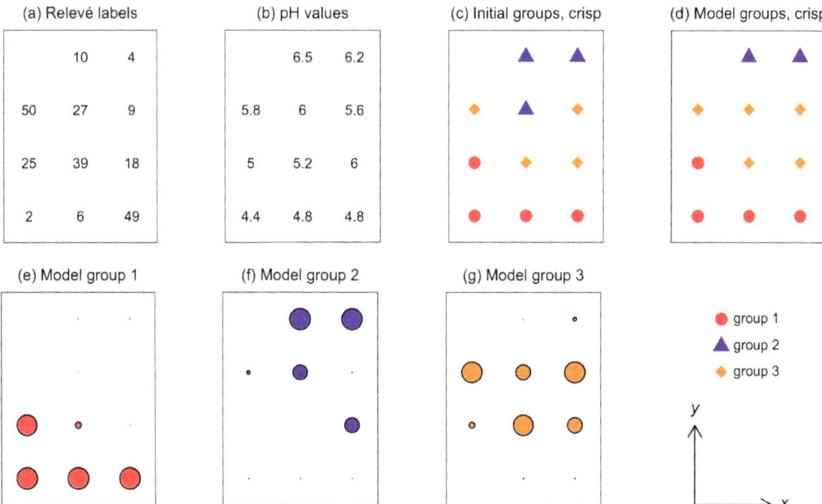

Figure 9.9: Multivariate logistic regression (see Table 9.1) with coordinates taken from Table 7.1.

In multinomial log-linear models the dependent input variable may be a nominal array with each state representing one vegetation type, that is, information we either get directly from a vegetation map or calculate from vegetation data using cluster analysis (Chapter 5). The states of this vector (the column labelled 'Group' in Table 9.1) define the membership of relevés to one of the vegetation types. Upon modelling, this vector is split up into 'dummy variables', denoted $g1$ through $g3$ in Table 9.1. This is why the method is called *multiple* logistic regression.

The interpretation of results is straightforward (Table 9.1, right-hand side). For relevé 4, as an example, the vector {0;1;0} after fitting with pH becomes {0.000;0.971;0.029}, an almost perfect agreement. To derive a discrete solution the maximum probability is set to one and the remaining to zero, that is {0;1;0}, in this case exactly as it is in the input. Model input and output in Table 9.1 is also presented spatially in Figure 9.9. Unlike in a discrete vegetation map the fitted model types have the potential to occur in all plots. The result of this very small example suggests that pH is a very good predictor for vegetation type as only relevé 27 changes membership, from type 2 to type 3.

When using function `multinom()` in the **nnet** package computation is simple. All we need is a classification of vegetation and one or several site factors as independent predictors. It is most convenient to take vegetation types from cluster analysis (Chapter 5), so we do this first:

```
vdm<- as.dist((1-cor(t(nveg^0.2)))/2)
o.hclr<- hclust(vdm,method="ward.D2")
o.grel<- as.factor(cutree(o.hclr,k=3))
o.mu<- nnet::multinom(o.grel ~ nsit$PH)
```

Variable `vdm` is the distance matrix of relevés [correlation as distance, Equation (4.6)] and setting `k=3` will form three groups, that is, vegetation types. With function `as.factor()` included, the result of function `cutree()` is interpreted as a vector of types, called 'factors' in $\mathcal{R}$. The model formula with the tilde means 'compute `o.grel` as a function of `nsit$PH`'. Results for further processing are now available in output list `o.mu`, as for instance the fitted values in Table 9.1:

```
round(o.mu$fitted.values,digits=3)
```

```
       1     2     3
1  1.000 0.000 0.000
2  0.000 0.971 0.029
3  0.999 0.000 0.001
4  0.000 0.001 0.999
5  0.000 1.000 0.000
6  0.000 0.484 0.516
7  0.926 0.000 0.074
8  0.000 0.484 0.516
9  0.082 0.000 0.918
10 0.999 0.000 0.001
11 0.000 0.026 0.974
```

Function `round()` facilitates the reading of the output. The residuals in `o.mu$residuals` are of interest too as they show in which relevé the fitting of groups was difficult to achieve.

Using now x- and y-axis in `nsit` allows the spatial distribution of all variables to be plotted as separate maps (Figure 9.9). Panels (a) and (b) are obtained as follows:

```
x<- nsit$X.AXIS ; y<- nsit$Y.AXIS
plot(x,y,type="n",asp=1) ; text(x,y,rownames(nveg))
plot(x,y,type="n",asp=1) ; text(x,y,nsit$PH)
```

Panel (c) uses the classification from cluster analysis, `o.grel`. The revised, crisp classification in panel (d) corresponds to the simulated vegetation type where occurrence probability is maximum. These two graphs are generated as follows:

```
newgr<- apply(o.mu$fitted.values,1,which.max)
plot(x,y,pch=as.integer(o.grel),cex=2.0,asp=1)
plot(x,y,pch=as.integer(newgr),cex=2.0,asp=1)
```

The probability of occurrence of the three types is displayed in Figure 9.9, panels (e), (f) and (g). The surface of the circles is proportional to the square root of the probabilities that `multinom()` yields:

```
fitval<- o.mu$fitted.values
for(i in 1:3) {
  plot(x,y,pch=16,cex=3*fitval[,i]^0.5,col="gray",asp=1)
  title(main=paste("Model group",i),cex.main=0.8,font.main=1)
}
```

Multiplication of the fitted values by 3 enlarges the circles when using small data sets. Function `title()` adds a title on top of the graphs.

As samples get larger, measuring the success of modelling is no longer trivial. There are many ways to do this, as for example evaluating model fit. Below we investigate how the real system, that is a vegetation sample, compares with the same when fitted. One can simply count the number of agreements (unchanged group numbers) and relate this to sample size. But as pointed out by Venables and Ripley (2010), the severity of error should be taken into account too. They suggest inspection of a confusion matrix, C, in which the allocation of relevés before and after modelling is compared, in the present case:

$$\mathbf{C} = \begin{matrix} 4 & 0 & 0 \\ 0 & 2 & 1 \\ 0 & 0 & 4 \end{matrix}$$

In this matrix the row index tells us from which group a data point originates and the column index where it went according to the model (here, one relevé comes from group 2 and changes to group 3). Data points not changing group membership are counted in the diagonal elements. A move of a relevé may be harmless when taking place among adjacent (that is, similar) vegetation types, but more serious in the case where the latter differ widely. Ripley (1996) suggests multiplying frequency by error cost. But what does cost mean in vegetation simulation?

Confusion matrix, C, can be weighted by distance among vegetation types, W, and these distances may serve as surrogate for cost: when distance is small, the difference between types is small and cost accordingly low, and vice versa. For computations, a matrix of group centroid scores is needed in which the three relevé groups are described by the relative frequencies of all species (see Table 7.10). This is an estimate for the occurrence probability of species within the relevé groups. A distance matrix of centroids may use correlation transformed to distance, for instance, serving as matrix of weight, W:

$$\mathbf{W} = \begin{matrix} 0.000 & 0.827 & 0.420 \\ 0.827 & 0.000 & 0.288 \\ 0.420 & 0.288 & 0.000 \end{matrix}$$

To derive the cost factor cf from that, the two matrices are multiplied and the sum of all elements taken:

$$cf = \sum_{i=1}^{ng}\sum_{j=1}^{ng} C * W \tag{9.6}$$

In this, $C * W$ is the element by element product of the two matrices and ng is the number of groups. The only non-zero element remaining is the second one in column 3, that is 0.288. Hence, cf is a relative measure allowing different models to be compared when based on the same data set.

Cost function [Equation (9.6)] is available through function `ccost()` in the `dave` package with the result found in `o.cc$ccost`:

```
o.cc<- ccost(veg=nveg,oldgr=o.grel,newgr=newgr,y=0.5)

Call:
ccost.default(veg = nveg, oldgr = o.grel, newgr = newgr, y = 0.5)
cf= 0.287552
```

This is cost factor cf from the output list `o.cc`. To get the confusion matrix, C, as well as weight matrix, W, one simply types `o.cc`.

9.8 Expected wetland vegetation (example)

Real world examples allow probing methodological and practical questions, two of which I address in this section:

1. Are two site factors, pH and average water level in this case, capable of reasonably explaining a fairly complex vegetation pattern?

2. Does simulation of vegetation types deliver a reliable prediction of species occurrence, just as reliable as GLMs (Section 9.3) and GAMs (Section 9.4) do?

Using the 'Schlaenggli' data set, the steps involved in computation are:

1. Clustering the 63 relevés (scalar transformation $x' = x^{0.2}$, minimum-variance clustering based on correlation coefficient, five groups formed). Plotting types according to their location in space (x- and y-axis).

2. Building a multinomial logistic regression model using two site factors, pH and average water level respectively.

3. Plotting the occurrence probabilities of all five vegetation types according to their location in the field (x- and y-axis).

4. For all plots, calculating species occurrence probabilities and comparing the resulting simulated maps with the originals.

Steps 1 through 3 are basically the same as in Section 9.7, only step 4, calculating species probabilities, is new. The data frame with vegetation data is `sveg` and the same with site data is `ssit`. Based on five vegetation types obtained by clustering the relevés, we select the two site factors pH and average water level as predicting variables of the model. This choice is the result of an evaluation of site factors, as shown in Table 9.2. Both site factors have distinct means for the five groups and fairly low standard deviations. The error probability in the last column stems from analysis of variance confirming high resolving power of the variables. This is all we need for multinomial regression [function `multinom()`]:

```
vdm<- as.dist((1-cor(t(sveg^0.2)))/2)
o.hclr<- hclust(vdm,method="ward.D2")
o.grel<- as.factor(cutree(o.hclr,k=5))
o.mu<- nnet::multinom(o.grel ~ pH.peat+Waterlev.av,data=ssit)
```

Again, variable `vdm` is the distance matrix of relevés and setting `k=5` will generate five groups. The means and standard deviations in Table 9.2 make use of the group structure `o.grel` too, as the second argument in function `tapply()`:

```
tapply(ssit$pH.peat,o.grel,mean)
tapply(ssit$pH.peat,o.grel,sd)
```

Plotting the fitted values of this model yields Figure 9.10, that is the fuzzy as well as the crisp model maps. It is done as shown in Section 9.7 with a five-fold loop for the five vegetation types:

Table 9.2: Means and standard deviations within groups of pH and water level in the model of the 'Schlaenggli' data set.

Parameter	Group	1	2	3	4	5	Error p
pH	Mean	4.40	5.04	6.10	5.40	6.07	<2e-16
	Standard deviation	0.417	0.544	0.346	0.370	0.364	
Water level (cm)	Mean	12.8	12.2	11.8	8.57	20.6	1.64e-09
	Standard deviation	3.72	1.83	3.61	2.64	2.07	

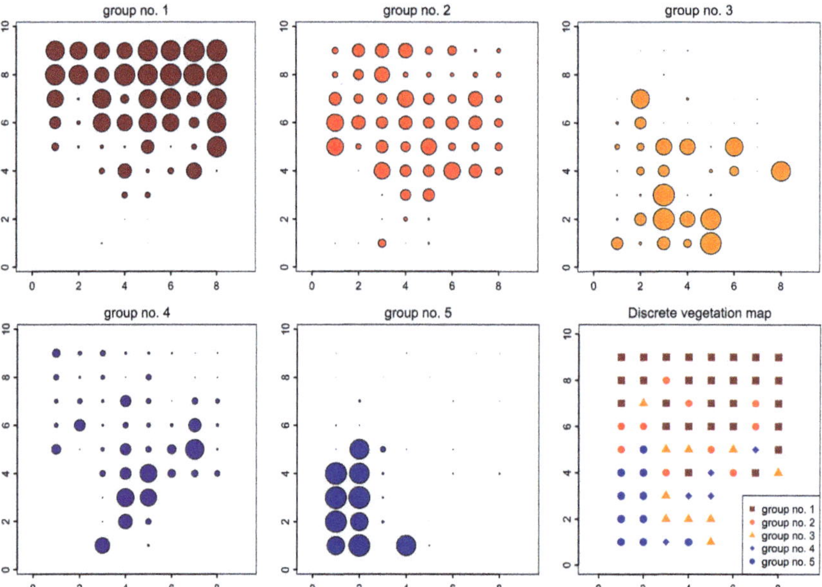

Figure 9.10: Simulated wetland vegetation. The 'Schlaenggli' data sets sveg and ssit are used. Probability of occurrence of all five vegetation types based on site conditions (pH and average water level) and the discrete vegetation map are shown. Circle size is proportional to probability.

```
x<- ssit$x.axis ; y<- ssit$y.axis ; fitval<- o.mu$fitted.values
for(i in 1:5) {
  plot(x,y,pch=16,cex=5*fitval[,i]^0.5,col="gray",asp=1)
  title(main=paste("group no.",i))
}
```

The plot() function displays the resulting five simulated vegetation types in separate maps, each with an individual title. But how do the maps agree with the real system? We first compute the simulated crisp classification newgr as done in Chapter 9.7 followed by calling function ccost() to provide answers to this question:

```
newgr<- apply(o.mu$fitted.values,1,which.max)
o.cc<- ccost(veg=sveg,oldgr=o.grel,newgr=newgr,y=0.5)
```

```
Call:
ccost.default(veg = sveg, oldgr =o.grel, newgr = newgr, y = 0.5)
cf= 2.838259
```

The cost factor, $cf = 2.838259$ is not bad for sample size of 63 relevés. The good fit also confirms the confusion matrix:

```
o.cc
```

```
Confusion matrix:
     [,1] [,2] [,3] [,4] [,5]
[1,]   22    2    0    0    0
[2,]    5    5    3    0    0
[3,]    0    2    7    0    1
[4,]    1    1    0    5    0
[5,]    0    0    0    0    9

Weight matrix:
          1         2         3         4         5
1 0.0000000 0.1186035 0.4122810 0.3086563 0.5273564
2 0.1186035 0.0000000 0.2781514 0.1609415 0.4270600
3 0.4122810 0.2781514 0.0000000 0.1650377 0.1476799
4 0.3086563 0.1609415 0.1650377 0.0000000 0.3821753
5 0.5273564 0.4270600 0.1476799 0.3821753 0.0000000
```

There are 17 out of 63 relevés changing group membership (the sum of all off-diagonal elements). While this appears to be many they mainly concern changes to closely related vegetation types only, as could be seen when inspecting the weight matrix. Because the new group membership is now also available we are ready to plot the discrete ('crisp') vegetation map shown in Figure 9.10:

```
plot(x,y,pch=newgr,cex=1.0,asp=1)
legend("bottomright",paste("gr.",c(1,2,3,4,5)),pch=c(1,2,3,4,5))
```

Parameter "bottomright" places the legend where there is sufficient space in this graph.

How is species probability derived? For this we need to know species *occurrence probability* within the types. An estimate of this is relative species frequency within the relevé groups, taken from a frequency table holding the centroids. Function centroid() uses the same classification o.grel as derived before. We denote the probability table by S and then have a look at the first species, *Vaccinium myrtillus*:

```
o.cent<- centroid(sveg,o.grel,y=0.2)
S<- o.cent$prob.table
colnames(S)[1] ; round(S[,1],digits=3)
```

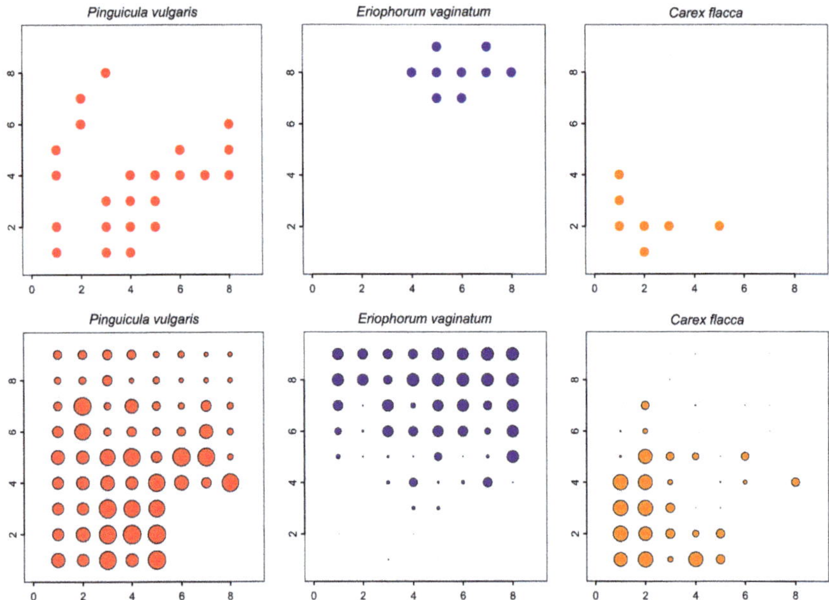

Figure 9.11: Occurrence probability of three selected species (lower panels). Occurrence of the same species in the field (upper panels).

```
[1] "Vaccinium.myrtillus"
    1     2     3     4     5
0.458 0.000 0.000 0.000 0.111
```

This is nothing else but the relative frequency of species 1 within the five vegetation types that were derived by clustering and we see that *Vaccinium myrtillus* mainly occurs in group number 1. It has to be compared to the 'fitted values' of function `multinom()`, via the occurrence probabilities of the vegetation types within the plots, denoted by M. For the first plot with label `"501"` this is the first row of M:

```
M<- o.mu$fitted.values
rownames(sveg)[1]
round(M[1,],digits=3)
```

```
[1] "501"
    1     2     3     4     5
0.964 0.034 0.000 0.001 0.000
```

Whereas from matrix S we displayed the first row, $s_{.1}$, from matrix M this was the first column, $m_{1.}$. The occurrence probability of *Vaccinium myrtillus* in the first plot, q_{11}, is the sum of the product of vectors m and s:

$$q_{11} = \sum_{i=1}^{k} s_{.i} m_{i.} \qquad (9.7)$$

In this, k is the number of groups, 5 in this case. Longhand calculation reveals occurrence probability of species 1 in plot 1 to be $q_{11} = 0.4415$, that is, $(0.458*0.964)+(0*0.034)+0+(0*0.001)+(0.111*0)$. Obviously, matrix Q is just the matrix product of M and S, $Q = MS$, and computation of the full matrix is obtained by:

```
Q<- as.data.frame(M %*% S)
```

In $\mathcal{R}$ the asterisk between the percentage signs is used for the inner product of vectors and matrices. Matrix Q is the entire sample of modelled vegetation data, with 63 rows and 119 columns in this example. Transforming this to a data frame as done above is most practical for further use.

Figure 9.11 is the spatial representation of both field data and simulated data. Species 19 (*Pinguicula vulgaris*), 15 (*Eriophorum vaginatum*) and 97 (*Carex flacca*) are chosen to illustrate different responses. In all cases the patterns of occurrence and expectations are rather similar. Species are frequently missing in the field data even though the ecological conditions, as suggested by simulated data, seem to be favourable. One possible explanation could be that plot size chosen for the survey (1 m^2) was just a bit too small, an issue discussed in more detail in Chapter 13.

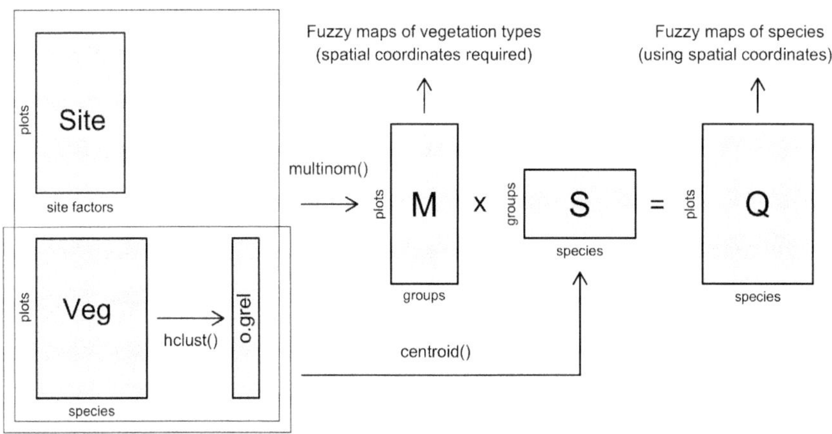

Figure 9.12: Steps of computation in multinomial logistic regression. 'Site' and 'Veg' are initial data, 'grel' is classification of relevés, M membership of plots to vegetation types, S is a relativized frequency table and Q is simulated vegetation data at species level.

To display field data as in the upper part of Figure 9.11 we need the spatial x- and y-coordinates in object `ssit` and the original vegetation data frame sveg from where we get presence–absence of species, *Pinguicula vulgaris*, species 19, for example:

```
plot(ssit$x.axis,ssit$y.axis,cex=sign(sveg[,19]),pch=16)
title(names(sveg[19]))
```

For the lower part of Figure 9.11 the model data frame Q is accessed instead of sveg and symbol size is set proportional to the square root of occurrence probability:

```
plot(ssit$x.axis,ssit$y.axis,cex=3*Q[,19]^0.5,pch=16,col="gray")
title(paste(names(sveg[19])))
```

Although not too abundant in the field, the simulation suggests that *Pinguicula vulgaris* has the potential to occur almost everywhere in the sample.

In conclusion, simulating vegetation types and extending the models to individual species offers more flexibility than the usual direct logistic regression approaches such as GLMs, GAMs and CART (Section 9.3, Section 9.4 and Section 9.5, respectively), because it requires a classification in which the number of vegetation types is chosen by the user. Unlike in any of the methods mentioned before there is no limitation in simulating rare species (Guisan *et al.* 2005) because regression always refers to vegetation types.

The entire procedure, although not difficult to understand, requires various steps, summarized in Figure 9.12. This shows the data sets needed, one holding vegetation and the other site factors, as well as the intermediate and final results. Arrows signify steps and their annotations are the functions used for these: function `hclust()` for group membership of relevés, `o.grel`, `multinom()` to compute matrix M of fitted values, `centroid()` for the frequency table S and matrix Q holding simulated vegetation data is just the matrix product of M and S.

10 Vegetation change in time

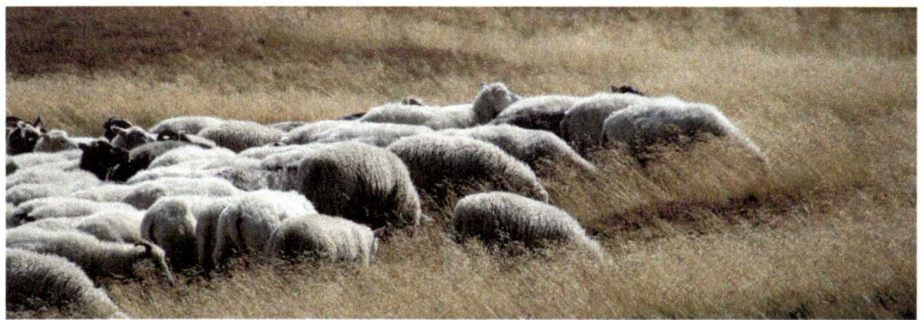

Although just another factor determining vegetation pattern, time has some unique properties. It is one-dimensional, proceeds in one direction only and as processes usually last longer than observed, time series data tend to underestimate temporal variation.

10.1 Coping with time

Assessing temporal change of vegetation as a multi-state system is a central issue in vegetation ecology (Wildi and Orlóci 2007). One could of course argue that time is but another attribute in a sample, as explained in Section 2.3.2, such that no specific treatment would be indicated. However, time has some unique properties. First of all it is one-dimensional, unlike space where direction is an issue and a decision may be needed when assessing order. Time always proceeds in the same direction, and even more importantly, it is transient. Once an event has taken place, there is no backtracking as can be done in space. This has consequences for investigating change in environmental systems, as discussed in detail by Green (1979). His hierarchical scheme of impact studies is shown in Figure 10.1. The most urgent question in the investigation of an impact is whether reference plots exist. Undoubtedly, striving for a reference is worthwhile, because once the impact has taken place there is no way to reverse the process. In experimental research, for instance, one may succeed in protecting plots

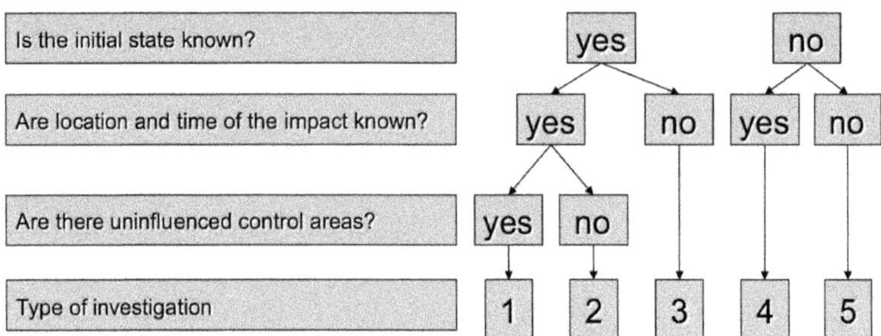

Figure 10.1: Type of environmental study needed to assess change. Adapted from Green (1979).

from an impact, but plots cannot be sheltered from time passing by. Further philosophical implications of time dependence are discussed in Legendre and Legendre (2012).

Time series theory is a specialized field of science (Beran 1994; Venables and Ripley 2010). In vegetation ecology the buzzword 'succession' is popular for addressing change in time (van der Maarel and Franklin 2013). In the sections below I discuss various issues of this. A first is detecting temporal autocorrelation (the dependence of temporal measurements). Assessing a trend (a temporal similarity pattern) is the subsequent topic which is related to the measurement of rate of change (the speed of the process) under various circumstances, including primary succession. I also explain a very basic idea apt to explain often observed continuous changes, the Markov models. Then, space-for-time substitution is introduced and demonstrated. The chapter concludes with applications of various methods to pollen data.

10.2 Temporal autocorrelation

When investigating a time series one has to be aware that this is usually just a fraction of a probably much longer lasting real process. Subsets drawn from measured time vectors most likely have different mean and variance than the entire process. When correlating two subsets originating from the same series, then these lack independence as would be required for statistical testing. The key to analysing this in a better way is by taking this fact into account.

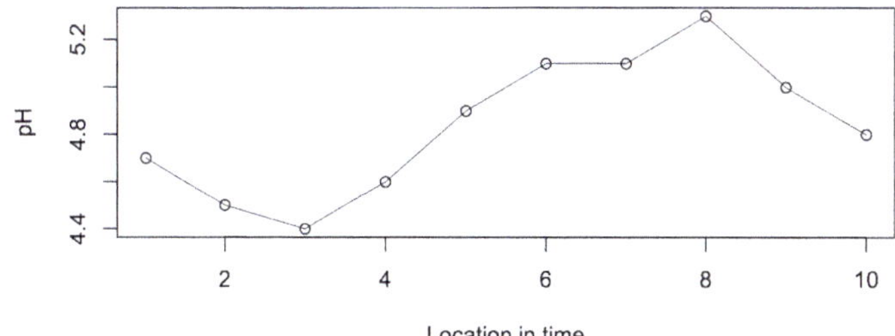

Figure 10.2: Temporal arrangement of measurements (pH) according to the columns of Table 10.1.

Table 10.1: Temporal autocorrelation in a time series (Figure 10.2). Measurements taken at equal step lengths, Δt, are correlated with neighbours 1–6 time steps apart. The two vectors used for correlation with time lag $\Delta t = 1$ are enhanced.

$\Delta t =$	0	1	2	3	4	5	6
vector	$x_{(l=0)}$	$x_{(l=1)}$	$x_{(l=2)}$	$x_{(l=3)}$	$x_{(l=4)}$	$x_{(l=5)}$	$x_{(l=6)}$
$n_{(l)}=$		9	8	7	6	5	4
$r_a=$	1.000	0.731	0.279	−0.137	−0.422	−0.482	−0.332
$x(1)$	4.7						
$x(2)$	4.5	4.7					
$x(3)$	4.4	4.5	4.7				
$x(4)$	4.6	4.4	4.5	4.7			
$x(5)$	4.9	4.6	4.4	4.5	4.7		
$x(6)$	5.1	4.9	4.6	4.4	4.5	4.7	
$x(7)$	5.1	5.1	4.9	4.6	4.4	4.5	4.7
$x(8)$	5.3	5.1	5.1	4.9	4.6	4.4	4.5
$x(9)$	5	5.3	5.1	5.1	4.9	4.6	4.4
$x(10)$	4.8	5	5.3	5.1	5.1	4.9	4.6
		4.8	5	5.3	5.1	5.1	4.9
			4.8	5	5.3	5.1	5.1
				4.8	5	5.3	5.1
					4.8	5	5.3
						4.8	5
							4.8

Measuring temporal autocorrelation is explained below using a univariate, artificial example (Figure 10.2, Table 10.1). Along a hypothetical gradient, 10 measurements x_t separated by equal step length are given. They are shifted by an increasing number of steps to mimic a time lag. The idea is to compare a vector to itself, but with its elements offset by a multiple of step length. If the elements of the series were independent, then correlation of the shifted series should approach zero. A consequence of shifting the series is that the first element in the first of two paired vectors and the last element in the second vector cannot be used any more in subsequent comparison and $n_{(l=1)}$ reduces to 9 when comparing the initial series with the second of lag $\Delta t = 1$. The first comparison starts with the paired values $\{4.5; 4.7\}$ and ends with $\{4.8; 5\}$. A straightforward idea is to correlate the two vectors in the usual way, but this would overestimate the relationship: the variance of the partly hidden process likely exceeds the one observed in the measurements. The formula for autocovariance addressing l time steps takes care of this (Beran 1994; Venables and Ripley 2010):

$$var_a(l) = \frac{1}{n_{(0)}} \sum_{t=1+l}^{n_{(0)}-l} (x_t - \overline{x}_{n_{(0)}})(x_{t+1} - \overline{x}_{n_{(0)}}) \tag{10.1}$$

Equation 10.1 resembles ordinary covariance, but with two distinct differences: division is always by $n_{(0)}$, that is the length of the total time series with lag $\Delta t = 0$, presently $n = 10$. This compensates for decreasing sample size when samples get smaller with growing lag. Then, the means $\overline{x}_{n_{(0)}}$ are always taken from the initial series and hence are the same for both vectors, x_t and $x_{t+\Delta t}$, respectively. Autocorrelation is obtained by dividing autocovariance by the variance of the entire series:

$$cor_a(\Delta t) = \frac{var_a(\Delta t)}{var_a(t = 0)} \tag{10.2}$$

$\mathcal{R}$ function `acf()` in package `stats` does all this in one single step, including graphical display:

```
pH <- c(4.7,4.5,4.4,4.6,4.9,5.1,5.1,5.3,5,4.8)
o.acf<- acf(pH)
o.acf
```

```
Autocorrelations of series pH, by lag
     0      1      2      3      4      5      6      7      8      9
 1.000  0.731  0.279 -0.137 -0.422 -0.482 -0.332 -0.132 -0.012  0.007
```

These are the autocorrelation values r_a in Table 10.1. They decrease with growing lag, become negative at lag 3, reach a minimum at lag 5 and then level off. The reason for this can be seen in Figure 10.2. The first few correlations result from monotonous change, but at large distance periodicity occurs causing the correlations to vanish when lag is sufficiently long. Function `acf()` does not print confidence intervals but these are plotted in a graph only. According to this $r_a = 0.731$ at lag 2 is significant at the 95% confidence level, the remaining are not.

10.3 Detecting trend

When sampling is repeated over time each new state will differ from the previous one, either as a consequence of limited precision of measurement or because change is going on in the system. If a lasting change occurs, it can still be blurred by noise and a trend may emerge only when sufficiently strong. Consequently, distinguishing randomness from trend is a mandatory prerequisite for any further step when looking at time series, and methods have to be found to investigate nature and possible causes of change.

Formally speaking, the question to be asked is whether the arrangement of plots in multivariate resemblance space reflects their order in time. If this is the case a trend exists and in the context of vegetation we would call it succession. An efficient way to recognize multivariate trends is to display the states in phase space; that is, to ordinate the relevés. In an ordination, trends inside a plot manifest as a point sequence in which the order accords with time. If this is the case then there is *temporal dependence* occurring, the classical case of succession.

To decide whether a trend is a local phenomenon only or more widespread, the use of replicate samples is required. This is the case in Figure 10.3(a,b) showing a subset of the data analysed in Wildi and Schütz (2000) with relevés documenting succession in 7 (preferentially selected out of 59) plots located in the Swiss National Park. The time steps are all 5-year intervals and the investigation period ranges from 1917 until 1996. The individual series are partly overlapping, forming a long, horseshoe-shaped gradient typical for PCA ordinations. This gradient is primarily a spatial one. But most series run into the same direction. Different types of temporal trends can be distinguished. Plot Pin3, as an example, shows a perfect trend in one distinct direction: the rate of change is almost constant. Plot N8, on the other hand, also shows a directed trend, but the rate of change varies considerably. FN2, in contrast, remains almost constant over time suggesting the absence of a trend.

The ordination axes in Figure 10.3(a,b) are computed for the entire data set. Isolating plot data and analysing these individually may reduce distortions caused by reducing the dimensions. This is done in Figure 10.3(c) for plot Tr6 and in Figure 10.3(d) for plot FN2. As expected, ordination axis 1 now approximately points into the direction of the main temporal trend.

Figure 10.3 is generated using function `pcaser()` in the `dave` package. Because data points may form dense clouds when change approaches zero, there are options implemented to draw lines connecting the points of series (`lines=TRUE`) or drawing arrows from the first to the last state in time (`arrows=TRUE`). Plot membership (the names of the plots) is found in variable `Plot.no` of data frame `sn7sit` and vegetation (the species) in `sn7veg`:

```
o.pcaser<- pcaser(sn7veg,sn7sit$Plot.no,y=0.25)
```

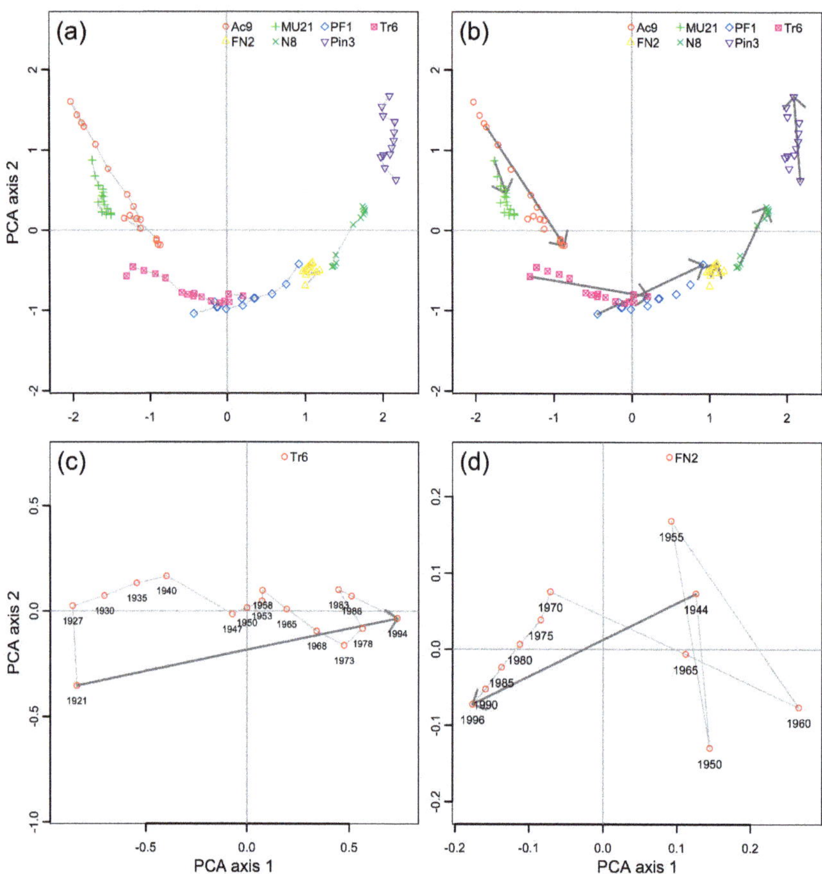

Figure 10.3: Ordination of data from seven plots in the Swiss National Park (Wildi and Schütz 2000). Subsequent states of plots connected by lines (a), arrows pointing from the initial to the final state (b). Panel (c): ordination restricted to plot Tr6. Panel (d): ordination restricted to plot FN2.

```
Call:
pcaser.default(veg = sn7veg, plotlabels = sn7sit$Plot.no, y = 0.25)

Eigenvalues (first 5): 64.541 18.421 11.114 4.002 1.179
```

As usual parameter y offers transformation according to $x' = x^{0.25}$. The first eigenvalue is much larger than in ordinary vegetation data because relevés in this data set are described by six variables only (addressing guilds rather than species). The plot method is used as follows:

```
plot(o.pcaser,lines=TRUE,arrows=FALSE)
```

This connects points of the same series by lines (a) whereas `lines=FALSE` and `arrows=TRUE` connects the first and the last state by arrows (b).

Figure 10.3(c) and (d) are done the same way, but using a subset of data only. For plot Tr6, for example, the rows we want to use, `sub`, are found using function `which()`:

```
sub<- which(sn7sit$Plot.no == "Tr6")
o.ps<- pcaser(sn7veg[sub,],factor(sn7sit$Plot.no[sub]),y=0.25)
plot(o.ps,lines=TRUE,arrows=TRUE,cex=0.6,lwd=0.5)
```

This reduces data frame `sn7veg` as well as `sn7sit` to the rows addressing plot Tr6 only.

10.4 Rate of change

Distance of multi-state observations per time step is nothing else but a measure for rate of change, that is, process velocity. In the simplest case a time series of vegetation consists of subsequent states separated by even time steps as illustrated in Figure 10.4(a). The five states documented by five relevés are then compared by calculating distances (see Section 4.2), which yields a 5 by 5 distance matrix [Figure 10.4(b)]. In the diagonal the self-comparisons are found; these are all zero (identity). All comparisons of states separated by one time step can be found in the first off-diagonal vector. I call these *rate of change of order* $o = 1$. The number of comparisons, c, is:

$$c(o) = n - o$$

where n is sample size (the number of time steps) and o is the order of change. For order 1 there are four possible comparisons, for order 2 only three and so on. The distance matrix can now be interpreted. Simple reasoning leads to the following considerations:

1. A change of any order can originate from random noise, trend or measurement errors.

2. If distances from the initial state monotonically increase with the number of time steps, then this accords with a constant rate of change. A trend will emerge.

Once we know that plots exhibit an overall trend, change in distance per time step reveals the dynamics of the process as explained in Figure 10.4. To demonstrate this a series of sufficient length is required enabling us also to vary the order (that is, time step length). Plot Tr6 from the Swiss National Park is suitable for that purpose. It encompasses 16 states ranging from 1921 until 1994. The resulting distance matrix has dimension 16 by 16.

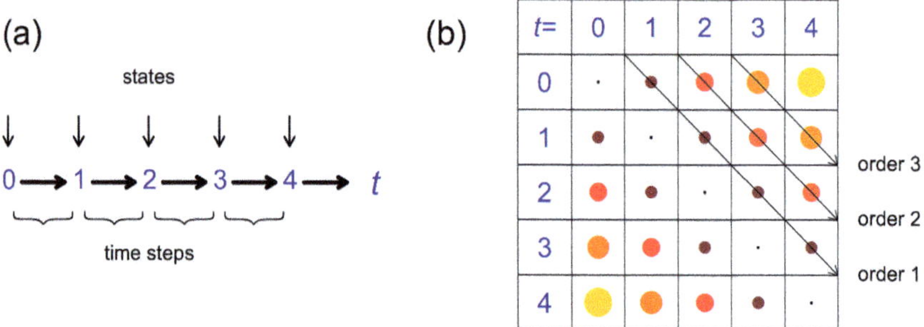

Figure 10.4: Measuring rate of change in time series analysis. (a) States and time steps in temporal systems. (b) Distance matrix depicting change in time.

To inspect this in more detail the x-axis forms the time scale, the off-diagonal vectors (elements along the arrows, Figure 10.4) of the distance matrix of Tr6 serve as y-scores in Figure 10.5. The lines accord with the different orders requested. This confirms that the fastest change occurred between 1940 and 1947, best seen in the rate of change of order one. Higher orders, encompassing longer time ranges, provide a smoother picture of dynamics and the number of states available decreases with increasing order.

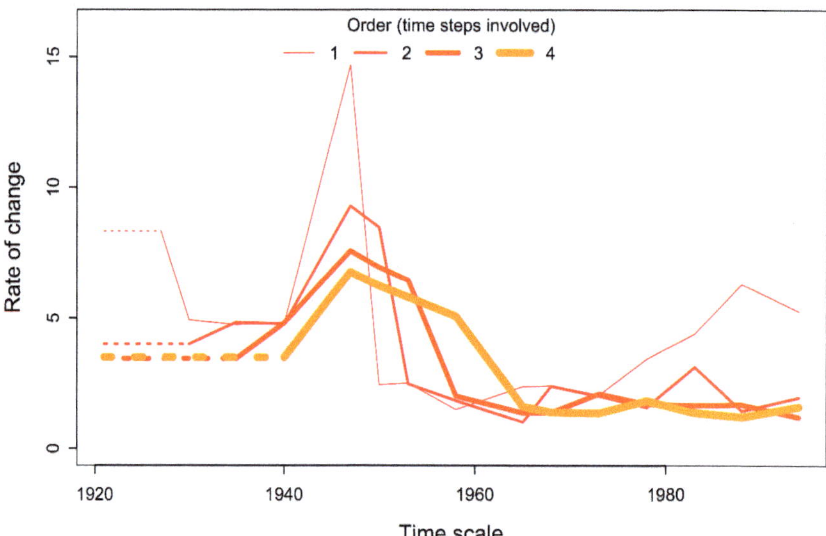

Figure 10.5: Rate of change in plot Tr6 in the Swiss National Park. Display of the first four off-diagonal vectors of the distance matrix, weighted by step length. See Figure 10.4 for selection of elements of the matrix.

The velocity profile in Figure 10.5 is obtained using function `speedprof()`. Two data frames are needed, one including vegetation data (all originating from the same plot), the other one the corresponding site information with time of sampling. We pick this from `sn7veg` and `sn7sit`, including rows belonging to plot Tr6 only:

```
sub<- which(sn7sit$Plot.no == "Tr6")
orders<- c(1,2,3,4) ; veg<- sn7veg[sub,] ; yr<- sn7sit$Year[sub]
o.spp<- speedprof(veg,yr,orders,y=1.0,adjust=TRUE)
plot(o.spp)
```

The `adjust` parameter sets the sum of cover values of all releveś to 100%. Because relevés in this data set already fulfil this condition the parameter has no effect. Although change in time (the distance between data points) in Figure 10.3(c) is similar to the same in Figure 10.5 (height of the speed profile), the speed profile only reflects the full dimensional rates of change.

10.5 Early succession: Vraconnaz revisited

Vegetation development starting from bare soil towards one or several equilibrium states is known as early succession (van der Valk 1992). It happens under various conditions, for instance on sediment deposits, in glacier forefields, after volcanic eruptions and in the course of various kinds of human impact. The challenge in data analysis is the fact that all species used as descriptors start with score zero. Hence, in terms of vegetation, the initial state is undefined. Probably the best way to overcome this is by including bare soil cover [as in the example of Lippe *et al.* (1985) used in Section 10.6] or even bare soil type as a variable. If cover percentage is used, the scores will sum up to 100%. This is the case in the example presented here, an investigation described in Feldmeyer-Christe *et al.* (2011) where early succession took place on bare soil after a bog burst occurring in the year 1987. The vegetation data set `vrveg` not only uses species cover as descriptors, but also bare soil, litter, and open water surface. The time series encompasses 11 plots from which yearly records are available starting in 1988 and ending in 2008, delivering as many as 21 time steps. Feldmeyer-Christe *et al.* (2011) are asking if succession converges towards one final equilibrium or if it ends up in different states. This question typically addresses direction of trend (Section 10.3). A second question is whether succession proceeds roughly at constant speed or if there is a process-specific velocity pattern. Although this is exactly what has been analysed in Section 10.4, each temporal state in this data set is documented by a sample of size 11 which also allows the interpretation of spatial variability.

The trend in the Vraconnaz data is shown in Figure 10.6 with lines connecting subsequent states in panel (a) and arrows pointing from the

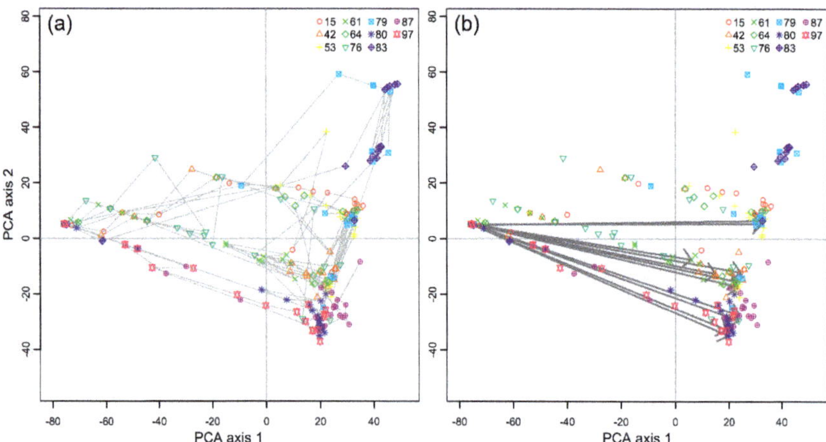

Figure 10.6: Ordination of data from bare peat plots of Vraconnaz (Feldmeyer-Christe *et al.* 2011). Subsequent states of plots connected by lines (a), arrows pointing from the initial to the final state (b).

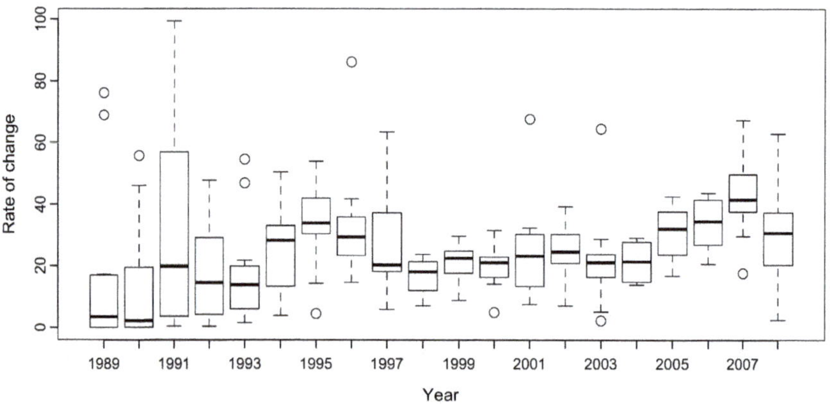

Figure 10.7: Boxplots (median, upper and lower quartile, minima and maxima; circles are outliers) of changes in vegetation for all years recorded in plots with bare peat (Feldmeyer-Christe *et al.* 2011).

initial to the final state in panel (b). Five distinct types emerge within a first phase, three of these persisting after 20 years (left side of ordinations). From Figure 10.7 we also learn that the first phase of succession is characterized by an increase of speed (an acceleration) up to about the year 1994, followed by a decrease. For the second phase acceleration resumes and it may even extend beyond the investigation period.

The velocity pattern in Figure 10.7 is a statistical representation of 11 speed profiles in the form of a boxplot. Unlike a single line (Figure 10.5, for example) this also illustrates the reliability of the pattern. Each bar of the plot depicts the distribution of the speed scores as found in a matrix with 11 plots as columns and the 20 time intervals as rows.

The ordinations revealing trend, Figure 10.6(a) and (b), can be reproduced the same way as shown in Section 10.3:

```
o.pcaser<- pcaser(vrveg,as.factor(vrsit$No_du_carre),y=1.0)
plot(o.pcaser,lines=TRUE,arrows=FALSE,cex=0.8)
```

These ordinations are based on the full cover-percentage scale (`y=1.0`) giving a quantitative view of the joint performance of the descriptors. The plot method of `pcaser()` accepts a graphical parameter for symbol size, `cex`. If `cex` is not specified it is assumed to be 1.0, the standard size. Reducing this is recommended for plots of large data sets.

To generate the boxplot of Figure 10.7 by function `boxplot()` the speed profiles of all 11 plots have to be merged into a single list, named `sp.list` below. Each time series in `vrveg` and `vrsit` is therefore isolated, the corresponding speedprofile derived and temporally merged in a matrix `spmat`. For technical reasons `spmat` is generated in advance and filled with empty elements (NAs). In order to split data frames `vrveg` and `vrsit` a list of plot names is also required. Function `levels()` extracts these from the vector of plot names in `vrsit`:

```
spmat<- matrix(rep(NA,220),nrow=11)
arr.lev<- levels(factor(vrsit$No_du_carre))
arr.lev
```

```
 [1] "15" "42" "53" "61" "64" "76" "79" "80" "83" "87" "97"
```

Within a loop of length 11 vegetation (`veg`) and site (`sit`) data frames are isolated for all individual plots with names taken from `arr.lev`. These are processed by function speedprof() and the resulting profiles are used to overwrite the as yet empty speed matrix `spmat` by the true values:

```
for (i in 1 : 11) {
  veg<- vrveg[vrsit$No_du_carre == arr.lev[i],]
  sit<- vrsit[vrsit$No_du_carre == arr.lev[i],]
  o.spp<- speedprof(veg,sit[,4],orders=1,y=1.0,adjust=TRUE)
  spmat[i,] <- o.spp$speed1
}
```

Matrix `spmat` is now split by columns into a list of 20 vectors corresponding to the 20 time intervals. Function `boxplot()` uses these to display the median, the upper and lower quartile, minima and maxima and also outliers (the circles):

```
sp.list<- split(t(spmat),o.spp$timescale[2:21])
boxplot(sp.list,xlab="Year",ylab="Rate of change")
```

The resulting boxplot can be customized in many ways (type `?boxplot` for detailed instructions).

10.6 Markov models

10.6.1 Method an example

Markov models are potentially able to reproduce a multivariate pattern of change. Very much like linear regression, they are fundamental in frequently successfully fitting locally observed processes. They ignore noise and nonlinearity and therefore fail to succeed when the rules of systems change, such as the hierarchy of competition among species. Fitting these models may be a first step in the evaluation of temporal patterns. Their functioning and use in evaluating vegetation process is explained below (Orlóci *et al.* 1993; Wildi 2001).

Changes in permanent plots can be interpreted as replacement processes. Several plant populations share the same resource, in the present case the physical space. Therefore, cover percentage is a straightforward (even though two-dimensional only) surrogate for measuring resource consumption. If gains and losses in space are in balance, so that any state of the system can be derived from the preceding one, then screening for a Markov process is worthwhile (Usher 1981).

Gain and loss of every species is assessed in a transition matrix P. This allows the computation of the state of a relevé vector, x, at time t from the previous step:

$$x_{t+1} = x_t P \tag{10.3}$$

In an ordinary permanent plot survey the vectors x are vegetation relevés. Unfortunately, such a kind of data does not allow estimation of the elements of the transition matrix: the gains and losses of the species (and this is the main reason why Markov models are not used routinely). If a species wins space then an ordinary relevé does not tell us which other species has lost it, and if a species loses ground, we do not know which one will profit from it. A Markov process, if present, remains undetected. Two plausible assumptions may help to overcome this situation in a method devised by Orlóci *et al.* (1993):

1. If a species loses part of the resource then any other dominating species will most likely profit (i.e. profit is proportional to the species cover).

2. If a species increases its occupation of the resource then the remaining dominant species will lose in proportion to their cover.

Both of these assumptions can be questioned. In succession, a change in abundance of a species may be caused by colonization of space by a newly arriving species, but a Markov model in its basic form does not foresee invasion. Moreover, a colonizing species may expand its cover at the expense of rare species. This is one indication why Markov models cannot be applied in isolation when invasion occurs.

As shown in Section 10.5 space as resource need not always be entirely occupied by vegetation. Therefore, it may make sense to add one more variable to the species list, quantifying the *open soil* (see Figure 10.8 for an example). Formally, open soil functions like any other ordinary species. Hence, the sum of all cover values, including open space, exactly amounts to 100%. This is achieved by the following transformation:

$$x'_{i,t} = \left(\frac{x_{i,t}}{\sum_{i=1}^{n} x_{i,t}}\right) * 100. \tag{10.4}$$

In this vector x contains the cover values of species i, t is the present time and n is the number of species. In the artificial example used for further explanations (Table 10.2), the sum of cover values is already 100%.

Table 10.2: Numerical example for demonstrating a Markov process (measured and modelled data). The elements are cover percentages.

	Measured data			Modelled data		
Time	$x_{t=1}$	$x_{t=2}$	$x_{t=3}$	$x_{t=1}$	$x_{t=2}$	$x_{t=3}$
Species 1	60	40	10	60	30	15
Species 2	25	35	55	25	42.5	51.23
Species 3	15	25	35	15	27.52	33.77
$\sum$	100	100	100	100	100	100

First of all the transition matrix for time step 1 to time step 2 is calculated (Orlóci *et al.* 1993). For each species i a difference in scores results, expressing change in time:

$$Diff(i) = x_{i,2} - x_{i,1} \tag{10.5}$$

Positive values of $Diff(i)$ signify a gain, negative ones a loss. The transition matrix contains all the losses of species i in row i, and the gains of the same in column i:

$$\mathbf{P} = \begin{pmatrix} p_{11} & \cdots & \downarrow & \cdots \\ \cdots & \cdots & \downarrow & \cdots \\ \rightarrow & \rightarrow & p_{ii} & \rightarrow \\ \cdots & \cdots & \downarrow & \cdots \end{pmatrix} \tag{10.6}$$

where $\downarrow$ is gain and $\rightarrow$ is loss of species i. The diagonal elements are the proportions each species covers at the end of the time step: $x_{i,t+1}$. The gains of species i at the expense of species h, as well as the losses of species i from species h, are given by:

$$Dev(h,i) = |Diff(i)|\frac{x_{h,t+1}}{\sum_i x_{i,t+1}} \tag{10.7}$$

This means that gains and losses occur in proportion to the resource (the space) each species occupies at the end of the actual time step. When processing the first species in our example, it can be seen that it loses 20% of the total resource from $t = 1$ to $t = 2$ ($Diff(1) = -20$). The new diagonal element is 40. From the lost 20% a portion of 35% goes to the second species, that is, 7%. To the third element go 20% of the 25% cover – 5% – completing the first row:

$$\mathbf{P}(t_1;t_2;spec.1) = \begin{bmatrix} 40 & 7 & 5 \\ 0 & 0 & 0 \\ 0 & 0 & 0 \end{bmatrix}$$

Species 2 achieves a win of 10% to be noted in column 2. This is again proportional to the covers of the species at time $t+1$:

$$\mathbf{P}(t_1;t_2;spec.1+2) = \begin{bmatrix} 40 & 7+4 & 5 \\ 0 & 35 & 0 \\ 0 & 2.5 & 0 \end{bmatrix}$$

The procedure is completed by processing species 3, winning 10%:

$$\mathbf{P}(t_1;t_2) = \begin{bmatrix} 40 & 7+4 & 5+4 \\ 0 & 35 & 3.5 \\ 0 & 2.5 & 25 \end{bmatrix}$$

After normalizing the rows (the sum adjusted to 1) the transition matrix is:

$$\mathbf{P}'(t_1;t_2) = \begin{bmatrix} 0.667 & 0.183 & 0.150 \\ 0 & 0.909 & 0.091 \\ 0 & 0.091 & 0.909 \end{bmatrix}$$

For each following time step the procedure resumes to yield one more transition matrix, as for time steps $t = 2$ to $t = 3$, where it is:

$$\mathbf{P}(t_2;t_3) = \begin{bmatrix} 10 & 10.5 & 11.5 \\ 0 & 55 & 5.5 \\ 0 & 7.0 & 35 \end{bmatrix}$$

For all time steps, all transition matrices, P, are averaged. The new transition matrix, $\overline{P}$, is assumed to be invariant for the entire time series. This means that it is kept constant over time and it will therefore outbalance fluctuations. In our example, after normalizing by rows, we get:

$$\overline{\mathbf{P}} = \begin{bmatrix} 0.500 & 0.2950 & 0.2050 \\ 0 & 0.9091 & 0.0909 \\ 0 & 0.1367 & 0.8633 \end{bmatrix}$$

Through simple matrix multiplication according to Equation (10.3) the modelled relevés are derived (Table 10.2, right-hand side). In these the first relevé is identical to the field data. It represents the initial state of the dynamic system; all subsequent states are modelled values.

Computation of this example in $\mathcal{R}$ starts with data input to generate the data frame that $\mathcal{R}$ function `fitmarkov()` of the `dave` package requires (Table 10.2):

```
x<- data.frame(matrix(c(60,40,10,25,35,55,15,25,35),ncol=3))
time<- c(1,2,3)
o.fm<- fitmarkov(x,time,adjust=FALSE)
plot(o.fm,colors=NULL,l.widths=NULL)
```

Parameter `adjust` does the transformation defined in Equation 10.4 when set to `TRUE`. The plot method of `fitmarkov()` generates a plot of raw and another of fitted data. Parameters `colors` and `widths` allow the lines in the graph to be customized. In the output list one finds the modelled (fitted) data and also the transition matrix:

```
o.fm$fitted.data

       X1        X2        X3
[1,] 0.60 0.2500000 0.1500000
[2,] 0.30 0.4247763 0.2752237
[3,] 0.15 0.5122808 0.3377192

o.fm$transition.matrix

     [,1]      [,2]      [,3]
[1,]  0.5 0.2950000 0.2050000
[2,]  0.0 0.9090909 0.0909091
[3,]  0.0 0.1366906 0.8633094
```

In real world examples successfully fitting a linear Markov process tells us that the succession involved most likely took place under constant rules of competition and facilitation. Orlóci *et al.* (1993) published a time series documenting recovery of a heathland after fire, using data from an investigation by Lippe *et al.* (1985). The example is presented here as a case where a linear Markov process successfully reproduces the temporal pattern.

The raw and the simulated data are shown in Figure 10.8, obtained from data frames `lveg` and `ltim`, the first holding the file of Lippe *et al.* (1985) vegetation data, the second the time steps involved:

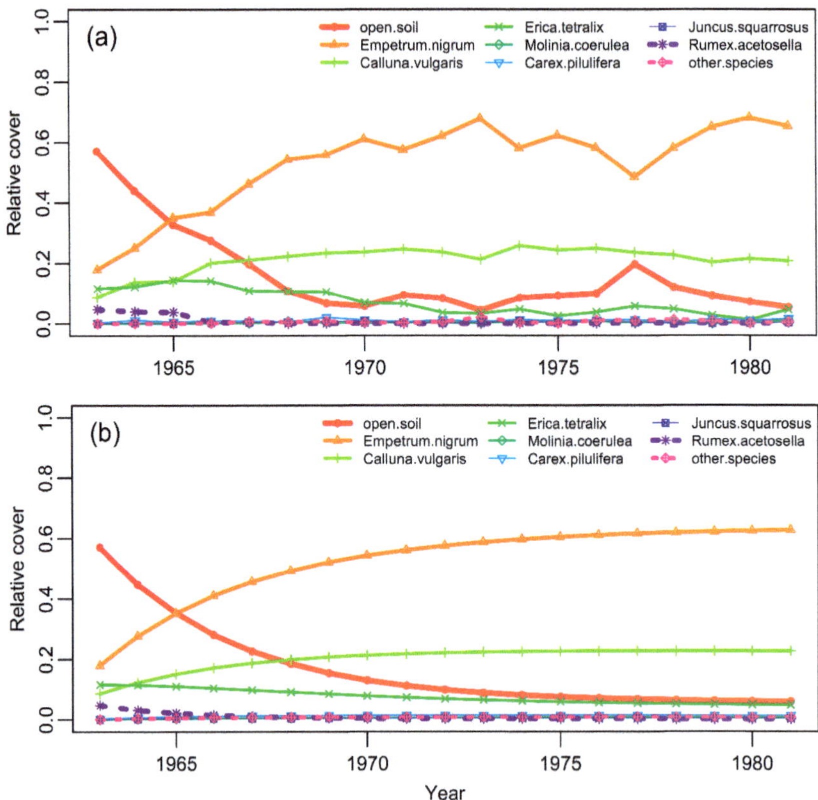

Figure 10.8: A Markov model of the Lippe *et al.* (1985) data set. (a) Field data; and (b) modelled data.

```
o.fm<- fitmarkov(lveg,ltim$Year,adjust=F)
plot(o.fm)
```

From the raw data [Figure 10.8(a)] it can be seen that in the first few years there is a rapid change. After about 8 years (≈1970), an equilibrium state is reached in which merely random oscillation occurs. One objective of the analysis is to determine the equilibrium state for which the Markov model is derived. The transition matrix is calculated as explained above; that is, from the 19 states of the system. It is the mean of 18 P matrices calculated for each time step. Then, beginning with the first field observation, 18 Markov relevés are derived through matrix multiplication. They are shown in Figure 10.8(b). After 19 years, the model has almost reached an equilibrium state. In the present example, it nicely fits the field data. Only the

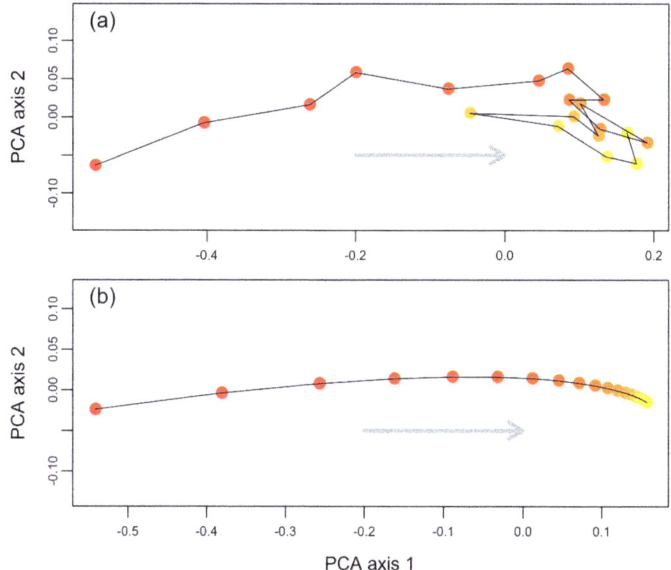

Figure 10.9: PCA ordination of the Lippe succession data. (a) Field data; and (b) Markov data. Arrows point in the direction of time.

deterministic part of the variation is reflected by the simulated time series; the random fluctuation is suppressed.

The Markov model explains multi-species change as long as this is linear. Linearity signifies that the transitions remain constant throughout the entire process. Successful fitting also means that the resulting similarity pattern does not depart too much from the observed, best seen in ordinations. Figure 10.9(a) projects the raw data of Lippe *et al.* (1985) in two-dimensional metric space extracted by PCA, whereas Figure 10.9(b) is the same of the fitted data. Figure 10.9(a) results from processing raw data taken from the output of function `fitmarkov()`.:

```
o.pcfit<- pca(o.fm$raw.data)
plot(o.pcfit)
lines(o.pcfit$scores[,1],o.pcfit$scores[,2])
```

The modelled process, Figure 9.7(b), uses fitted data:

```
o.pcfit<-pca(o.fm$fitted.data)
plot(o.pcfit)
lines(o.pcfit$scores[,1],o.pcfit$scores[,2])
```

The result illustrates the effect of fitting, suppressing random fluctuation, best seen near the final part of process, on the right-hand side of the ordinations.

10.6.2 Limitations and practice

The strength of Markov models is in revealing change in the dominance of species. This is why cover scores should be used that closely reflect consumption of resource, cover percentage for example. Furthermore, the interpretation of a graph, such as the one in Figure 10.9(b), becomes overburdened if too many species are involved. Very many species with rather low cover hardly ever add to the understanding of dynamics as conceived in the Markov process. In the 11 plots included in the Vraconnaz data `vrveg` (Section 10.5), there are as many as 154 species involved and most of these are rare. It is therefore recommended to limit their number as far as possible. A method worth looking at is the identification of orthogonal components, that is, the RANK method explained in Section 6.7.1. This carries the risk that some species with high cover scores are suppressed - if they correlate with other equally important ones. A criterion of selection that may be preferable to control is the use of overall cover, the sum of cover values across all time steps.

To demonstrate this with the Vraconnaz data an individual plot has to be retrieved first from data frames `vrveg` and `vrsit` as shown in Section 10.5:

```
veg<- vrveg[vrsit$No_du_carre == "15",]
sit<- vrsit[vrsit$No_du_carre == "15",]
```

This isolates the plot named `"15"` from vector `No_du_carre` (French). Vegetation data `vrveg` now still holds all the 154 species variables, mostly empty vectors. Those can be eliminated as explained in Section 7.6.3, but it is not really needed because the following step will do it as well.

To identify the dominant species, vector `f.sc` is derived, intended to hold the total sum of cover values throughout the entire range of succession. Its content is then ordered and the five species with highest cover are selected as descriptors. These serve as input for function `fitmarkov()`:

```
f.sc<- apply(veg,2,sum)
veg<- veg[,order(f.sc,decreasing=TRUE)[1:5]]
o.fm<- fitmarkov(veg,sit$Jahr,adjust=TRUE)
plot(o.fm)
```

This is how to get the panels (a) and (b) in Figure 10.10. The example demonstrates how the model representation can simplify the interpretation of time series. However, change occurring in early succession is delayed here by about five years and this exceeds the descriptive capability of a linear Markov model. It also fails to handle Gaussian-shaped response curves such as `litiere_seche` (dry litter) in panel (a) and any of the strong fluctuations occurring towards the end of the series. In conclusion, Figure 10.10 is a typical example where linear Markov models miss the capture of important features of dynamics: applications of Markov models should be limited to time series with moderate change and minor fluctuations only.

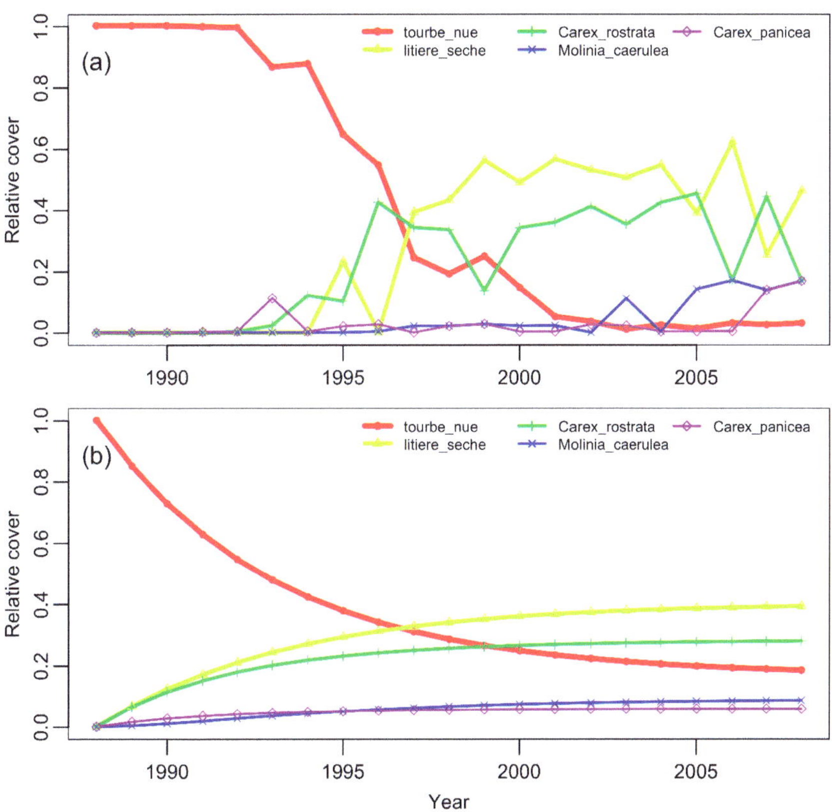

Figure 10.10: A Markov model of plot 15 of the Vraconnaz data. (a) Raw data of plot 15, (b) modelled data.

10.7 Space-for-time substitution

10.7.1 Principle and method

In long-term investigations species eventually exhibit a characteristic pattern of change: constancy, increase, decrease, random fluctuation, periodicity, and so on (Huismann *et al.* 1993). If the observation time is sufficiently long, many of these patterns turn out to be fragments of a bell-shaped response function. In space-for-time substitution, one assumes that several different fragments of the same response curve can be found, but occurring in different locations. If these fragments overlap, the entire response curve can be restored. This principle is sketched schematically in Figure 10.11. The thick lines show hypothetical response curves of the same species over eight time steps. At first glance they seem to be different in nature – sometimes with a tendency to increase and sometimes decreasing. However, when the curves of plots 2 and 3 are properly shifted (thin lines), a single response evolves encompassing 14 time steps. This is of course just an interpretation, because such curves never fit exactly. It is therefore essential that they overlap sufficiently, as is the case in Figure 10.11. In real situations, there are many species involved, and an overlap in vegetation data should yield a meaningful result for a majority of them.

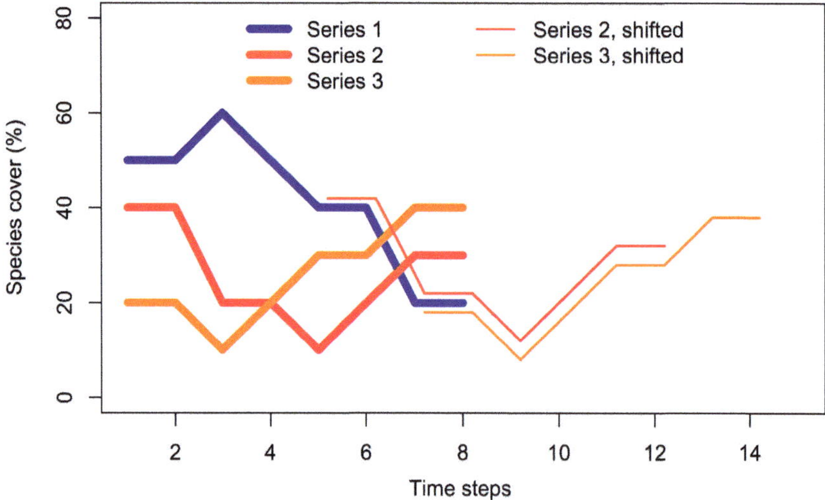

Figure 10.11: The principle of space-for-time substitution in the univariate case (Ghosh and Wildi 2007).

The fact that a succession trend sometimes ends in one plot but continues in a different one has been known for a long time. It has added much to the monoclimax theory strongly debated in the first half of the 20th century (Clements 1916, see also in van der Maarel and Franklin 2013). Although there has never been a doubt that such phenomena exist, the handling and interpretation has led to controversy. Much of this can be read in the review by Pickett (1989): he warns against the uncritical use of such data. Although I give an example of a really successful application below, some of the most frequently mentioned shortcomings and pitfalls are the following:

- Superimposing time series data is always hampered by stochastic noise and disturbance. The resulting synthetic series therefore suffers from some uncertainty.

- Vegetation change within two different plots will never be identical as the plots will most likely differ in site condition as well as in the species pool.

- Succession may not proceed at the same speed in different plots, prohibiting a perfect fit of series.
- The overlay of series mimics monoclimax. Alternative paths as in polyclimax can hardly be identified by this approach alone.

I am unaware of any experiments that would test the potentials and failures of space-for-time substitution, simply because these would last too long. All examples stem from surveys. Often, it was not even intended that long-term temporal patterns would be sought, just the data suggest this in the course of the analysis, as was the case in the succession data from the Swiss National Park shown below. The method applied here has been developed specifically to screen the more than 100 time series available for a general temporal trend (Wildi and Schütz 2000). The aim of the method is to find an unequivocal solution to the superposition of several (that is, more than two) multivariate time series rather than having to search for an iterative solution by trial and error.

Plot AC1																	
Year 19..	*30*	*35*	*42*	*47*	*53*	*57*	*60*	*65*	*68*	*74*	*81*	*85*	*90*	*94*			
"Aconitum"	72	71	70	56	44	26	48	40	34	31	24	23	22	22			
"Deschampsia"	18	19	21	27	38	46	40	42	44	46	51	54	58	60			
"Trisetum"	8	7	6	9	7	14	7	9	11	13	12	11	9	8			
Plot AC9																	
Year 19..				*17*	*22*	*25*	*32*	*35*	*40*	*47*	*50*	*53*	*59*	*65*	*68*	*74*	*82*
"Aconitum"				85	86	86	85	72	57	39	33	25	23	16	29	24	16
"Deschampsia"				12	10	12	14	22	32	40	43	46	55	57	47	49	46
"Trisetum"				3	3	1	0	2	5	7	9	10	8	11	11	11	15
Plot AC1/9																	
Age (yr)	*0*	*5*	*10*	*15*	*20*	*25*	*30*	*35*	*40*	*45*	*50*	*55*	*60*	*65*	*70*	*75*	*80*
"Aconitum"	72.0	71.0	70.0	70.5	65.0	56.0	66.5	56.0	45.5	35.0	28.5	24.0	22.5	19.0	29.0	24.0	16.0
"Deschampsia"	18.0	19.0	21.0	19.5	24.0	29.0	27.0	32.0	38.0	43.0	47.0	50.0	56.5	58.5	47.0	49.0	46.0
"Trisetum"	8.0	7.0	6.0	6.0	5.0	7.5	3.5	5.5	8.0	10.0	10.5	10.5	8.5	9.5	11.0	11.0	15.0

Figure 10.12: The similarity of time series AC1 and AC9 accords with comparison of relevés taken in 1990 (AC1) and 1959 (AC9), respectively.

The problem with finding the best solution when many time series are available is in identifying the most suitable pairs of response curves for fusion. The best candidates are those where the overlapping observations are the most similar compared with all other time series. In the following the similarity of time series is defined as the similarity of the two most similar observations in any two time series. This is shown in Figure 10.12, an example from two plots in the Swiss National Park. The species set is reduced to three for simplicity. The 'real' plots, AC1 and AC9, were surveyed in 'real' years: AC1 from 1930 until 1994 and AC9 from 1917 until 1982. When comparing the species composition (using Euclidean distance) the most similar observations are those from 1990 in AC1 and from 1959 in AC9. The series are now shifted until these two observations are located in the same column. The new series AC1/9 encompasses 80 years: now just in terms of age without any specific dates. It can also be seen that the method requires time steps of equal length. In the present case minor deviations from the standard time step of 5 years were corrected by interpolations. The steps leading to an unequivocal solution when fusing three or more time series are the following:

- Compute a resemblance matrix of time series. Resemblance (distance) is defined as shown in Figure 10.12.
- Derive the minimum spanning tree of time series (Gower and Ross 1969). This is a graph showing the nearest neighbours of all time series in the form of a tree.
- Position the observations by overlapping the time series according to the order given in the minimum spanning tree. This yields the relative age of each series.
- Compute the mean composition of the new synthetic time series by averaging all scores pertaining to the same time step.

The minimum spanning tree yields a unique solution to the problem. This is not necessarily the 'true' one, but it is the one delivering the shortest possible series based on the data used.

10.7.2 Swiss National Park succession (example)

The results shown below are from Wildi and Schütz (2000). The original time series are of exceptional length and the first observations date back to the year 1917, when J. Braun-Blanquet established the first permanent plots in the Park with the intention to document future reforestation of pastures (Figure 10.13).

It took as many as 80 years to detect that the data could be interpreted according to the idea of space-for-time substitution. The plots do not constitute a statistical sampling design but they are dispersed all over the previous pastures of the park. The data set used below addresses 59 of them, consisting of 751 relevés. The species are merged into six groups, carrying the genus names of 'dominants'.

Data preparation is explained in more detail in Wildi and Schütz (2000). The steps involved in the analysis are the same as those shown in Section 10.7.1. In Figure 10.14 the minimum spanning tree for the 59 time series is shown, embedded into a PCOA ordination. This is not just a single line as an ordering principle but a complex tree. Processing this by fusing time series pairwise yields the arrangement in Figure 10.15. The resulting synthetic time steps (83) minus one are multiplied by step length of 5 years, yielding a model time span of 410 years (Figure 10.16).

Figure 10.13: *Pinus mugo* on a former pasture in the Swiss National Park, just escaping the reach of browsing red deer.

All steps described above are implemented in $\mathcal{R}$ function `overly()` of the `dave` package. Data frame `sn59veg` is the original as used in the paper of Wildi and Schütz (2000) and plot labels and year of sampling are located in `sn59sit`:

```
names(sn59sit)
```

```
[1] "Plot.no"      "Year"
```

Relevés with identical plot label, variable `Plot.no` in `sn59sit`, belong to the same time series. Labels may consist of any character string. The year of sampling is not really needed, because the algorithm generates its own artificial time scale. Function `overly()` offers data transformation for the computations according to $x' = x^y$ for which y has to be chosen (Section 3.2). Not surprisingly, Wildi and Schütz (2007) show that this transformation affects the final result too. For the final average scores in Figure 10.15 untransformed raw data are used. The sampling interval `sint` in this project is 5 years, line widths `l.widths` and `colors` are set to `NULL` forcing the plot method to apply default settings:

```
o.over<- overly(sn59veg,sn59sit$Plot.no,y=0.5,sint=5)
plot(o.over,colors=NULL,l.widths=NULL)
```

```
Call:
overly.default(veg = sn59veg, Plot.no = sn59sit$Plot.no, y = 0.5,
    sint = 5)
Number of time steps in new time series:   83
Time span of the new time series:          0 - 410
```

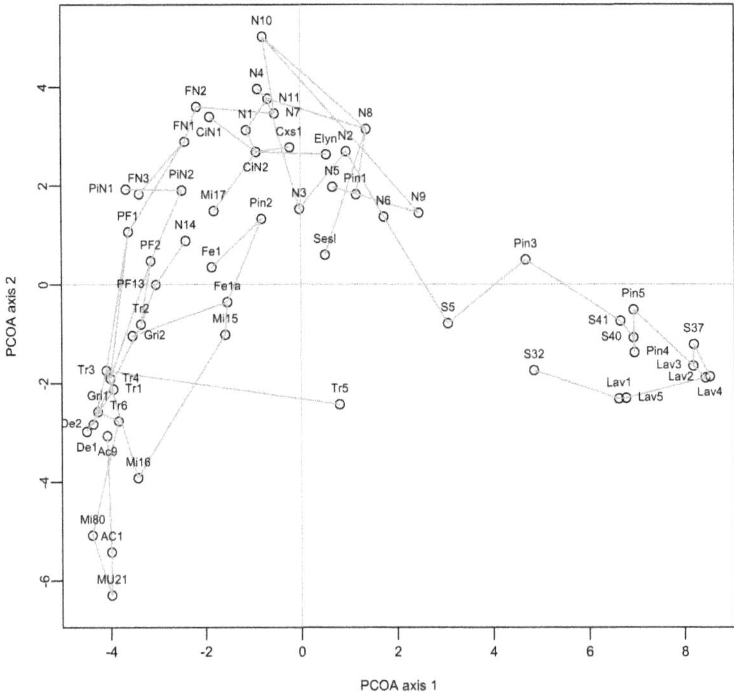

Figure 10.14: Minimum spanning tree of 59 time series from the Swiss National Park. Each dot represents a plot labelled by plot name.

The plot method generates the graphs shown in Figure 10.14, Figure 10.15 and Figure 10.16. The list `o.over` contains among other results the new, synthetic time series for further processing.

The overall trend (Figure 10.16) can be interpreted as follows: an initial *Aconitum* phase, resulting from livestock grazing and fertilization, dominates for about 50 years after the cessation of grazing by livestock. A *Deschampsia* phase then emerges and remains dominant for about 15 years. A later transition to a grassland dominated by *Festuca rubra* is most likely caused by grazing activity of red deer (Krüsi *et al.* 1998; Achermann *et al.* 2000). This is followed by a *Carex sempervirens* phase that may last 150 years. Finally, *Pinus montana* seedlings begin to establish, slowly initiating the reforestation phase (Figure 10.13).

The pattern revealed in this example must be strongly nonlinear, as the bell-shaped response curves in Figure 10.16 suggest. It has been shown by Wildi and Schütz (2007) that computed process length, a result of the analysis, varies to some extent depending on the transformation chosen for the species scores. Unlike in static modelling (Chapter 9), no reference

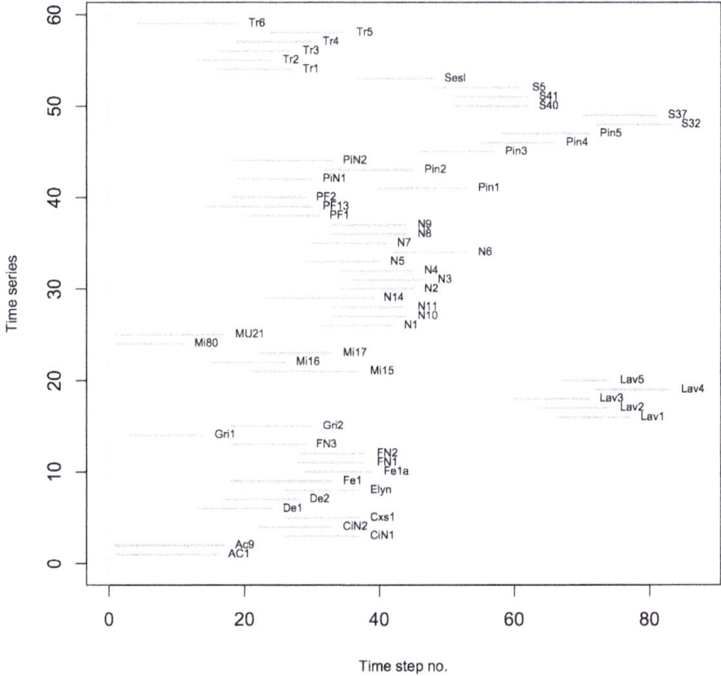

Figure 10.15: Order of 59 time series from the Swiss National Park.

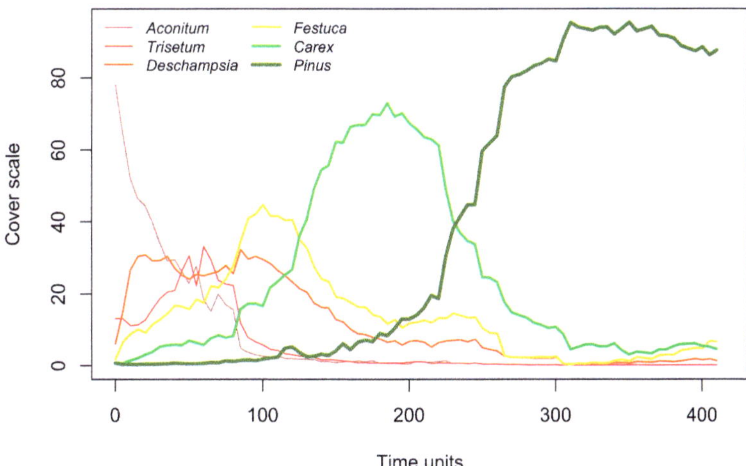

Figure 10.16: Succession in abandoned pastures of the Swiss National Park, derived by space-for-time substitution.

measurement exists to tell us which of the transformations is best, leaving us with some uncertainty about succession velocity.

A sample in space and time and therefore a candidate for an attempt at space-for-time substitution as explained in Section 10.7.2 is the Vraconnaz data frame, `vrveg` (Section 10.5). Because bare soil was created simultaneously in all plots one would not expect that spatial variation mimics a temporal shift. Or, vice-versa, the method used should not suggest this. Probing for space-for-time substitution with Vraconnaz data is done as follows:

```
o.ov<- overly(vrveg[,1:60],vrsit$No_du_carre,y=0.5,sint=1)
plot(o.ov,colors=NULL,l.widths=NULL)
```

The result (not shown here) is as trivial as can be expected. Unlike in Figure 10.15 the 11 time series are considered temporally aligned and none of these are shifted.

10.8 Dynamics in pollen diagrams

Vegetation series encompassing time spans of thousands of years can only be found in fossil records, for example in pollen diagrams (Lischke 2005). In these, extreme nonlinearity can be expected because very long time spans increase the chance that changes in the functional role of species within the vegetation cover will occur, caused mainly by invasions and extinctions. I demonstrate typical patterns using pollen of tree species of the Soppensee profile from Lotter (1999) (see also Lischke *et al.* 2002), documenting the change in tree species on the Swiss Plateau from about 13,000 BP until 5700 BP (Figure 10.17), without considering the changes in the technology of ^{14}C dating that have taken place in the meantime.

First we look at the velocity of the process, defined as the rate of change in total species composition per time unit:

$$V = \frac{d}{\delta t} \tag{10.8}$$

where d is the Euclidean distance (a measure of dissimilarity) between any two consecutive states in the pollen diagram. This type of calculation allows the derivation of a velocity profile of the change processes over the period of measurement (Figure 10.18). As explained in Section 10.4, different time steps lengths can be used for this, the shortest of 50 years given by the temporal resolution of data, whereas the longest steps of 800 years show the long-term trends. The 50-year time step is also used in the ordination of Figure 10.19, panel (a). The states in time are the circles and their diameter is proportional to velocity. There are linear phases where velocity is high and others where it is low.

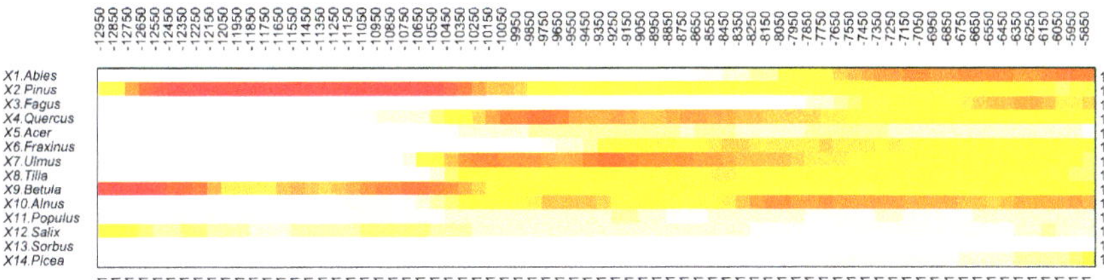

Figure 10.17: Tree species in a pollen diagram covering about 7000 years (Lotter 1999). Synoptic table with every second sample displayed from data frame pveg. The x-axis is year BP.

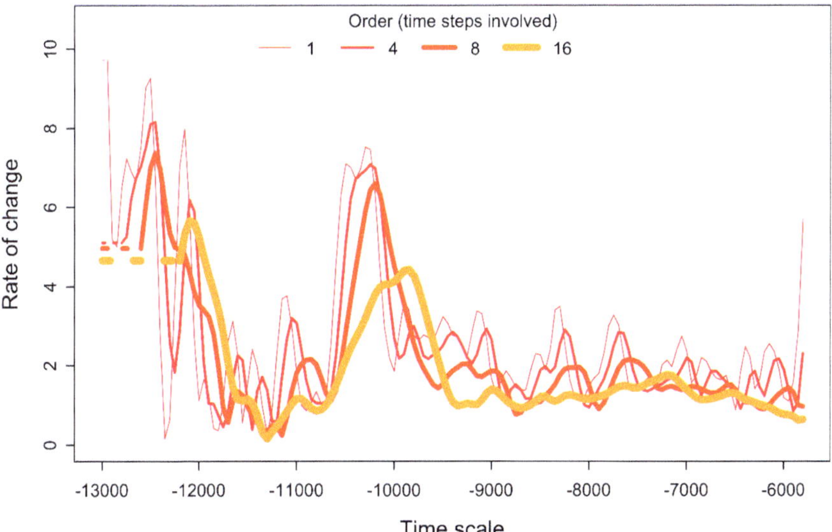

Figure 10.18: Velocity profile of the Soppensee pollen diagram (Lotter 1999). The time step length in the data is 50 years.

Even when inspecting the velocity profiles in detail [Figure 10.19, panel (a)], no evidence is found for velocity being related to nonlinearity, such as that happening around 12,000 and 10,000 BP. But there is yet another interesting way of looking at similarity, when taking its second derivative, acceleration in the change process, A:

$$A = \frac{V_{t+\delta t} - V_t}{\delta t} \tag{10.9}$$

When A is positive, the velocity increases; when it is negative, the velocity decreases. In Figure 10.19, panel (b), acceleration is used to interpret the same time trajectory as in Figure 10.19, panel (a): circles are now propor-

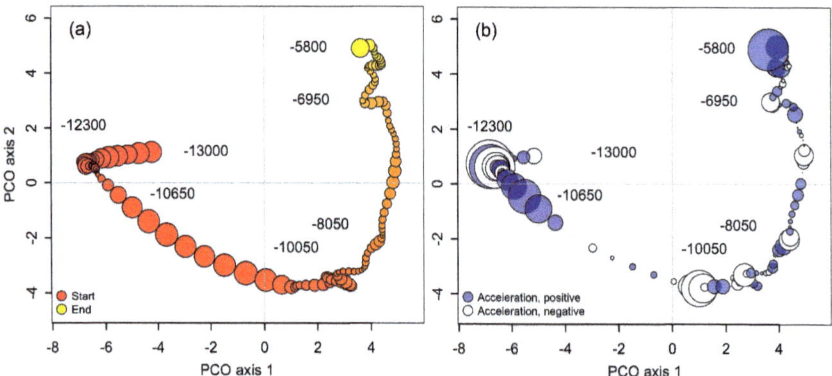

Figure 10.19: Time trajectory of the Soppensee pollen diagram (Lotter 1999). Panel (a): diameter of circles are proportional to the velocity of change. Panel (b): diameter of circles proportional to change of velocity (positive or negative acceleration).

tional to acceleration. The result is striking: linear phases, whether fast or slow, show low acceleration. Nonlinear phases, on the other hand, are characterized by pronounced positive as well as negative accelerations. Hence, strong fluctuations in the dynamics of the systems distinguish nonlinear from smooth, predictable linear phases. The large circles at the beginning and at the end of the series, respectively, may be caused by disturbance effects of sampling rather than vegetation dynamics.

Is this finding a generally valid rule? The answer is, probably not. First of all, the observation is restricted to one profile only and in this there are just two phases of strong nonlinearity. Secondly, fluctuation manifesting in velocity and acceleration can be caused by disturbance in the profile. Sedimentation is dependent on many factors and cannot be expected to be constant over millennia. Hence, investigation of many more profiles would be needed to approve the velocity pattern.

It is also worth considering not only the quantitative component in the data, but also the qualitative one. In the analyses shown so far the quantitative view is adopted; that is, the species scores are percentages of total pollen abundance. In these analyses the most abundant types of pollen dominate the result. Rare species, including the invaders in their first years of arrival, do not contribute much to the rate of change. However, species scores can be transformed to presence–absence (or any intermediate scale, see Section 3.4) so that the velocity expresses change in species presence. This is shown in Figure 10.20, where panel (a) depicts change in the quantity of pollen composition while panel (b) is change in quality; that is, the emergence or disappearance of pollen of a specific species. Obviously, these two types of change happen at different points in time. There are phases in which several species emerge [panel (b)] but no considerable change in quantity can be observed [panel (a)] and vice versa. When comparing the profile in panel (b) with the data in data frame `pveg` we can even identify the origin of the 11 peaks found. These are, from left to right:

1 Invasion of *Quercus*.
2 Invasion of *Alnus*.
3 Disappearance and re-emergence of *Quercus*.
4 *Ulmus* invading.
5 Joint invasion of *Tilia* and *Acer*.
6 Invasion of *Fraxinus*.
7 Disappearance of *Fraxinus*.
8 Re-emergence of *Fraxinus*.
9 *Abies* invading.
10 *Fagus* and *Picea* arriving simultaneously.
11 *Sorbus* appearing.

The rather high temporal resolution given in Figure 10.20 also emits a signal when a species disappears temporarily, but re-emerges in a subsequent time step. This can be a sampling artefact and the inspection of the raw data is inevitable for a careful interpretation.

The presentation of the raw data as shown in Figure 10.17 is a graphical display of a classical synoptical table where the sampling units are the columns and the descriptive variables (pollen types) the rows (Section 7.6). It is drawn using every second sample only. In $\mathcal{R}$ a vector of sequential row numbers `ri` is generated first and within function Mtabs() only every second element is taken. The double-percentage sign as it is used here is the modulo function in which elements are considered only if division by 2, in this case, delivers zero. Hence, to get every fourth element only, one would write `ri[ri%%4==0]`, and so forth:

```
ri<- c(1:nrow(pveg))
o.Mt<- Mtabs(pveg[ri[ri%%2==0],],method="raw")
plot(o.Mt)
```

The velocity profile in Figure 10.18 is obtained as shown in Section 10.4, that is, from function `speedprof()` in package `dave`:

```
orders<- c(1,4,8,16)
o.spp<- speedprof(pveg,psit$Years.B.P,orders,adjust=FALSE)
plot(o.spp)
```

The ordinations in Figure 10.20 (a) and (b) are generated by function `vvelocity()`, using the same data sources `pveg` and `psit` as above:

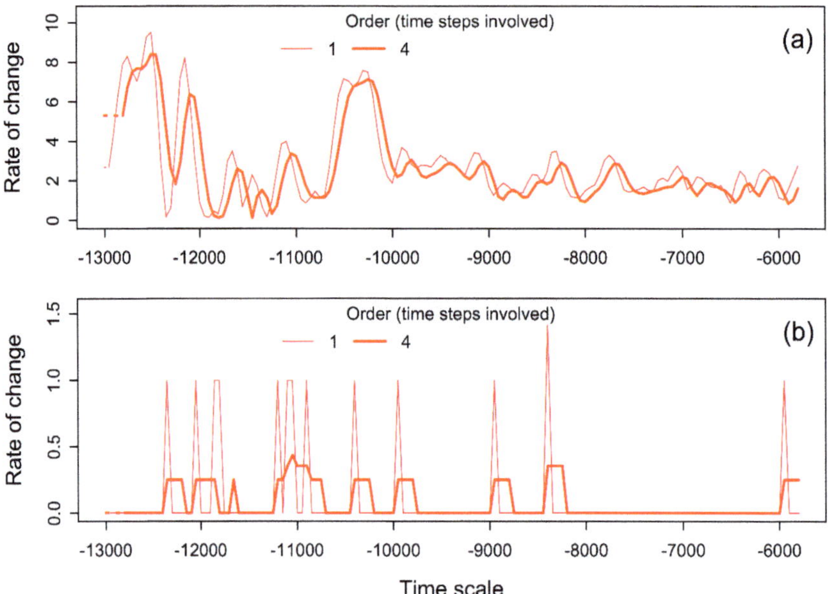

Figure 10.20: Velocity profiles of quantitative (a) and qualitative content (b). The peaks in (a) stem from dominance change of species, those in (b) from species invasions and extinctions (see details in the main text).

```
tlabs<- c(1,15,48,60,100,122,145)
o.vvel<- vvelocity(pveg,psit$Years.B.P,y=0.5)
plot(o.vvel,tlabs=tlabs,scal=1)
```

In this, `tlabs` denotes time steps to be labelled in the graph. Parameter `scal` allows for adjustment of circle diameter and `y` is defining the transformation of species scores as usual.

The speedprofiles in Figure 10.20 are generated by function `speedprof()` again. For panel (a) this is done as follows:

```
o.spp<- speedprof(pveg,psit$Years.B.P,orders=c(1,4),adjust=TRUE)
plot(o.spp)
```

For panel (b) `pveg` is transformed to presence–absence, using `sign(pveg)`.

The analysis of pollen data reminds us to interpret short time series with great care. Investigations of human beings may easily fall into a phase where change is slow and continuous, sometimes lasting centuries. In the Soppensee data, change in direction and velocity tend to happen unexpectedly and quickly.

11 Dynamic modelling

Given the state of an ecosystem, mathematically formulated rules of change mimic dynamic system behaviour. Dynamic modelling is a perfect tool to evaluate the outcome of assumed or observed sets of mechanisms.

11.1 Principles of systems

It may come as a surprise to see dynamic modelling in a book on vegetation ecology. But this topic perfectly fits the general framework outlined in Chapter 1 by implementing investigations inside the model world. Probably the first dynamic models of the type used here served the investigation of systems other than ecological, mainly economic and industrial. Forrester (1968), in his pioneering book *Principles of Systems*, gives a simple definition of the subject: 'As used here, a "system" means a grouping of parts that operate together for a common purpose.' In models of such systems, 'parts' are described by state variables like weight of plant biomass, percentage cover of vegetation, plant nutrients per cubic decimetre of soil, population size of a species in a plot and so on. Hence, dynamic models potentially serve the analysis of natural systems. But what is the meaning of the buzzword 'model' in the present context? Again, Forrester (1968) gives a simple explanation: 'A model is a substitute for an object or a system.' When modelling we are working with this substitute, being a system by itself, but likely less complex than the real system it describes. When modelling we investigate the substitute, for example by performing test runs and studying how it succeeds or fails without doing harm to the real system – not even damaging the computer we use. In the early days of electronic computing everyone was fascinated by the apparently unlimited possibilities of simulation, culminating in the world model of Dennis L. Meadows, by which he justified his *Limits to Growth* (Meadows *et al.* 1972). Limits to modelling were experienced later because the computer models proved difficult to handle when complexity increased, as illustrated in an early attempt shown in Figure 11.1. Even simple models may be difficult to handle and to understand; small is often more beautiful.

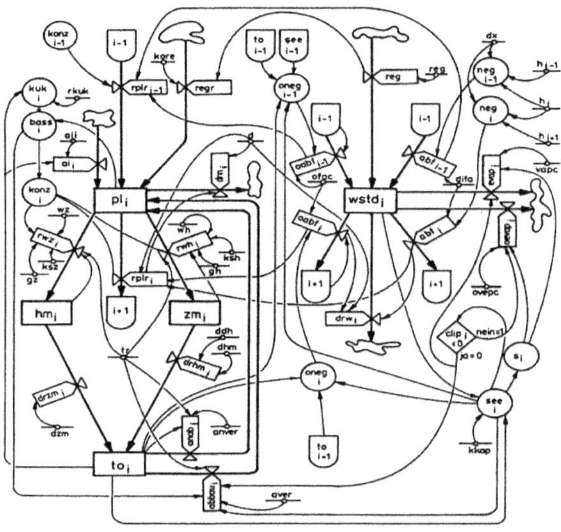

Figure 11.1: Attempt to get a dynamic model under control (Wildi 1976).

Dynamic modelling is an easy way to play with hypotheses, assumptions and parameters. The rules how state variables change are described by one or several differential equations. Starting with initial conditions given by the modeller the resulting change of the entire system in time is derived through numerical integration, carried out automatically. When introducing the method I start with the simplest systems, comprising one state variable only. These are exponential and logistic growth equations, followed by an evaluation of the well-known Lotka–Volterra equations addressing the interaction of two species populations, a system suitable to demonstrate the crucial role of parameter selection. The final example, a succession study, extends the principle from merely temporal to spatially explicit systems.

11.2 Simulating exponential growth

Time has only one dimension and the simplest types of models do not consider space. The temporal change of a system is described in the form of differential equations. Numerical integration of these equations yields a state vector describing the state of the system in the future. An example is the exponential growth equation:

$$\frac{\delta X}{\delta t} = rX \tag{11.1}$$

where X (the state variable) is, for example, the number of individuals at time t, and r is the rate of population increase (Maynard Smith 1974). This equation has the well-known deterministic solution:

$$X = X_0 e^{at} \tag{11.2}$$

This assumes that within a short time span, δt, each individual will give rise to a fraction, $r\delta t$, of new individuals. It is of course more adequate to reason in stochastic terms and so $r\delta t$ is the probability of an individual having an offspring within δt, or no offspring with the probability of $1 - r\delta t$. As shown by Maynard Smith (1974), the formula for the average population size is:

$$\hat{X} = X_0 e^{at} \tag{11.3}$$

and the variance of X is:

$$var(X) = X_0 e^{2at}(1 - e^{-at}) \tag{11.4}$$

Integrating more complex differential equations is difficult, and frequently impossible. This is where numerical integration comes into play: the principle shown in Figure 11.2. In numerical integration, one specific outcome is calculated based on one assumed initial state of the system. The example starts with a population size of $X_{t=0} = 10$. The growth rate is $r = 0.1$, meaning that any given individual has this probability of having an offspring within a single time unit t. For each time step, the new state of the population size is determined based on the previous one:

$$X_{t=1} = X_{t=0} + \frac{\Delta X}{\Delta t} \tag{11.5}$$

This kind of calculation is also known as Euler's rule. After one time step, $X = 11.0$, and after two steps we get $X = 12.1$ (if time step length is equal to one time unit, see below). In numerical integration the symbol Δ is used instead of δ because the time step has a finite length, chosen by the user. This yields just an approximation of the real process, which is continuous and not discrete. An ideal population of large size will grow from the very beginning of the process, leading to a somewhat faster growth than in a discrete case. In numerical integration the approximation is improved when recalculating X in smaller time step lengths. The effect is shown in Table 11.1. Obviously, precision increases with reduced time step length for the calculation. Setting the time step too short, however, may lead to computational errors due to the limited precision current computers provide.

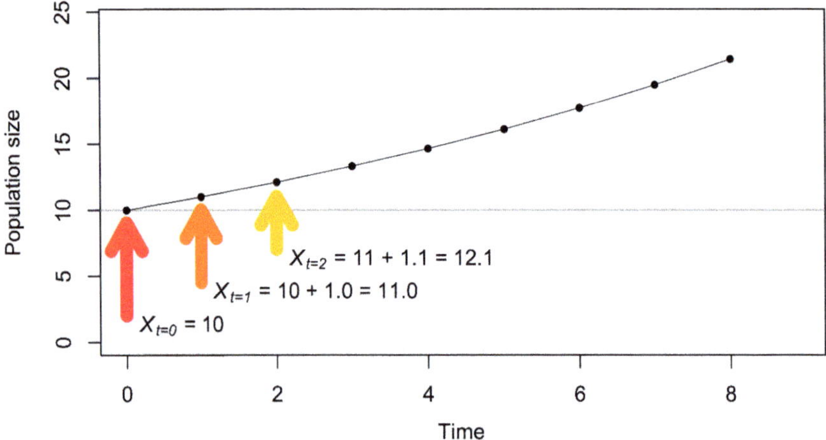

Figure 11.2: Numerical integration of the exponential growth equation [Equation (10.1)] using Euler's rule. Time step length Δt is 1.0.

Table 11.1: The effect of time step length in numerical integration. Column 2 holds the exact solutions, columns 3 and 4 are the approximations.

Time	$\Delta t \to 0.$	$\Delta t = 0.5$	$\Delta t = 1.0$
$X_{t=1}$	11.0517	11.025	11.000
$X_{t=2}$	12.214	12.155	12.100
$X_{t=3}$	13.498	13.401	13.310
$X_{t=4}$	14.918	14.774	14.641
$X_{t=5}$	16.487	16.289	16.105

11.3 Logistic growth

Exponential growth is widespread in natural systems, but only at the beginning of growth processes. Any positive growth rate ultimately ends up in a population size tending towards infinity. As early as 1825 the mathematician Pierre François Verhulst published an equation known today as the logistic equation which takes care of the limiting role of resources as exponential growth proceeds. In this the new element introduced is an upper limit for population size, C, the carrying capacity:

$$\frac{\delta X1}{\delta t} = rX\frac{C-X}{C} \qquad (11.6)$$

As the state variable X approaches the carrying capacity, C, growth becomes zero and the population stops growing. Equation 11.6 does not explain why the carrying capacity occurs. But one could think, for example, of an open water pond which is overgrown by *Lemna* (duckweed). At the beginning there is no limiting resource (open space, in this case) but once the pond is overgrown growth must stop.

As for exponential growth the explicit solution to the differential equation (Equation 11.6) is known (Maynard Smith 1974). But logistic growth is another good example to demonstrate numerical integration. Although there are numerous functions available in $\mathcal{R}$ to support this I explain how explicit calculation is done. The three basic steps involved remain the same for even very complex models:

- Set up the time frame (beginning, end, time step length), define parameters (growth rates, carrying capacity, other) and state variables at the begin of simulation (initial conditions).
- Program a time loop (a `for` statement in $\mathcal{R}$, see below). Embedded in this are the differential equations and functions for numerical integration.
- Establish model output. This may include text, but frequently graphical display.

For the logistic growth equation we simply write this down in $\mathcal{R}$:

```
tmax <- 100                     # time limit of simulation
tstep<- 1.0                     # time step length
nt<- tmax/tstep                 # number of time steps
t<- seq(0,tmax-tstep,tstep)     # vector of time steps
x<- rep(0,nt)                   # vector storing the state variable
x[1]<- 1.0                      # initial state of x
r<- 0.1                         # growth rate
c<- 20                          # carrying capacity
```

For time `t` a vector is prepared using function `seq()` with three parameters: the first giving the first element of the vector, the second the last element and the third is the increment used for the elements in between. The content of this vector then is $t = (0, 1, 2, \ldots, 99)$. Function `rep()` is used to initialize the state vector `x` and it repeats 0 `nt` times.

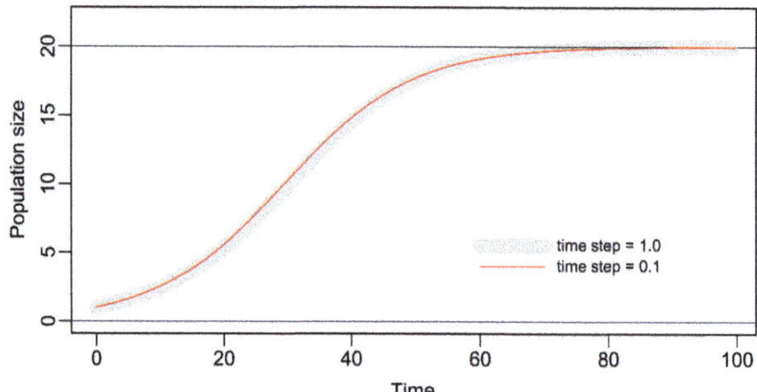

Figure 11.3: Simulating logistic growth using two different time step lengths. Carrying capacity is $c = 20$.

Then follows the second step, assessing the time loop (using function `for()`) with `nt` steps of computation. Inside the curly braces is the differential equation (Equation 11.6) and a line for numerical integration (Equation 11.5):

```
for (i in 2:nt) {
   dx<- r*x[i-1]*(c- x[i-1])/c
   x[i] <- x[i-1]+(dx*tstep)
}
```

Finally the state variable, x, is displayed as a function of time to yield the graph of Figure 11.3:

```
plot(t,x,type="n",xlab="Time",ylab="Population size")
lines(t,x,pch=1,lwd=6,col=gray(0.8))
abline(h=c(0,c),lwd=0.4)
```

Function `plot()` is called first to open a plotting window, followed by `lines()` to draw the resulting function. In Figure 11.3 there are two of these lines illustrating the effect of time step length chosen. The shorter the time step the more precise is the approximation, but the overall gain in precision is minor in the case of simple logistic growth. It is also noteworthy that the simulation starts at population size $x[1] = 1.0$ as the model does not consider invasion and a value of zero would result in no growth at all. At low population size the latter is close to exponential growth, when approaching the carrying capacity this slows down yielding the S-shaped curve typical for logistic growth.

11.4 The Lotka–Volterra model

The Lotka–Volterra equations, independently devised by Alfred J. Lotka (Lotka 1925) and Vito Volterra (Volterra 1926) are an excellent example of a model describing a dynamic system with two interacting state variables. It is also known as predator–prey system for which it was originally designed, but it can equally be used for guilds instead of species populations including organisms other than animals (plants, algae or bacteria, for example). In Watkinson's (1987, p. 178) view it is an appropriate way to model plant-herbivore interactions. Describing the rather complex relationships in a parsimonious way made it most appealing to theoretical biologists. It gained much attention when MacLulich (1936) published statistics of lynx (a predator) and hare (a prey) pelts traded in North America. This time series of about 125 years shows the synchronous oscillations of both populations with peaks of the predator always showing a lag behind those of prey species (Figure 11.4).

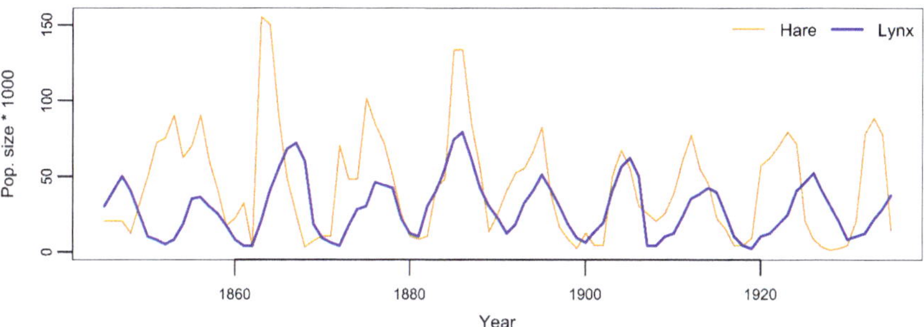

Figure 11.4: Abundance of lynx and snowshoe hare, approximated by the number of pelts received by the Hudson Bay Company (MacLulich 1937). Redrawn after Odum (1971, p. 191).

The model of Lotka and Volterra turned out to be full of not always easily recognized assumptions making interpretation a true challenge. For the sake of simplicity we assume that a first state variable, $X1$, is a prey population and a second, $X2$, the same of a predator. Using the notation shown in Table 11.2 the two differential equations can be written as:

$$\begin{aligned} \frac{\delta X1}{dt} &= X1(g1 - d1X2) \\ \frac{\delta X2}{dt} &= -X2(d2 - g2X1) \end{aligned} \tag{11.7}$$

Having a closer look at Equations 11.7 reveals that interaction between two populations is built-in whenever two state variables occur within the

Table 11.2: List of the parameters of the Lotka–Volterra model

Term	Role in predator–prey system	Explications
$X1$	Population size of prey	Number of individuals
$X2$	Population size of predator	Number of individuals
$g1$	Growth rate of prey without predators	Unlimited nutrient supply assumed
$g2$	Growth rate of predator per prey caught	Growth dependent on consumption
$d1$	Death rate of prey per predator	Consumption rate of predator per prey
$d2$	Death rate of predators	In absence of prey

same differential equation. Hence, we set up the time loop for simulation as shown in Section 11.3, and try to find some meaningful assumptions for the parameters including the initial conditions for state variables $X1$ and $X2$:

```
tmax <- 400 ; tstep<- 0.1
nt<- tmax/tstep ; t<- seq(0,tmax-tstep,tstep)
X1<- rep(0,nt) ;  X2<- rep(0,nt)
X1[1]<- 300 ; X2[1]<- 20
g1<- 0.025 ; g2<- 0.0005
d1<- 0.0005 ; d2<- 0.15
```

Simulation encompasses the computation of the temporal increments given by Equations 11.7 and subsequent numerical integration of the two state variables:

```
for (i in 2:nt) {
   dx1<- X1[i-1]*(g1-d1*X2[i-1])
   dx2<- -X2[i-1]*(d2-g2* X1[i-1])
   X1[i]<- X1[i-1]+(dx1*tstep)
   X2[i]<- X2[i-1]+(dx2*tstep)
}
```

Printing of the state variables reproduces the results given in Figure 11.5, panel (a):

```
par(mfrow=c(2,1))
 plot(c(0,tmax),c(0,max(X1)),type="n",xlab="Time",ylab="Pop. size")
 lines(t,X1,pch=1,lwd=2,col=gray(0.7))
 lines(t,X2,pch=1,lwd=6,col=gray(0.7))
```

There exists a rich literature about the Lotka–Volterra equations (see, for instance, Begon *et al.* 2005). Most authors agree that the assumptions involved in the model usually over-simplify the complexity of real systems. However, already in this parsimonious form two-species interactions reveal some unexpected properties, many mentioned in the book of Lotka (1925) already. Using numerical integration these so-called Lotka–Volterra rules can be illustrated as shown in Figure 11.5.

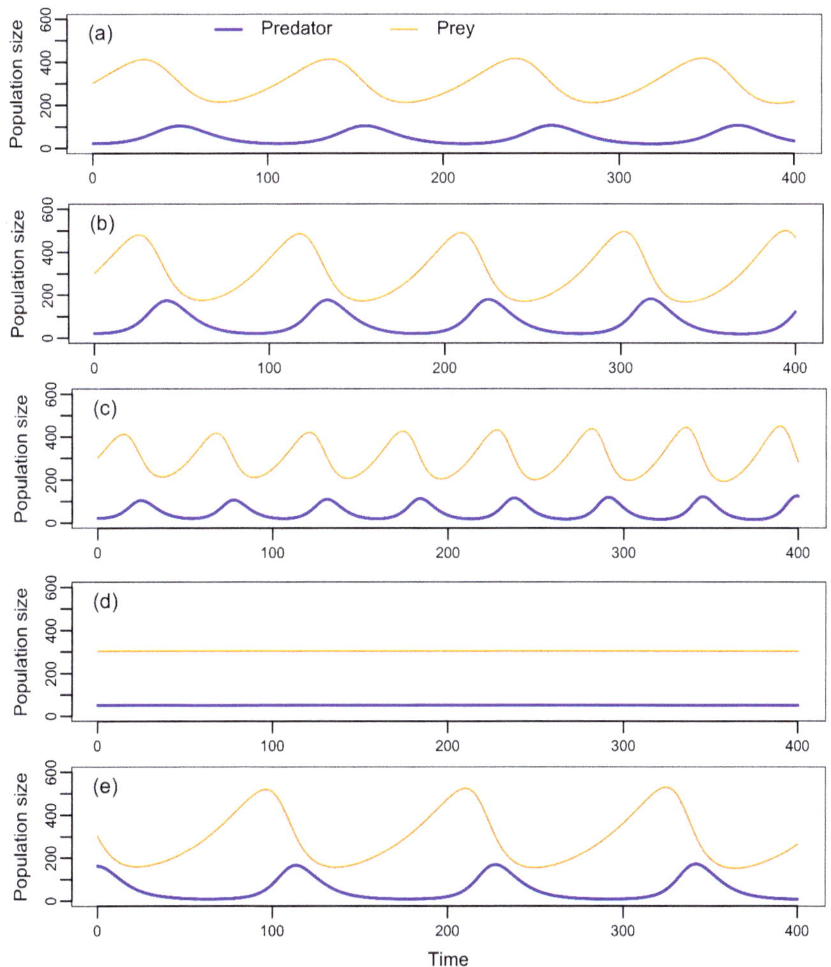

Figure 11.5: Response of the Lotka–Volterra model. (a) Parameters set as given in the main text. (b) Increased growth rate of prey. (c) All growth and death rates multiplied by 2. (d) Initial population size of predator increased to 50. (e) Initial population size of predator set to 160.

Setting the initial conditions and the paramters intuitively may easily end up in a situation where either predators or prey rapidly become extinct. In the numerical example shown in panel (a) of Figure 11.5 with parameters given above predators and prey coexist. The changes of population sizes in time reveal some rules:

(1) Population size of predators and prey oscillate at the same frequency.

(2) Predator peaks always lag behind those of the prey species.

(3) Mean population sizes remain constant over various cycles.

That the prey population is larger than the predator population is intuitive. But what must happen to get a larger predator population? This surprisingly happens when the survival conditions for the prey are improved. In panel (b) of Figure 11.5 the respective growth rate, $g1$, is increased from 0.025 to 0.035, giving rise to the next general rule:

(4) Predator population is increased by improving survival conditions of the prey population (and vice versa).

This also increases the amplitude of the oscillation. But what determines its frequency? A straightforward assumption would be process velocity. To check for this all parameters (growth rates and interaction coefficients) are doubled to generate panel (c). And in fact this also causes frequency to increase accordingly, suggesting the following rule:

(5) Frequency of population fluctuation is determined by the overall velocity of processes, that is, growth rate and interaction coefficients.

One more question is why oscillation occurs at all. Apparently this is caused by the lack of equilibrium at the beginning of the process. Such an equilibrium can be found, in the present case when setting the initial population size of the predators to $X2 = 50$. The outcome is shown in Figure 11.5, panel (d), meaning that

(6) Oscillations are caused by lack of an equilibrium at the outset of the process.

The fluctuations are carried over from cycle to cycle. If there is an equilibrium from the very beginning, no fluctuation occurs. In the real world, however, one would expect that a disturbance event would happen eventually, bringing oscillation back.

In all examples so far the initial population size of predators is below the equilibrium state – except in panel (d). To evaluate the opposite case the initial size of $X2$ is set to 150 and the result displayed in panel (e). As can be seen there the shape of the functions remains unchanged, only a shift along the time axis can be observed.

The Lotka–Volterra model is a good example to demonstrate the potential of differential equations in revealing unexpected system properties. But it can also be used to consider pitfalls typical for models of the kind. One is the absence of resource dependence, that is, the prey population is allowed to grow without limitation unlike in the case of logistic growth. In a set of differential equations energy or matter can enter the system from nowhere or vanish if not controlled explicitly. Furthermore, it is rather unlikely that a predator would rely on a single prey species only. In practice it is rather difficult to even assess the proper population sizes and almost impossible to measure the interaction coefficients, $g1$ and $d1$. In Figure 11.5 these result from trial and error. Another abstraction concerns population size which in the equations is assumed to be continuous. Lotka (1925) was already aware of this but he argued that discrete handling is not really needed if the populations are sufficiently large. Finally, space is also lacking in this model, an issue addressed next.

11.5 Simulating space processes

In space–time modelling, space is assumed to be discrete, just like time in numerical integration (Section 11.2). For simplicity, two-dimensional spatial systems are frequently designed as square, systematic grids. Spacing grid cells regularly and assuming finite extension facilitates computations. The state variables, which in time models imply no spatial extent, account for the content of these cells. Spatial interactions proceed through exchange between cells; this may involve matter, energy or information. Exchange is of course also a function of time and the model has to express how much is moved from one cell to the next per time unit. This movement is either directed or undirected (diffuse).

In ecosystems diffusion will hardly happen in isolation. Simultaneously, a temporal process is taking place inside all grid cells and these therefore contain their own temporal models. The entire model claims to describe a space–time process.

Assumptions have to be made about the exchange process between cells. In Figure 11.6(a) exchange takes place in the horizontal and vertical direction only. An alternative would be to also allow fluxes in the direction of the diagonals. In Figure 11.6(a) two time steps are needed to reach the closest diagonal cell. When designing the exchange process, the balance of matter, energy and information has to be maintained. What leaves one cell has to arrive either in the next or in a controlled sink. An example of a directed process is water flow induced by gravity, for which orientation of the slope determines direction.

In the case of diffusion no specific direction is specified. An assumption of this kind is used in the succession model of the Swiss National Park (Wildi 2002, see Section 10.7.2), where diffusion applies to propagation of plant species and these potentially spread in any direction [Figure 11.6(b)]. I assume that a small fraction of the content of any two adjacent cells is exchanged at each time step. Some of the species newly arriving in a cell will successfully establish and spread, while others will disappear; this depends on the local conditions, set by the temporal model within each cell.

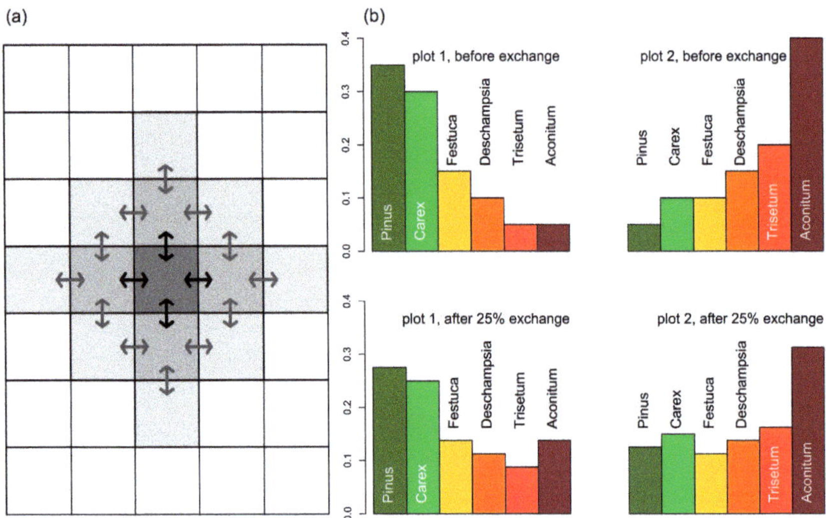

Figure 11.6: The mechanism of spatial exchange. (a) Model design in which exchange proceeds horizontally and vertically only. (b) Example of partial exchange of total content of neighbouring cells to simulate diffusion.

11.6 Processes in the Swiss National Park

This is an example of the application of dynamic modelling techniques presented in the previous sections. The outset of the exercise is the temporal pattern of succession in the Swiss National Park, revealed by space-for-time substitution as explained in Section 10.7.2. The description of the model follows Wildi (2002).

11.6.1 The temporal model

To keep complexity under control, species are grouped into six guilds (assemblages of species) (Wildi and Schütz 2000): these are the state variables. The basic process considered is thus colonization of plots and subsequent species interactions. The plots (the cells in the model) accord with the research grid established in the Swiss National Park for the purpose of investigation. Plot size is 20 m by 20 m, the number of plots within the unforested investigation area, Alp Stabelchod, is 286 (Achermann *et al.* 2000, see Figure 11.9). For simplicity, it is assumed that the total surface occupied by the species guilds never exceeds 100% of the plot. The model plot is eventually overgrown by one or several species guilds, so that in the end no open soil is left.

Next, the objective is to quantify overgrowth and also species replacement. In the original time series from the permanent plots it can be observed that overgrowth always starts slowly (Wildi and Schütz 2000). With

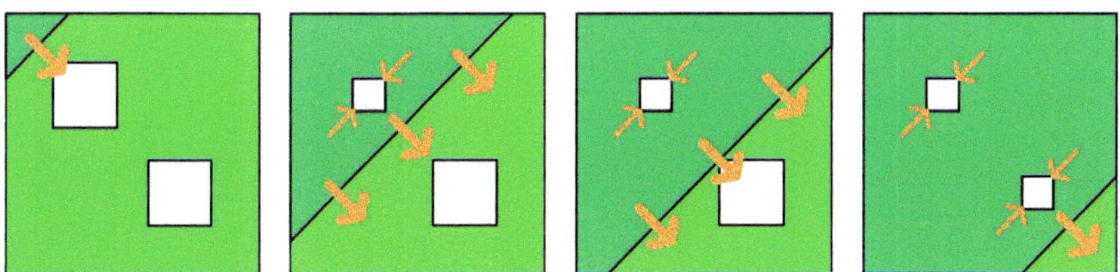

Figure 11.7: Overgrowth of a plot by a new guild. The white squares indicate patches inappropriate for growth, causing spread to slow down towards the end of the invasion process.

increased cover of the guilds, spread accelerates. When approaching 100% of the surface, overgrowth slows down. This finding is sketched in Figure 11.7. A function that mimics this behaviour is the logistic growth equation; in the case of only one guild, it has the general form:

$$\frac{dX}{dt} = Xr\frac{100\% - X}{K} \tag{11.8}$$

(Wissel 1989, see also Equation 11.6). Here, r is the growth rate of guild X and K is the carrying capacity; that is, 100% of the plot surface. As X is also measured in terms of percentage, the space not yet occupied is $100\% - X$. Colonization stops when X reaches 100%. The growth is regulated by X itself, as a result of intra-specific competition. Logistic growth in a temporal model requires all guilds X_i to be present in a minimum quantity at the beginning of any simulation run. No growth would occur otherwise.

Competition comes into play because of two assumptions. First, the gain in cover of guild X_i is at the expense of any other guild with lower competition power (or open ground). In order to keep the cover percentages balanced, the growth equation will be composed of two components: a gain by population growth and a loss to stronger competing guilds. Second, $100\% - X_i$ is the available space for the best competing guild, i, only. If there is another, stronger competing guild, X_j, then the space reduces to $100\% - X_i - X_j$. As will be seen in the description of the model, the mechanism has to make provision for many more competing guilds; six in the present case. Based on previous findings (Wildi and Schütz 2000) the following order of competition power was assessed:

Pinus (1) > *Carex* (2) > *Festuca* (3) > *Deschampsia* (4) > *Trisetum* (5) > *Aconitum* (6)

The logistic growth equation for *Carex* (X_2), which can be outcompeted by *Pinus* (X_1) only, is given by:

$$\begin{aligned}\frac{dX_2}{dt} &= X_2 r_2 \frac{100\% - X_1 - X_2}{K} \\ &- (\frac{dX_1}{dt})\frac{X_2}{\sum_{i=2}^{6} X_i}\end{aligned} \quad (11.9)$$

In this, K is again carrying capacity. *Carex* is growing according to the logistic growth, Equation 11.8. But then a portion of the surface that *Pinus* (X_1) is winning is subtracted from the growth of *Carex*. For *Festuca* (X_3) there is additional proportional loss to *Pinus* and *Carex*:

$$\begin{aligned}\frac{dX_3}{dt} &= X_3 r_3 \frac{100\% - X_1 - X_2 - X_3}{K} \\ &- (\frac{dX_1}{dt} + \frac{dX_2}{dt})\frac{X_3}{\sum_{i=3}^{6} X_i}\end{aligned} \quad (11.10)$$

The growth equations for all subsequent guilds are built accordingly. Trampling by grazing animals and also subsequent recolonization are very fast processes able to accelerate succession markedly. However, in $\mathcal{R}$ function `SNPtm()` of the `dave` package, the use of which is explained below, this is omitted for the sake of simplicity.

Numerical integration of the differential equation requires an initial state of the system to be set, preferably based on real world measurements. Under present circumstances it is best to take this from the result of space-for-time substitution, the initial state of succession determined in Section 10.7.2 and displayed in Figure 10.16. This, of course, may not be the ideal choice, because field data are always affected by random noise. Despite this, Table 11.3 shows that cover taken in the model is close to the same in Figure 10.16. Only the initial state of *Pinus* is assumed to be even lower, because in the real system this is probably absent at the beginning of succession and the model assumption of presence is an abstraction.

Table 11.3: Initial values in the Swiss National Park temporal model compared with the state observed in the field (Figure 10.16) and growth rates used in the temporal model.

Guild	Model state (%)	Field state (%)	Growth rate
Pinus	0.50	0.66	0.025
Carex	0.70	0.70	0.040
Festuca	1.70	1.65	0.045
Deschampsia	6.1	6.0	0.045
Trisetum	13.0	13.0	0.045
Aconitum	78.0	78.0	0.045

Estimating growth rates of species guilds is far more difficult as there is probably no way of measuring. The values listed in Table 11.3 result from trial and error in the course of simulation. In striving for simplicity all are assumed identical ($r = 0.045$) except for the two guilds *Carex* and *Pinus* arriving last in succession.

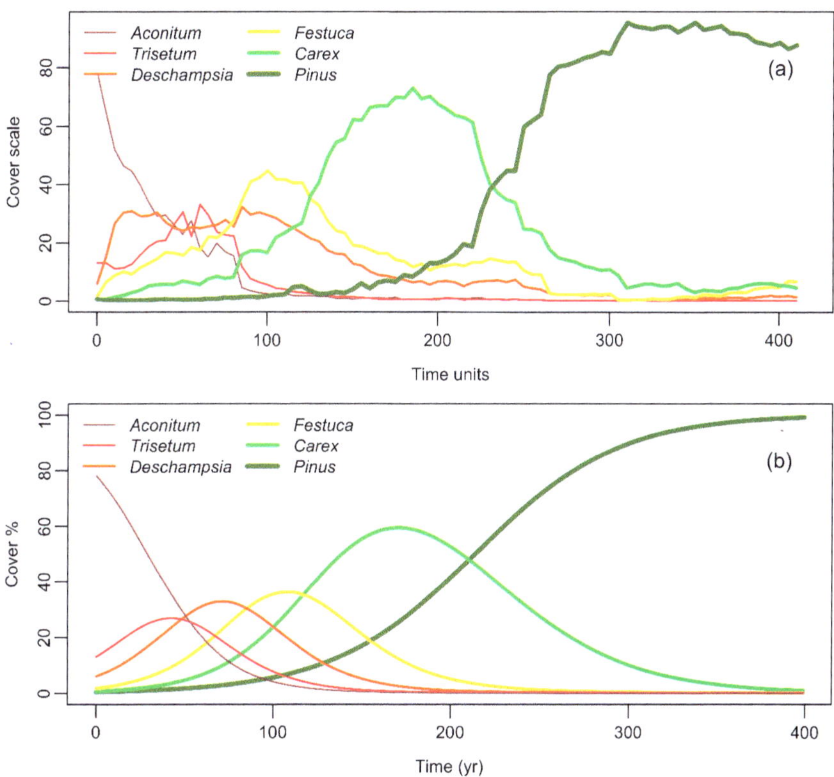

Figure 11.8: Original (a) and simulated (b) temporal succession in former pastures of the Swiss National Park. Panel (a) is identical to Figure 10.16.

Figure 11.8(b) is the simulation result when using parameters from Table 11.3, whereas Figure 11.8 (a) depicts the real world reference, that is, the same as Figure 10.16. What we learn from the result is that the model assumptions result in a bell-shaped response of guilds. The overall shape of processes is rather well reproduced by the model. The latter seems to fail near the end of succession, where there is a decrease in *Pinus* not reflected by the model.

The objective of modelling is not only to imitate processes, but also to learn about variation in the model response. The Swiss National Park temporal model is packed into $\mathcal{R}$ function `SNPtm()` included in the `dave` package. Figure 11.8(b) is obtained by using default settings, according to:

```
o.SNPtm<- SNPtm(trange=400,tsl=1.0,x6=NULL,r6=NULL)
```

```
plot(o.SNPtm)
```

In this, **trange** is temporal range of the simulation and **tsl** is time step length used in simulation. The default values for the initial state **x6** and the growth rates **r6** are the same as in Table 11.3. These can be set individually as follows:

```
x6<- c(0.50,0.70,1.7,6.1,13.0,78.0)
r6<- c(0.025,0.040,0.045,0.045,0.045,0.045)
o.SNPtm<- SNPtm(trange=400,tsl=1.0,x6,r6)
plot(o.SNPtm)
```

```
Call:
SNPtm(trange = 400, tsl = 1, x6 = NULL, r6 = NULL)

Time range:         0 - 400
Time step length: 1
Species:          Initial state:   Final state:  Growth rate:
 Pinus                      0.5           99.1          0.025
 Carex                      0.7            0.9          0.040
 Festuca                    1.7            0.0          0.045
 Deschampsia                6.0            0.0          0.045
 Trisetum                  13.0            0.0          0.045
 Aconitum                  78.0            0.0          0.045
```

The numerical result of simulation is provided in list **o.SNPtm** including other simulation parameters used.

11.6.2 The spatial model

A major drawback recognized in the temporal model is the lack of realistic assumptions for species invasion. This is the motivation to add two more dimensions, x-axis and y-axis in space. For the full space–time model the following notation is used:

$$\vec{x}_{i,x,y,t}|i = 1, \ldots, 6; x = 1, \ldots, 30; y = 1, \ldots, 40; t = 1, \ldots, 400| \qquad (11.11)$$

where i stands for guild, x and y are the spatial coordinates and t is time in years. The model space is a grid of 30 by 40 plots (Figure 11.9). Not only the pasture but also the adjacent forest stands fit into this rectangle. The spread of any one guild happens by spatial exchange. A portion of the content of any plot is transferred yearly to the neighbouring plots, as shown in Figure 11.6. The gains, g, and losses, l, are balanced:

$$\begin{aligned}
\vec{x}_{i,x,y,t+1} &= \vec{x}_{i,x,y,t} + \vec{g}_{i,x,y,t} - \vec{l}_{i,x,y,t} & (11.12)\\
\vec{g}_{i,x,y,t} &= d(\vec{x}_{i,x-1,y,t} + \vec{x}_{i,x+1,y,t} + \vec{x}_{i,x,y-1,t} + \vec{x}_{i,x,y+1,t}) & (11.13)\\
\vec{l}_{i,x,y,t} &= 4d(\vec{l}_{i,x,y,t}) & (11.14)
\end{aligned}$$

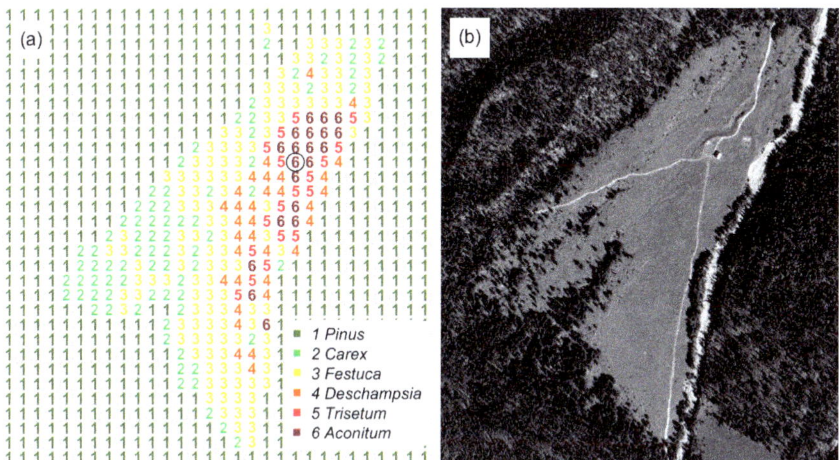

Figure 11.9: (a) Spatial design of the Swiss National Park model with the initial state (see Table 11.4 for contents of the cells). (b) Aerial image of Alp Stabelchod (Aerial Image 2000 ©Swiss National Park).

Table 11.4: Six discrete vegetation states used as initial conditions of the six state variables (guilds) in spatial modelling, cover percentage. The map in Figure 11.9(a) as well as the same in the first column ($t = 0$) in Figure 11.10 are composed of these states.

Guild	No.	State 1	State 2	State 3	State 4	State 5	State 6	$\sum$
Pinus	1	0.00	0.00	1.00	1.00	8.00	90.00	100.00
Carex	2	1.00	1.00	10.00	15.00	65.00	8.00	100.00
Festuca	3	2.00	3.00	30.00	42.00	20.00	3.00	100.00
Deschampsia	4	7.00	15.00	35.00	35.00	6.00	2.00	100.00
Trisetum	5	10.00	35.00	35.00	15.00	5.00	0.00	100.00
Aconitum	6	50.00	17.50	17.50	10.00	5.00	0.00	100.00

From Equation 11.13 we see that the gain always comes from all four directions as illustrated in Figure 11.6(a). The losses in all four directions (Equation 11.14) are the same as they are proportional to the composition of the central plot. The velocity of exchange is given by factor d. This is assumed constant, even though spatial processes may be faster where more animals are browsing. Having no measurements of exchange at hand, I keep it at the very low level of $d = 0.001$, that is, one tenth of a per cent per year.

Along the edges of the system, outside the meadow, the exchange is mirrored. All these plots are covered by *Pinus mugo* forest, the final state of succession considered in the model.

Including spatial extent in the model poses problems with the initial condition of the meadow; that is, the state of all non-forested 268 plots in the year 1917 (outset of succession), which is not known to us precisely.

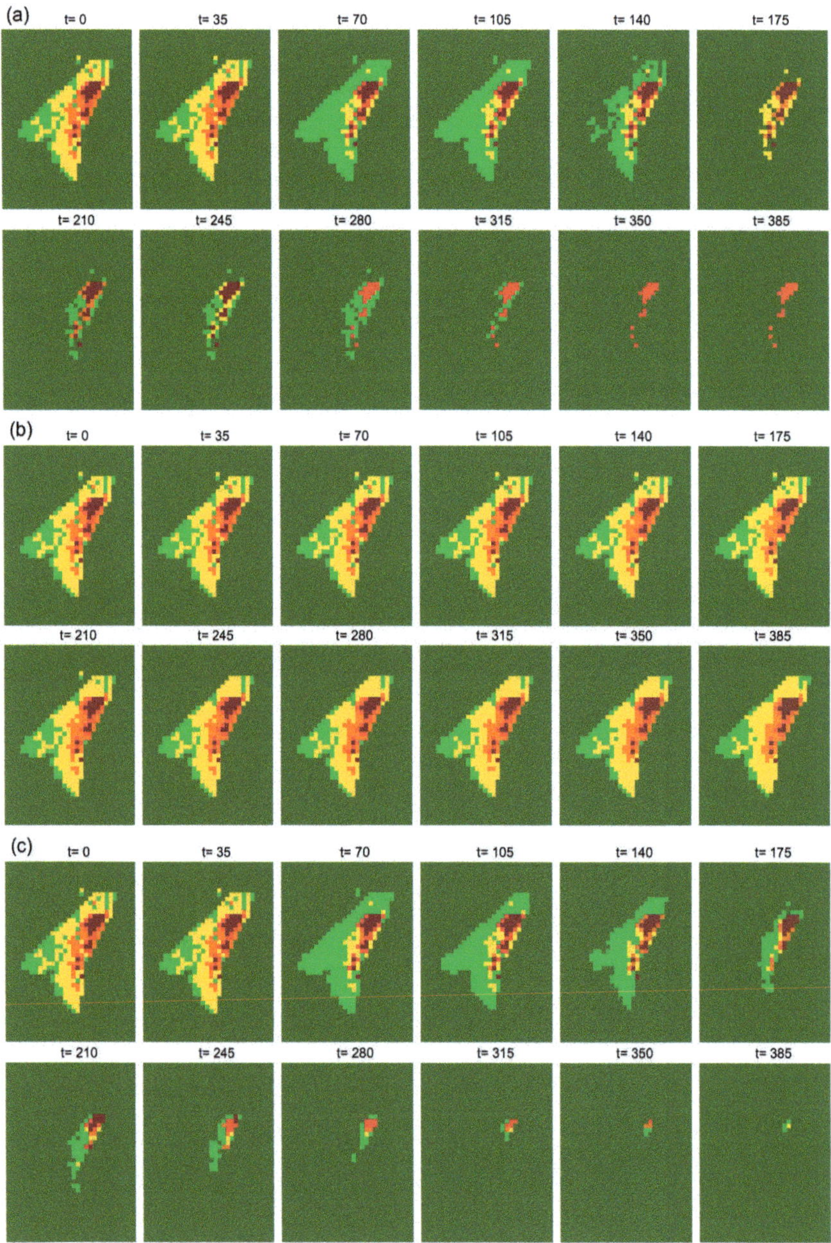

Figure 11.10: Results of the spatial simulation of succession, Alp Stabelchod. Series (a): temporal model only. Series (b): spatial diffusion only. Series (c): spatial and temporal processes combined. Colours accord with succession states [see Figure 11.9(a)].

From the present state of the meadow and the direction and rate of change observed in many similar plots we are able to suggest a simplified state using identical composition for all cells belonging to the same vegetation type as initial conditions in the year 1917 (Table 11.4, Figure 11.9). Hence, at the beginning of simulation, all plots of the same succession stage have identical guild scores and the system consists of a limited number of discrete states, whereas in reality the vegetation forms a continuum. As soon as the simulation begins, diffusion causes differentiation of cells and all maps become continuous.

The persistence of the meadow through subsequent successional guilds and finally *Pinus* forests lasts about 500 years in simulations using the temporal model. Because of the lack of spatial interactions, vegetation boundaries do not move in temporal simulation and only vegetation composition changes. As a result, the boundaries of vegetation types remain unchanged over the entire simulation time. This can be observed in all states of the simulation run shown in Figure 11.10(a). Vegetation boundaries finally only vanish because all plots reach the final state of succession where they have identical composition (that is, forest).

The effect of spatial exchange among pixels can be simulated in isolation. Assuming an extremely low rate of exchange of $d = 0.001$ and all growth rates set to zero, the state of the system after 400 years is shown in Figure 11.10(b). Overall composition is almost the same as at the beginning, but the composition pattern of the maps has become more homogeneous compared with the initial state. The *Aconitum* stage has slightly increased in surface. The spatial process changes the state of the meadow very slowly and it does not explain the results obtained from the permanent plot survey.

Finally, the spatial and temporal processes are run simultaneously. This accelerates the simulation of succession considerably and the meadow is almost entirely covered by *Pinus mugo* after 400 years [Figure 11.10(c)]. It can now be seen that vegetation boundaries have moved and differ from their initial state. The diffusion process causes *Pinus* to invade the meadow from the edges towards the centre.

The lesson to be learned from this simulation exercise is that neither the temporal model alone nor the spatial one is suited to explain succession. In order to understand the major processes invasion has to be allowed, suggesting a spatial component be added to the temporal model. While simple models may be easy to handle and understand, extra complexity in this case is indispensable.

This space–time model is also written in $\mathcal{R}$ available as function `SNPsm()` included in the `dave` package. The calls below generate series (a), that is, a temporal model only with diffusion `diff=0.000` and the 6 growth rates set to the defaults (`r6=NULL`):

```
r6=NULL
o.tSNP<- SNPsm(trange=400,tsl=5.0,diff=0.000,r6)
plot(o.tSNP,out.seq=35,col=TRUE)
```

For series (b) in Figure 11.10, all growth rates must be set to 0, but spatial diffusion turned on (`diff=0.001`):

```
r6=c(0,0,0,0,0,0)
o.sSNP<- SNPsm(trange=400,tsl=5.0,diff=0.001,r6)
plot(o.sSNP,out.seq=35,col=TRUE)
```

Finally, growth rates and diffusion are activated to yield series (c):

```
r6=NULL
o.stSNP<- SNPsm(trange=400,tsl=5.0,diff=0.001,r6)
plot(o.stSNP,out.seq=35,col=TRUE)
```

The very long time step length of `tsl=5.0` will speed up the computations, but lead to low precision results. Parameter `out.seq` defines the interval for printing the vegetation maps. Function `SNPsm()` also draws the map of initial conditions in Figure 11.9.

12 Revising classifications

Classifications relate vegetation ecology to practical demands of society. Revising classifications aims to improve vegetation description for applied use while enhancing their power of predicting environmental conditions. This case study illustrates a path towards revision, performance testing and presentation.

12.1 Beyond statistical analysis

Classifications are means of communication and they are indispensable for various practical applications, such as drawing vegetation maps, assessing endangered vegetation types for legal protection or as a reference in monitoring projects (Mucina 1997). This kind of application contradicts the nature of vegetation, which is in perpetual change. Progress in scientific recognition too adds to the transience of established classifications because species definitions and names change, driven by progress in genetics, for instance. What can be done under these circumstances?

De Cáceres *et al.* (2015) reconcile this conflict of interests as follows: 'a vegetation classification may be understood as a set of vegetation types where new types may be added if needed, but where previously defined types may be modified or discarded only after careful reflection'. This perception is based on the assumption that vegetation consists of a finite number of discrete types where new ones are allowed to evolve and old ones to vanish.

But as we know from succession studies (Feldmeyer-Christe *et al.* 2011, Section 10.5) changes are mostly continuous: single species drop out of plots and new ones invade eventually. Therefore classifications are not really an appropriate means to analyse change. What, then, is the justification for revising classifications?

The goal of revision is to improve on existing knowledge and the most important distinction from ordinary pattern recognition lies in the sources of information. Rather than striving for new high-quality samples (Section 2.3) existing data comes into play. A motivation can be the availability of vegetation databases (see Dengler *et al.* 2011 for a global overview on databases). Using databases such as TURBOVEG (Hennekens and Schaminée 2001) or VEGEDAZ (Küchler 2009), and related processing systems like JUICE (Tichý 2002), is a technical requirement. These trigger the dream of many vegetation scientists to extend present knowledge about vegetation by combining samples of different origin. To do so a method would be needed assimilating samples to the extent that they can be analysed statistically, as for instance proposed by Lengyel *et al.* (2011) or Tichý *et al.* (2014). No solution to this problem has been found to date. A less ambitious way is devised by De Cáceres *et al.* (2015). They propose splitting classification systems into portions, each of these representing a 'consistent classification section' (CCS), that is, 'a subset of classification system where vegetation types are defined using the same criteria and procedures.' Hence, in their terminology, the example I present below concerns a single CCS for which I strive for an improved classification. Extending a given CCS is hardly feasible without obliterating the sampling design, but dropping part of the sample is a justified operation in a revision as we expect outliers (single members of a vegetation type, for instance) and artefacts (under-represented vegetation types along the border of the investigation area) to hamper classification in retrieved data.

One more reason why vegetation samples may have to be reduced prior to analysis is a statistical one. Testing a group pattern is only feasible if groups are sufficiently large. And as soon as alternative classifications (unrevised versus revised, for instance) have to be evaluated, testing for quality is indispensable. In the context of revision this requires measurable quality criteria. While it is easy to apply methods like k-means reallocation to a predefined group structure (Section 5.7) we still do not know if the result is any 'better' for our purpose. Up to now we have met two useful quality criteria. An intrinsic one is the proportion of variance explained by the classification (Section 7.5.3). This is high if the variance between groups (that is, between the group centroids) is high and the same within groups (group heterogeneity) is low, resulting in crisp, easy to recognize classifications. An external criterion offers analysis of variance (Section 7.2.1). It is a means to test whether a group structure found in vegetation can be confirmed by an external independent variable, a site factor in this case. It measures the ability of the classification to predict the state of a site factor. 'Good' classifications result in a low error probability of corresponding ecological models.

Evaluating revisions also means that the final classification should share properties with the initial one, for instance, having the same sample size. This allows the results of ecological models to be compared directly without getting in conflict with degrees of freedom. The questions posed by revisions are straightforward. (1) Can a simple re-allocation of sampling units considerably improve the quality of an existing classification? (2) Would a replacement of the existing classification by a new, unsupervised solution improve the results achieved by re-allocation? (3) Is there a way to retain an existing nomenclature despite some changes of classification? The first two questions address competing paradigms, one relying on a pre-defined (expert-based) classification, the other one looking for a new solution without considering past knowledge.

12.2 Wetland data

The data used in this chapter is taken from the first survey of the Swiss mire-monitoring project (Grünig *et al.* 2005; Feldmeyer-Christe *et al.* 2007) known as 'Swiss wetland vegetation data'. Sampling has taken place in the years 1997 through 2002 capturing a comprehensive sample of wetlands, ranging from ombrotrophic peat bogs through fens to open-water reed vegetation at altitudes from about 300 to 2400 m a.s.l. The cover-abundance scale used is logarithmic with ranks from 0 (0% cover) to 4 (up to 100% cover). The full sample consists of 17,608 relevés of which I analyse a random sample of size 1496. This is organized in two data frames, `wetveg` for vegetation records and `wetsit` for complementary site information. The latter also includes a phytosociological classification according to the system of Braun-Blanquet (1932). Although this system is sometimes considered comprehensive, robust, inert against spatial and temporal variability, and so on (Ewald 2003; Dengler *et al.* 2008; Mucina *et al.* 2016), in this example it is expected to have flaws for various reasons (Graf *et al.* 2010), one being the restricted species list used for assignment. It is the goal of the revision to evaluate and to improve the classification if feasible.

As mentioned in Section 12.1 the evaluation of a revision is most meaningful if there exist independent reference data to be used in model context. For the example given below the data frame of site factors, already named `wetsit` in the second edition of this book (Wildi 2013), has been extended by some climate variables suitable for testing the predictive power of classifications. These are listed in Table 12.1. Some new soil variables, also included in the current release, are not used as there are too many missing values.

Table 12.1: Site factors in data frame `wetsit` used for testing classifications.

Site factor	Name	Description of variable
Location	`X, Y, Z`	Coordinates of the Swiss coordinate system. Variable `Z` is elevation in metres a.s.l.
Degree days	`ddeg300`	Annual integrated degree days above 3°C
Precipitation	`precyy`	Annual sum of the monthly mean precipitation sum (1961-1990) in 1/10 mm
Frost days	`sfroyy`	Annual average number of frost days during the growing season
Min. temperature	`tminall`	Annual minimum of monthly mean of minimum temperature
Water balance	`swb`	Annual average site water balance in 1/10 mm (precipitation minus evaporation)
Radiation	`sradyy`	Mean yearly global direct and diffuse radiation

12.3 Preprocessing data

Preprocessing and even more so revising classifications may require various data adaptations, some of which are highly suggested, others not always needed. To facilitate overview the instructions required for data editing and analysis with $\mathcal{R}$ are presented separately in Section 12.6 including references given to explanations in preceding chapters.

12.3.1 Suppressing outliers

A first step in the adaptation of a sample is the elimination of possible outliers. A sampling unit is deemed an outlier when it differs considerably

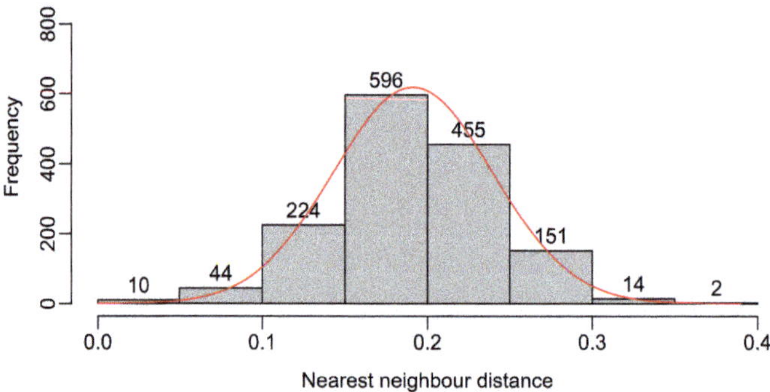

Figure 12.1: Frequency distribution of nearest-neighbour distances of relevés in the entire sample ($n = 1496$) of the Swiss wetland vegetation data. Correlation as distance [Equation (4.6)] used. Superimposed is a fitted normal distribution.

in content from all others in the same sample. There are various reasons why this can happen: a sampling area frequently includes locations that do not really conform with the aim of the investigation (a wet patch in a dry area, for instance), a technical measurement error caused by malfunction of an instrument or by a trivial typing error. Irrespective of cause, outliers can impose a detrimental effect on any statistical analysis and omitting these is highly recommended.

Identification of outliers proceeds through screening for nearest neighbours. The nearest neighbour of any one relevé can be found in the distance matrix by searching for the smallest distance in the row or column pertaining to this relevé (the diagonal elements notwithstanding). How isolated a sampling unit really is, can be quantified: it is the distance to the most similar relevé in the sample. This most similar relevé is its 'nearest neighbour'. To assess and erase outliers a threshold is needed. This can be chosen preferentially: one can decide that any neighbour distance of, say, $d \geq 0.3$, is deemed an outlier situation. However, before doing so it is good practice to find out what a 'normal' nearest-neighbour situation is in any one data set. Inspecting the distribution function of nearest-neighbour distances allows for this. In the present example the bar chart in Figure 12.1 documents the issue. Obviously, the nearest-neighbour distances in the data set `wetveg` are almost normally distributed. A real outlier would have to be above the range of these values and there are no relevés really fulfilling this condition (see right-hand side of the histogram). Figure 12.1 is generated by function `outlier()` in which a threshold is set for the minimum distance where a sampling unit is reported to be an outlier (Section 12.6). When this is set to 0.25, for example, then 151+14+2=167 relevés are deemed outliers.

12.3.2 Selecting groups

As pointed out by Lengyel *et al.* (2011) data retrieved from databases will typically show unbalanced group size: some types will be over-represented and many others be represented by one or very few sampling units only. The idea is to analyse groups with sufficient size only, for which a threshold is set. This will allow statistical properties of entire groups to be evaluated.

In the wetland example the belongings of relevés to associations are found in the data frame `wetsit` and therefore group number and size can be determined. In Figure 12.2 all 172 associations occurring are ordered and displayed by decreasing frequency. Almost half of these have one or two members only. A vertical line marks a threshold of group size 10. Associations to the right are now excluded from revision – in the hope that the remaining sample will still reflect the major content and exhibit the major pattern of the full sample. There are 36 associations with 10 or more members remaining in the reduced data set `wetveg.r` and the sum of group sizes is 1091, the size of the new sample to be used for revision. Finally, the data frame of site information, `wetsit`, is reduced to `wetsit.r` to include the same plots as does vegetation data.

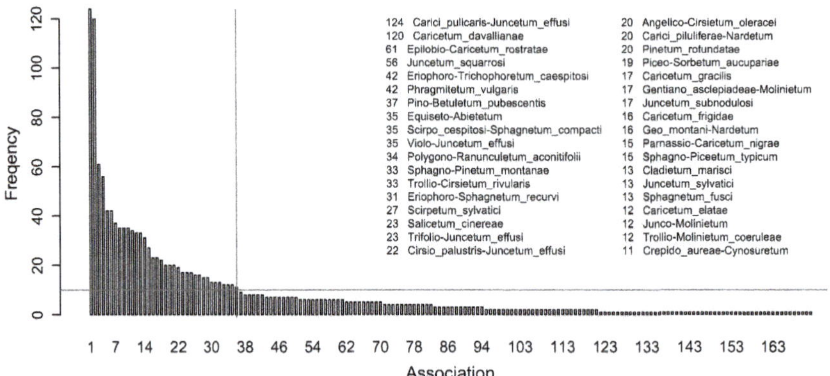

Figure 12.2: Group sizes in the sample of wetland data. Frequency of associations in decreasing order. The horizontal and vertical lines mark an optional threshold of 10 for suppressing low-frequency groups.

Having data selection completed ordination is a means to visualize a possible change in resemblance pattern caused by data reduction. In Figure 12.3 three alliances are highlighted forming the edges of the triangular point cloud. Not surprisingly the overall shape of the full [panel (a), $n = 1496$] compared to the reduced sample with minimum group size 10 [panel (b), $n = 1091$] and the same with group size 20 [panel (c), $n = 873$] is rather similar, confirming that we are still revising the core body of the sample.

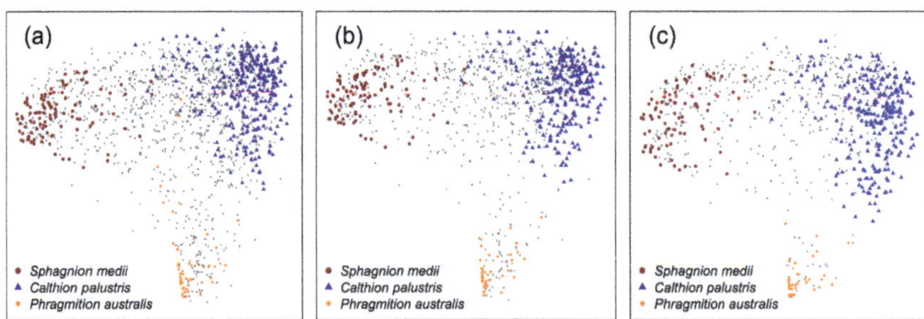

Figure 12.3: Comparison of the full [$n = 1496$, panel (a)] versus the reduced wetland sample with minimum groups size 10 [$n = 1091$, panel (b)] and minimum group size 20 [$n = 873$, panel (c)] respectively.

To generate Figure 12.3 (a) we first derive a distance matrix `ddr` for which we use correlation as distance according to Equation (4.6). This is then subjected to principal coordinates analysis by function `pco()` yielding the x- and y-coordinates for plotting, located in `o.pco$points`:

```
ddr<- as.dist((1-cor(t(wetveg)))/2)
o.pco<- pco(ddr)
plot(o.pco$points,pch=16,cex=0.2,asp=1)
```

In the `plot()` function we use dots (`pch=16`) of rather small size (`cex=0.2`). In the subsequent step these are over-plotted by larger points – but only if these belong to the alliance *Sphagnion medii*, for instance. Alliance numbers are stored in column `wetsit$Verband_Nr` (German for alliance number):

```
x.c<- o.pco$points[wetsit$Verband_Nr == 809.02,1]
y.c<- o.pco$points[wetsit$Verband_Nr == 809.02,2]
points(x.c,y.c,pch=16,cex=0.6,col="black")
```

To visualize the other alliances 809.02 is replaced by 813.04 for *Calthion palustris* and 806.01 for *Phragmition australis*, for instance.

12.4 Evaluating classification revisions

In this case study classification by associations is taken from site data `wetsit.r` where it is located in a column named `"Assoziation1_ek1"`. This is the outset for reallocation by the k-means method (Section 5.7) generating an updated classification whose quality has to be compared with the initial one. The common ground of quality assessment is the distance matrix of relevés, named `ddr` in Section 12.6. Function `adonis()` (NP-MANOVA, Section 7.5) determines the between groups as well as the total sum of squares as shown in Section 7.5.3. Whereas the total always remains unchanged, between groups sum of squares varies with the classification used. The results of comparisons are summarized in the first column of Table 12.2.

When running NP-MANOVA we note that in the phytosociological classification between groups, sum of squares explains about 44% of a total of 99.947. But as we now use the revised classification the model is improving. As shown in Section 12.6 we first derive the centroids of the phytosociological classification and feed these to function `kmeans()` for updating. According to function `adonis()` the variance explained by the revised classification is now up to 50%, a clear improvement (Table 12.2).

Unfortunately there is no statistical test possible to directly compare the two models. Hence, we focus on the second criterion of quality, the predicting power of the classifications with respect to an independent site factor. In this rather mountainous sampling area elevation, variable `Z` in `wetsit.r`, promises to be a good predictor. Analysis of variance of elevation as a

Table 12.2: Evaluation of classifications derived by different methods. Quality is measured as variance explained by classification and by F-values of analysis of variance based on six independent variables.

Classification	NP_MANOVA Variance	Analysis of variance F-value					
	% explained	Elevation	Degree days	Precipitation	Min. temp.	Water balance	Radiation
Phytosociological	44	43.2	51.9	19.6	31.4	50.7	4.9
Phytosociological, revised	50	73.2	74.4	23.3	59.6	45.9	9.9
Ward's method	50	107.5	99.0	29.4	83.2	65.2	8.2
Complete linkage	47	67.5	79.4	26.4	54.5	48.2	5.9
Average linkage	39	55.8	65.7	28.3	48.2	67.7	7.2

function of the phytosociological classification (see Section 12.6 for technical details) yields the following output:

```
                          Df    Sum Sq Mean Sq F value Pr(>F)
as.factor(Assoziation1_ek1)   35 112958174 3227376    43.2 <2e-16 ***
Residuals                 1055  78807008   74699
---
Signif. codes:  0 *** 0.001 ** 0.01 * 0.05 . 0.1   1
```

The total sum of squares is $112'958'174 + 78'807'008 = 191'765'182$; the classification explains about 59% of this. The F-value reaches an impressive 43.2. The model is (as all in the sequel) highly significant. We now repeat this with the revised classification found in `o.kmeans$cluster`:

```
                          Df    Sum Sq Mean Sq F value Pr(>F)
as.factor(o.kmeans$cluster)   35 135847919 3881369   73.23 <2e-16 ***
Residuals                 1055  55917264   53002
---
Signif. codes:  0 *** 0.001 ** 0.01 * 0.05 . 0.1   1
```

The classification now explains about 71% of total variance and the F-value rises to 73.23. Hence, the revised classification has clearly improved the predicting power with respect to elevation.

Having concluded that it is worth revising this classification (if the goal is ecological modelling) we would like to know how this compares to an entirely new classification generated by unsupervised clustering (Ward's method, Section 5.3.3). For the sake of comparability of results this is done with the same data frame (`wetveg.r`), based on the same distance measure (correlation as distance) and generating the same number of groups, $k = 36$. When now running `adonis()` we observe that the sum of squares explained is almost the same as in the revised classification, that is, about 50% (Table 12.2). However, the ecological model (analysis of variance) using elevation again performs even better:

```
              Df    Sum Sq Mean Sq F value Pr(>F)
class3        35 149827204 4280777   107.7 <2e-16 ***
Residuals   1055  41937978   39752
---
Signif. codes:  0 *** 0.001 ** 0.01 * 0.05 . 0.1   1
```

In this console output we see that the F-value of 107.7 achieved by this solution by far exceeds the previous classifications. Although k-means reallocation improved the phytosociological classification, it apparently got stuck in a local optimum. The progress, of course, depends on the clustering method too and "ward.D2" in this case proved to be the best choice. Still better than the revised classification are "average" and "complete". In terms of the internal criterion (NP-MANOVA, sums of squares explained by between group variance), the results differ from the external criteria (Table 12.2). Sum of squares explained by the classification can be rather low (39% when using average linkage clustering, for example) while the predictive power is still quite high ($F = 55.8$). The predictive power achieved with respect to some more environmental factors is finally shown in graphical form in Figure 12.4. The results confirm that differences in performance emerge when using predictable factors only (elevation, for example) but not when predictability is low (radiation).

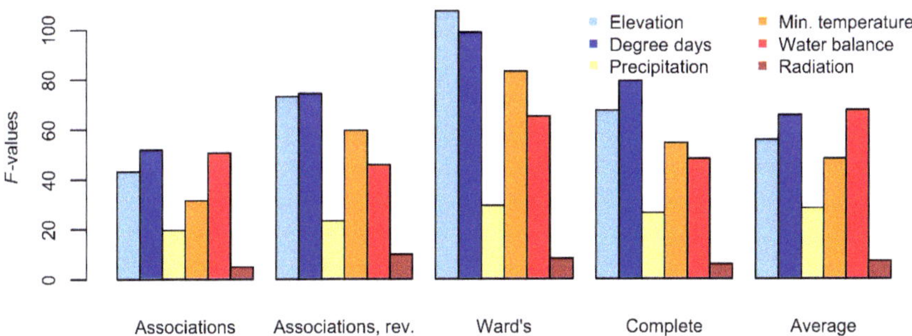

Figure 12.4: F-values achieved in environmental models based on different classifications

Predictability achieved by different classifications can be illustrated graphically using boxplots. This is shown in Figure 12.5 with elevation displayed as a function of classification. Variance of elevation within groups is by far largest in the phytosociological classification, panel (a), followed by the same after reallocation, graph (b) and Ward's clustering, panel (c). In all three cases the groups are ordered by elevation (median) to facilitate the interpretation, a feature that is not part of the analysis.

Although the verdict is clear, it is important to keep in mind that this is a single case study only and the methodological findings might differ in other examples. The main factors that determine the results (and the interpretation) are the classification methods involved and the independent

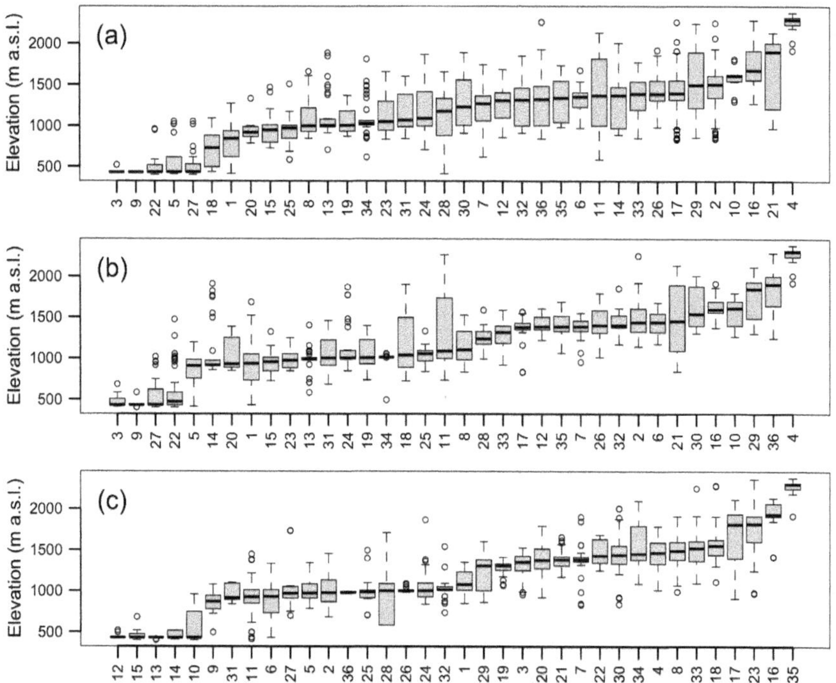

Figure 12.5: Boxplots showing the response of three classifications to elevation. Panel (a) uses the initial, phytosociological classification, panel (b) the same after revision by the k-means method and panel (c) the result of clustering by Ward's method.

site factors used. The variables best explaining classifications in the present case are elevation and degree days, both closely related to the consumption of warmth by plants (Table 12.2). Whatever site factor is used to measure the predictive power of classification, a revision of phytosociological classification by re-allocation is recommended. But a clustering solution may likely find an even better performing classification. Ward's method, in this case, allows for the best models. It generates groups with well-balanced group size, displayed when typing `table(o.grel)` immediately after clustering.

12.5 Carry-over nomenclature?

A strength of phytosociological classification is the nomenclature it uses (Weber *et al.* 2000). In the example of the wetland data each of the 172 associations carries a name given in the past in published classifications (Mucina 1997; Mucina *et al.* 2016). After reducing the sample as explained

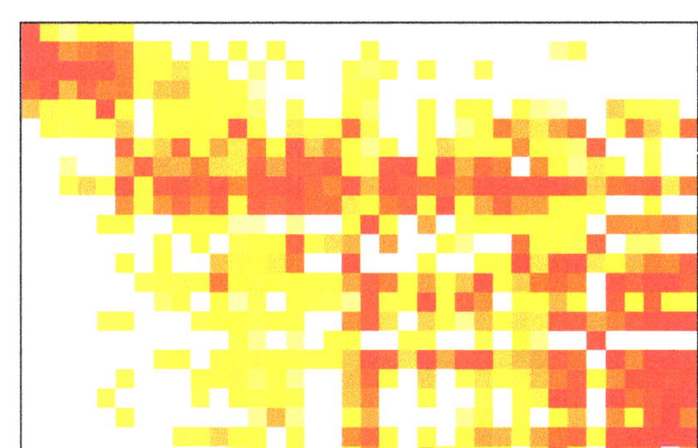

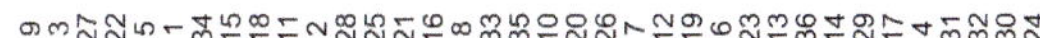

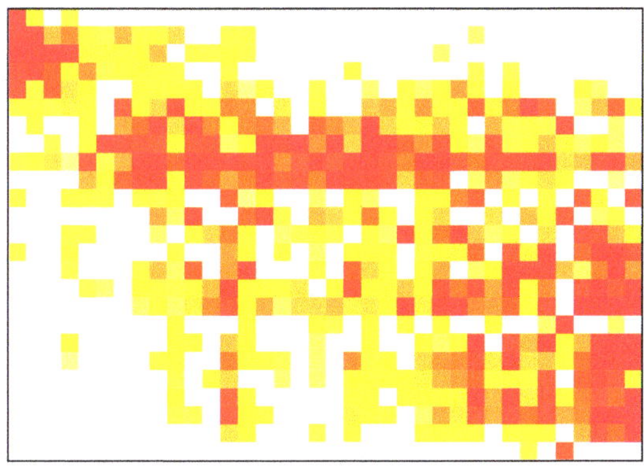

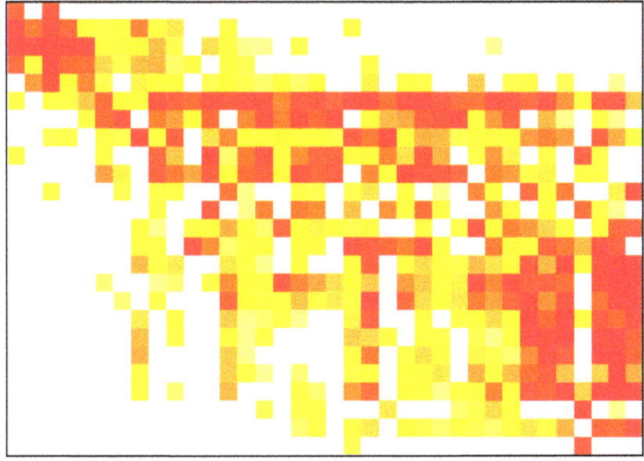

Figure 12.6: Summary tables (centroids) of three alternative classifications. Panel (a) uses the phytosociological classification and the 25 species ranked first within this, panel (b) the same after revision by the k-means method and panel (c) the result of clustering by Ward's method.

Table 12.3: Jancey's ranking applied to three alternative classifications.

Rank	No.	Species	F-value	error p
Phytosociological classification				
1	653	*Phragmites_australis*	89.559	1.26e-287
2	256	*Cladium_mariscus*	84.835	1.68e-278
3	364	*Eriophorum_vaginatum*	61.528	2.78e-227
4	186	*Carex_elata*	32.741	9.51e-143
5	901	*Sphagnum_magellanicum*	31.504	2.48e-138
6	508	*Juncus_triglumis*	31.224	2.57e-137
7	983	*Vaccinium_uliginosum*	30.300	5.96e-134
8	981	*Vaccinium_myrtillus*	29.639	1.65e-131
9	888	*Sorbus_aucuparia*	27.488	2.16e-123
10	192	*Carex_frigida*	26.067	7.17e-118
Phytosociological classification, revised				
1	364	*Eriophorum_vaginatum*	78.917	1.55e-266
2	653	*Phragmites_australis*	60.053	1.26e-223
3	981	*Vaccinium_myrtillus*	45.921	2.77e-185
4	983	*Vaccinium_uliginosum*	45.119	6.64e-183
5	256	*Cladium_mariscus*	43.888	3.31e-179
6	713	*Polytrichum_juniperinum*	39.313	6.33e-165
7	955	*Trichophorum_cespitosum*	37.236	3.86e-158
8	963	*Trifolium_pratense*	37.201	5.04e-158
9	892	*Sphagnum_capillifolium*	36.328	4.12e-155
10	508	*Juncus_triglumis*	36.055	3.41e-154
Ward's sum of squares clustering				
1	653	*Phragmites_australis*	88.424	1.82e-285
2	364	*Eriophorum_vaginatum*	75.222	9.8e-259
3	508	*Juncus_triglumis*	51.185	2.57e-200
4	256	*Cladium_mariscus*	39.688	3.98e-166
5	983	*Vaccinium_uliginosum*	39.157	2.01e-164
6	901	*Sphagnum_magellanicum*	37.708	1.06e-159
7	713	*Polytrichum_juniperinum*	36.986	2.62e-157
8	981	*Vaccinium_myrtillus*	34.695	1.46e-149
9	186	*Carex_elata*	32.374	1.9e-141
10	892	*Sphagnum_capillifolium*	29.436	9.36e-131

in Section 12.3 there are 36 types remaining (Figure 12.2), the outset of the efforts of revision.

In phytosociology diagnostic species are often inherited from previous investigations and they are reused in subsequent classification projects, a way to carry-over expert knowledge. In our revised classification, and even more so when using unsupervised classification, prior lists of such species do not exist. Hence, we are forced to determine these in a posterior approach (Bruelheide 1995, 1997). Applying Jancey's ranking to the different classifications (Section 7.2.2) is one such possibility (outperforming IndVal, see Wildi and Feldmeyer-Christe 2013) identifying the best differentiating species (Table 12.3) by their corresponding F-values. Species lists of this kind are specific to classifications and therefore not providing an all-purpose nomenclature. However, they can still be used to establish a context dependent nomenclature. Context dependent means that the diagnostic species and also their F-values are used in one specific classification only (Table 12.3).

Lists of the best ranked species according to Jancey's ranking are useful for an overview of the content of any given classification. This is shown in Figure 12.6, a summary of the three classifications. The columns of the matrices are the centroids of the respective groups, derived by function `aggregate()` previously used to prepare reallocation (Section 5.7). The species lists are reduced according to Jancey's ranking applied to the full data sets (see Section 12.6 for technical details).

In Figure 12.6 species as well as group centroids are ordered according to correspondence analysis to facilitate the interpretation, as explained in Section 7.6.2. The three tables carry very much the same information: the three classifications identify the same dominating floristic trend in the sample, that is, from basic towards acidic soil conditions. In practice one would probably increase the number of diagnostic species, from 25 to 50 or even more.

12.6 Step by step in $\mathcal{R}$

As in any other numerical analysis data must be checked and screened for irregularities and possibly reduced prior to revision:

Outlier detection, Figure 12.1:

```
o.outl<- outlier(wetveg, thresh=0.25,y=0.5)
plot(o.outl)
```

```
Call:
outlier.default(veg = wetveg, thresh = 0.25, y = 0.5)

Threshold value:                        0.25
Original number of releves, species:   1496 1163
Remaining number of releves, species:  1329 1094
```

If deemed necessary (which is not the case here) the data set reduced accordingly, `o.outl$new.data`, can be used in the sequel.

Plotting group frequency. This shows how to display the size of phytosociological groups graphically in decreasing order in the form of a barplot, Figure 12.2:

```
ass.tab <- table(wetsit[,"Assoziation1_ek1"])
ass.o <- order(ass.tab, decreasing=TRUE)
nam<- seq(1,length(ass.o),1)
barplot(ass.tab[ass.o],xlab="Groups",ylab="Frequency",names.arg=nam)
```

We can check in advance what the size of data frames would be when reducing data to minimal group size of 10, for example:

```
ass.n<-sum(ifelse(ass.tab >= 10,1,0))
ass.n ; sum(ass.tab[ass.o[1:ass.n]])
```

```
[1] 36
[1] 1091
```

Deleting low frequency groups. This is how to reduce vegetation data to the groups with minimum group size, 10 in this example. Reductions of this kind are needed if tests are intended to compare revisions of existing classifications. First, a complete list of association names is isolated in a vector `ass.nam` which is subsequently transformed into a table (in terms of $\mathcal{R}$), `ass.tab`. This is then reduced to the elements with frequency of at least threshold 10. The final vegetation data frame `wetveg.r` will contain rows found in this list only. The operator `%in%` signifies 'occurring in':

```
ass.nam<- wetsit[,"Assoziation1_ek1"]
ass.tab<- as.data.frame(table(ass.nam))
ass.tab<- as.vector(ass.tab[ass.tab[,2] >= 10,])
wetveg.r<- wetveg[(ass.nam %in% ass.tab[,1]),]
wetveg.r<- wetveg.r[,apply(sign(wetveg.r),2,sum) > 0]
```

It likely happens that in the reduced data frame some species vectors will be empty. These are erased in the last line (see Section 7.6.3).

Adjusting site data. The rows encountered in the vegetation data frame `wetveg.r` are selected to get a compatible data frame `wetsit.r`:

```
a<-rownames(wetveg) ; b<-rownames(wetveg.r)
wetsit.r<- droplevels(wetsit[a[a %in% b],])
```

Function `droplevels()` internally erases association names no longer used (a peculiarity of $\mathcal{R}$).

Reallocation. Preparing a revised classification through k-means reallocation of the phytosociological version (see Section 5.7). For k-means reallocation the centroids `cent` of the initial classification must be provided:

```
class1<- as.factor(wetsit.r[,"Assoziation1_ek1"])
cent<- aggregate(wetveg.r^0.5,list(class1),mean)
o.kmeans<- kmeans(wetveg.r^0.5, cent[,-1])
```

In this `cent` is the centroid of associations. The negative index removes unwanted group names. The new group membership of relevés is found in `o.kmeans$cluster`.

Classification through clustering. While still using the same sample this creates a new classification free of expert knowledge. For the sake of comparison the same number of groups is chosen as in the phytosociological classification, `ass.n`$= k = 36$:

```
ddr<- as.dist((1-cor(t(wetveg.r^0.5)))/2)
o.hclr<- hclust(ddr,method="ward.D2")
o.grel<- cutree(o.hclr,k=ass.n)
```

The new classification is in `o.grel`. The method parameter in function `hclust` can be changed to use various other clustering algorithms. In terms of $\mathcal{R}$ all three classifications are now available as factors:

```
class1<- as.factor(wetsit.r[,"Assoziation1_ek1"])
class2<- as.factor(o.kmeans$cluster)
class3<- as.factor(o.grel)
```

Explained variance. All three classifications, the phytosociological before and after revision and the one obtained by unsupervised clustering are evaluated for the proportion of variance explained, using function `adonis()`:

```
o.ado1<- adonis(ddr~class1)
o.ado1$aov.tab$SumsOfSqs[1] ; o.ado1$aov.tab$SumsOfSqs[3]
o.ado2<- adonis(ddr~class2)
o.ado2$aov.tab$SumsOfSqs[1] ; o.ado2$aov.tab$SumsOfSqs[3]
o.ado3<- adonis(ddr~class3)
o.ado3$aov.tab$SumsOfSqs[1] ; o.ado3$aov.tab$SumsOfSqs[3]
```

This displays variance explained by classification followed by the total in all three cases (Table 12.2).

Predictive power. All three classifications are tested for predicting elevation, Z, by analysis of variance:

```
o.aov1<- aov(Z~class1,data=wetsit.r)
summary(o.aov1)
o.aov2<- aov(Z~class2,data=wetsit.r)
summary(o.aov2)
o.aov3<- aov(Z~class3,data=wetsit.r)
summary(o.aov3)
```

This displays the summary of all three models respectively (Table 12.2). The F-values can directly be compared because the degrees of freedom are identical. As the altitudinal trend is strong, all three models are highly significant.

Jancey's ranking. This is a method to a posteriori identify diagnostic species. The complete list of species is displayed according to decreasing F-value when typing `print(o.srank1)`, for instance:

```
o.srank1<- srank(wetveg.r,class1,method="jancey",y=0.5)
o.srank2<- srank(wetveg.r,class2,method="jancey",y=0.5)
o.srank3<- srank(wetveg.r,class3,method="jancey",y=0.5)
```

The first 10 ranks of these lists are given in Table 12.3.
Boxplots. This shows elevation, `wetsit.r$Z`, as a function of the respective classification (Figure 12.5). In a first step group means of elevation are computed using function `tapply()`. These are subsequently ordered according to increasing elevation to facilitate the interpretation of the boxplot:

```
o.Z<- order(tapply(wetsit.r$Z,list(class1),median))
o.nl<- levels(class1)[o.Z]
o.class<- factor(class1,levels=o.nl)
par(mfrow=c(3,1))
boxplot(wetsit.r$Z~o.class,las=2,lwd=0.5,ylab="Elevation (m)")
```

This prints the association names along the x-axis of the boxplot. Function `par()` and parameters `las` and `lwd` in function `boxplot()` are attempts to embellish the resulting graphs. For the remaining boxplots `class1` is replaced by `class2` and `class3` respectively (three times!).

Reducing samples. The full data frame `wetveg.r` can be reduced to include a set of differentiating species only, those devised by Jancey's ranking:

```
wetveg.r1<- wetveg.r[,o.srank1$species[1:100]]
wetveg.r2<- wetveg.r[,o.srank2$species[1:100]]
wetveg.r3<- wetveg.r[,o.srank3$species[1:100]]
```

Centroid tables. In order to get an overview of the content of the classifications (Figure 12.6) we calculate the centroids again, then reduce columns (the species) to the 25 ranked highest by Jancey's method and order the rows and columns according to the first axis of correspondence analysis, an operation provided in function `Mtabs()`:

```
wetcent<- aggregate(wetveg.r^0.5,list(class1),mean)[,-1]
wetcent1.r<- wetcent[,o.srank1$species.no[1:25]]
o.Mt<-Mtabs(wetcent1.r,method="ca")
plot(o.Mt)
```

In practice one would probably like to inspect some more species, 50 for example, or even 100. Using `class2` provides a table of the revised classification and `class3` the version with Ward's clustering used. There is a possibility that the tables will occur in mirrored order as the sign of the coordinates in correspondence analysis is set by chance.

12.7 Revising classification - or data?

As observed in the present example vegetation response along ecological gradients can be strong, causing nonlinearity in patterns (Section 6.5). The good news is that even an expert-based, subjective classification will likely reflect the major trends in the environment. It may be sufficient to use all-purpose classifications based on a small number of character and differential species to get ecological models that are highly significant. But this comes with the risk of underexploiting the predictive power of vegetation types. And in fact revisions shown in this chapter using k-means reallocation improve models, most likely because the k-means method draws on complete species composition. This finding supports the key hypothesis of similarity theory (Section 1.2; Feoli and Orlóci 2011), telling that similarity in biological space runs parallel with similarity in environmental space. All-purpose classifications reflecting agreement among experts have the advantage of setting landmarks in the vast similarity space of vegetation samples. But in ecological modelling they seem to underperform and apparently deserve improvement by methods relying on similarity theory. Unfortunately, in any new classification context, the original nomenclature that the system of Braun-Blanquet (1932) provides is no longer applicable because species composition of vegetation types change.

Whether using numerically updated classifications or new ones from unsupervised clustering, the goal remains the same: trying to get a maximum of benefit out of existing data. But classifications can also be weak due to the underlying data, if these poorly describe the investigation object or because change occurred in vegetation or in the environment. In either case new data are needed which then may lead to new classifications. This is in fact the idea of the new vegetation survey under development in North America (Jennings *et al.* 2003, 2009) in which relevé data form the basis of vegetation description and classification is derived from this by the user. Hence, classification is becoming a subordinate, more flexible tool in vegetation description.

13 Swiss forests: a case study

Vegetation ecology performs best in the context of comprehensive statistical samples. In the present case study the sampling plan is a square grid constrained to the forested area of Switzerland. It uses variable plot size for vegetation data and complements this with a set of environmental variables. This allows a stunning variety of questions to be answered.

13.1 Aim of the study

This case study illustrates some applications of the methods explained in preceding chapters. It is a real world example assessing ecological and also methodological questions. The vegetation data originate from a survey across Switzerland (Wohlgemuth *et al.* 2008) which aimed to reveal relationships between vegetation composition and the growth rate of tree species. Just as in an experimental approach, questions have been posed prior to investigation and a sampling design chosen to obtain answers. But as is often the case in large surveys the order of steps can be reversed. In some examples the data are the outset of the exercise and the questions that they could potentially answer are identified later. This imposes restrictions on the analytical methods used, but it allows an exploration of the variable set far beyond its initial scope.

The analyses shown below concentrate on species composition and species spatial distribution. An alternative would be to focus on species richness, as done by Wohlgemuth *et al.* (2008), who found that there is a high correlation between diversity (the number of species per plot) and the canopy cover of trees (a surrogate for light availability inside the forest stands). However, they also detected that the spatial resolution of the survey is probably not optimal when investigating biodiversity, as their correlations increased after clumping the sampling units into landscape patches of about 100 km^2 in size.

Not surprisingly, the scale as well as the variables chosen in sampling design determine the potentials and restrictions of application. As will be shown below, the sampling area is the territory of Switzerland (~41,000 km^2), restricted to its forested area (~30% of the surface). The strong elevation gradient is the main cause of spatial variation in climate, and climatic relationships are therefore a focus. Concentrating the study on forests was possible because human influence is weak and controlled, a consequence of the strong regulations imposed on forest management by public law. The question of the strength of human disturbance will be raised too. Due to the extent of the study area, traces of the post-glacial history of the vegetation can be expected, as well as patterns related to the diversity in the parent material for soil formation. There is no explicit temporal information included in this data set. Yet, the fact that young trees (seedling and shrub layer) are distinguished from full-grown trees opens a window on change in time. Variables describing the soil conditions and the geological pattern are as yet scarce. The statistical analysis should reveal whether these are needed to disclose the relationship between vegetation and site conditions.

13.2 Structure of the data set

The vegetation data are organized in the tradition of forest ecology: each tree species is recorded three times (in the herb, the shrub and the tree layer). Shrubs are recorded twice (herb and shrub layer) and herbs and mosses just once. Furthermore, all locations are sampled three times within concentric round plots of 30, 200 and 500 m^2, respectively. The majority of the examples shown below use the 200 m^2 data set.

In the environmental data set three categories of variables are found. First, variables recorded or verified in the field: elevation, slope and pH of the upper soil layer. Secondly, climate data interpolated from meteorological stations, as described by Zimmermann and Kienast (1999). Thirdly, 'azonal variables' taken from the Swiss Soil Quality Map, interpolating properties of soil types to yield a rank scale with range 1–6. The quality of the latter is limited due to the low resolution of the original map (scale 1:200,000) and the arbitrarily chosen soil types. These variables are labelled as 'soil map'. Environmental factors included in the site factor data set are the following (order according to field records, climatic data and data from soil map):

Sampling plan (`x.coord, y.coord`). The sample is a subset of the grid used in the Swiss National Forest Inventory (NFI; Brassel and Brändli 1999). Sampling units are located at each intersecting point of the 4 km×4 km coordinate grid of Switzerland. Only forest stands are taken (definition of NFI). The resulting sample size is $n = 726$ and it represents an unbiased state of the Swiss forests. True locations of plots deviate by 5–30 m from the grid: the precision of the navigation tools used at the time was rather limited.

Elevation, m a.s.l. (`hoehe, elev`). Two variables are given for technical considerations. The first stems from the field survey, the other is derived from a digital terrain model (DTM). The range of the elevation scores does not reflect the Swiss topography as plots above the timber line are not part of the sample.

Slope, deg (`slp`). This is derived from the digital terrain model and therefore it is affected by errors inherent in this.

pH upper soil layer (`pH.LFI`). Samples of the upper soil layer were taken in the first survey of the Swiss NFI (Brassel and Brändli 1999). I replaced 43 missing values with the mean of the sample (pH 5.1095).

Degree days, °C days (`ddeg.0`). Daily temperature is interpolated from a set of climatic stations (Zimmermann and Kienast 1999). Degree days are the integral of the daily mean temperature curve above the zero line.

Yearly precipitation, mm (`prcp.yy`). Precipitation exhibits higher yearly spatial variation than, for example, temperature. Accordingly, interpolated variables are also less reliable.

Frost days during growing season (`sfro.yy`). On frost days the night temperature drops below the freezing point. The growing season lasts from March until the end of September.

Coldest mean monthly temperature, °C (`tave.cc`). This is a proxy for the risk of frost drought. Trees are exposed to this, whereas plants below the snow cover are protected. The coldest month usually is January.

Mean yearly global radiation (`srad.yy`). This is the sum of direct and diffuse radiation over the entire year.

Yearly water balance, mm yr^{-1} (`swb`). The monthly water balance – water gain by precipitation minus water loss by evapotranspiration – is integrated over the whole year.

Moisture index, i.e. water balance in July, mm yr^{-1} (`min7`). This is the water balance as explained above, but for July only. July is the warmest month in Switzerland.

Soil depth, soil map (`s.depth`). This is estimated on a 1–6 scale addressing suitability for agriculture.

Nutrients, soil map (`s.nutr`). A 1–6 scale expresses availability of main plant nutrients.

Water capacity, soil map (`s.wcap`). This is the maximum possible water holding capacity estimated on a 1–6 scale.

Water permeability, soil map (`s.wperm`). Permeability estimated on a 1–6 scale.

Soil wetness, soil map (`s.wetn`). Excess average water content to limit agricultural use, estimated on a 1–6 scale.

Table 13.1: Data sets used in Chapter 13. See also Appendix 14

Data frame	Rows; columns	Contents
ws30	726; 1262	Swiss forest grid, 30 m^2 plots, vegetation
ws200	726; 1262	Swiss forest grid, 200 m^2 plots, vegetation
ws500	726; 1262	Swiss forest grid, 500 m^2 plots, vegetation
wssit	726; 20	Swiss forest grid, site (for all plots)

The vegetation survey, taking place in the years around 1995, used the Braun-Blanquet code (Braun-Blanquet 1964) transformed to a rank scale which can be further adapted as shown in Table 3.4. Sample size is $n = 726$. Separate data sets were generated for plot sizes of 30, 200 and 500 m^2. VEG-EDAZ (Küchler 2009) was used to update taxonomy where needed. The analyses finally require four data files, listed in Table 13.1 and Appendix 14.

13.3 Selected questions

13.3.1 Is the similarity pattern discrete or continuous?

The worst case scenario in terms of pattern recognition is disclosing patterns caused by the sampling plan itself, as may happen in preferential sampling.

Table 13.2: Composition of eight vegetation types in terms of tree layer and some site factors. Species scores are percentage frequencies.

Group no.	1	2	3	4	5	6	7	8
Group size	164	125	95	92	63	48	105	34
Elevation (mean, stdv.)	788	630	865	1341	1256	1469	1753	702
	228	172	265	191	335	271	263	252
Degree days (mean, stdv.)	2718	2964	2607	2039	2201	1847	1618	3455
	346	285	407	355	552	349	333	496
Precipitation (mean, stdv.)	1354	1208	1416	1629	1394	1417	1356	1780
	245	211	279	267	423	354	338	147
pH (mean, stdv.)	6.16	5.16	4.36	4.90	5.22	5.89	4.17	4.16
	1.00	1.32	1.24	1.17	1.06	1.25	1.12	0.56
Fraxinus excelsior	37	43	9	7	8	2	0	47
Quercus robur	5	17	5	0	0	0	0	6
Fagus sylvatica	85	70	75	28	6	2	3	32
Pinus sylvestris	15	14	6	0	13	10	8	0
Abies alba	35	38	71	16	10	17	2	0
Picea abies	50	66	87	73	63	96	64	0
Larix decidua	4	7	6	1	19	15	67	3
Alnus incana	2	3	0	11	6	2	1	6
Castanea sativa	1	2	1	0	0	0	1	76

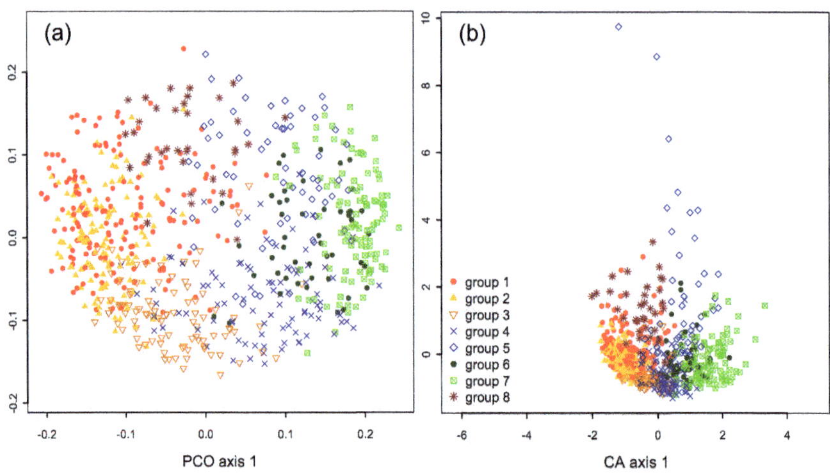

Figure 13.1: Principal coordinates analysis (a) and correspondence analysis (b) with eight vegetation types (Table 13.2) of the Swiss forest data set overlaid.

This, however, is unlikely to occur in the systematic sampling plan used here. One could argue that discrete patterns should not occur because the territory encompasses a huge altitudinal gradient, which is continuous by nature (Austin 2013b). But there are other factors with the potential to generate discontinuity, such as bedrock type and forest management. In

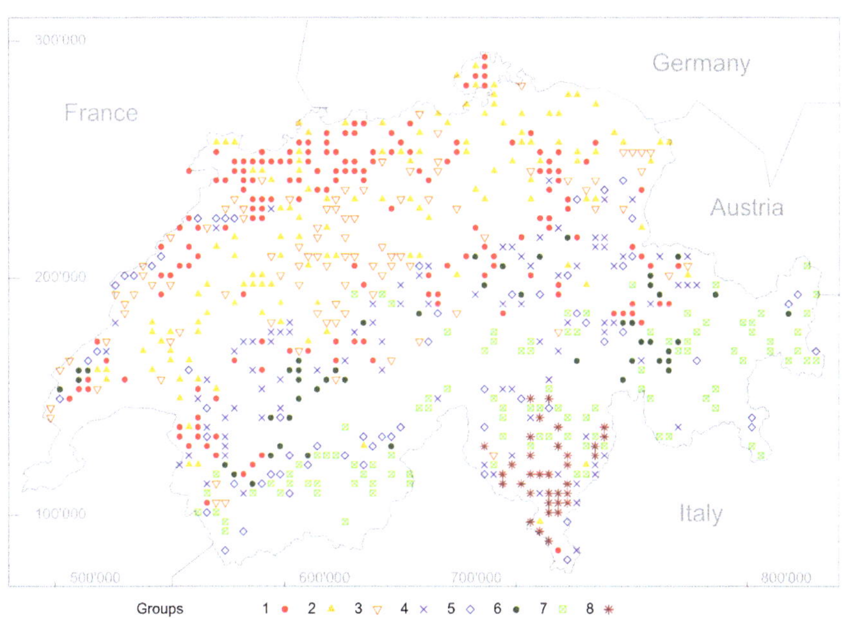

Figure 13.2: Vegetation map of Swiss forests (eight groups from cluster analysis, Table 13.2).

order to explore the species resemblance pattern in detail I use classification and ordination applied to the entire data set.

In Figure 13.1 classification (eight types derived by cluster analysis) is superimposed onto ordinations of PCOA [Figure 13.1(a)] and CA [Figure 13.1(b)], both confirming continuity of pattern. Unlike in the analysis of small data sets (Figure 6.5 and Figure 6.7, for example), differences between the two ordination methods are striking. In PCOA strong nonlinearity is responsible for a roundish point cloud, whereas CA generates a horseshoe with outliers appearing in the upper part of Figure 13.1(b) – a hallmark of the limitations of CA when applied to large data sets. Strong resemblance gradients prevail in any case whereas even faint traces of discontinuities are lacking.

The classification used is the result of cluster analysis of the vegetation data frame `ws200`:

```
vdm<- as.dist((1 - cor(t(ws200^0.25)))/2)
o.hclr<- hclust(vdm,method="ward.D2")
o.grel<- as.factor(cutree(o.hclr,k=8))
```

In this `vdm` is the distance matrix (correlation used as distance) and function `cutree()` generates eight relevé groups stored in `o.grel`. The cluster analysis used above does not tell us anything about the composition of these groups. One attempt to gain some insight is shown in the upper part of Table 13.2 presenting group means and standard deviations of selected site factors. Among these, elevation seemingly is a rather group specific parameter with distinct means and fairly low variation. As shown earlier (Chapter 9.8) means and standard deviations in Table 13.2 are most easily obtained when using function `tapply()`:

```
tapply(wssit$elev,o.grel,mean)
```

The first parameter in `tapply()` is the variable to be analysed, in the present case elevation. The second is group membership and the third is the function to be used. To calculate the standard deviation, `mean` is replaced by `sd`. To get the same for the degree days variable `wssit$ddeg.0` is used (see Section 13.2), for precipitation it is `wssit$prcp.yy` and for pH it is `wssit$pH.LFI`.

In cases where the investigator has some preliminary knowledge about vegetation, a frequency table with a selection of well-known species roughly reveals the composition of groups. Because the present data set stems from Central European forests, a lot is known about dominating tree species and we may simply select a few for display after computing group centroids with function `centroid()`:

```
o.centroid<- centroid(ws200,o.grel,y=1)
mytrees<- c(346,83,77,34,25,28,31,69,702)
round(t(o.centroid$prob.table[,mytrees]),digits=2)
```

The third line accesses the matrix of centroids restricted to nine species. This is displayed in transposed form [function `t()`] with species now in rows. When multiplying by 100, we get the content shown in the lower portion of Table 13.2. The first three vegetation types are typical *Fagus sylvatica* forests whereas the last encompasses the *Castanea sativa* forests of the southern Alps, for example.

As shown in Chapter 12 the predicting power of environmental factors can be visualized using boxplots. This is done in Figure 13.3. Panel (a) confirms that cluster analysis generates a classification reasonably well reflecting elevation. The pH value in panel (b), on the other hand, is of 'azonal' nature as it is related to geological conditions, that is, the parent material from which different soil types have involved. Since classification, `o.grel`, is available already from previous steps, the boxplot is obtained directly by function `boxplot()`:

```
boxplot(wssit$elev~o.grel,ylab="Elevation (m)",col="gray")
```

To get the same for pH we choose the corresponding site factor, `wssit$pH.LFI`, instead of elevation. Clearly, elevation better reflects the vegetation pattern

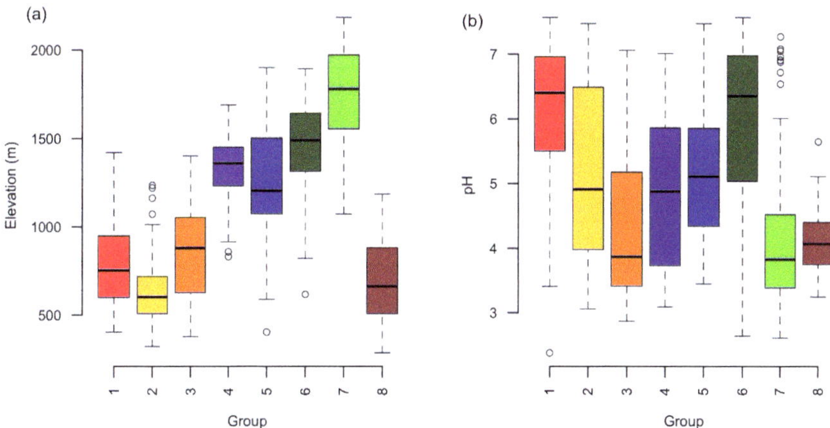

Figure 13.3: Boxplot of Swiss forest types (eight groups). Panel (a): Elevation as response. Panel (b): pH as response.

than pH. But in both graphs the boxes overlap confirming that the environmental factors form continuous gradients.

A different way to get an impression of the vegetation types is when drawing a map (Figure 13.2). This requires access to the geographical coordinates from data frame `wssit`, and group membership `o.grel` is used to select different plot symbols for vegetation types (parameter `pch`):

```
plot(wssit$x.coord,wssit$y.coord,pch=as.integer(o.grel),cex=0.5,asp=1)
```

The parameter `asp=1` is needed to avoid distortion of the map. In this the types stretch in south-westerly to north-easterly direction to form vegetation zones parallel to the Alps. Hence, there is evidence for a continuous pattern prevailing in space.

The same distance matrix `vdm` is also used in PCOA ordination, another way to strive for discontinuity [Figure 13.1(a)]:

```
o.pco<- pco(vdm) ; x<- o.pco$points[,1] ; y<- o.pco$points[,2]
plot(x,y,pch=as.integer(o.grel),asp=1,cex=0.6)
```

Again, group membership `o.grel` accords with different symbols and `asp=1` avoids distortion of the ordination. CA in Figure 13.1(b) is obtained as follows:

```
o.ca<- cca(ws200^0.25)
x<- o.ca$CA$u[,1] ; y<- o.ca$CA$u[,2]
plot(x,y,asp=1,pch=as.integer(o.grel),cex=0.6)
```

Obviously, the overall similarity pattern is perfectly continuous. In comparison CA ordination is seriously hampered by outliers and PCOA in this case may be the safer choice.

13.3.2 Is there a scale effect from plot size?

All locations have been surveyed threefold using concentric plots of size 30, 200 and 500 m^2, respectively. When plot size is increased, the number of species tends to increase as well. This is rapidly checked by counting species occurrence in the three different data sets, for example those shown in Table 13.2:

```
mytrees<- c(346,83,77,34,25,28,31,69,702)
for(i in 1:9) {s<- sum(sign(ws30[,mytrees[i]])) ; cat(s,"\n")}
```

In this the vector sums of signum-transformed data are written to console using function cat(). When we repeat this with objects ws200 and ws500, respectively, we get the frequencies in Table 13.3.

Table 13.3: Frequencies of tree species in data sets with plot sizes of 30, 200 and 500 m^2.

	Plot size		
Species	30 m^2	200 m^2	500 m^2
Fraxinus excelsior	90	152	190
Quercus robur	15	37	54
Fagus sylvatica	281	343	377
Pinus sylvestris	49	69	81
Abies alba	142	203	244
Picea abies	352	468	515
Larix decidua	72	113	141
Alnus incana	13	26	40
Castanea sativa	25	32	33

Do additional species found in large plots contribute to the distinction of relevé groups (vegetation types)? Could different plot sizes yield an alternative classification of forest types? A joint analysis of plots of different size is required to explore these questions, for which data sets ws30, ws200 and ws500 have to be merged.

A convenient way to merge vegetation data is using $\mathcal{R}$ function merge(). Because merge() generates 'automatic' row names, the originals have to be saved first and restored after merging:

```
rn.all<- c(rownames(ws30),rownames(ws200),rownames(ws500))
```

Merging proceeds by processing ws30 and ws200 first and then adding ws500:

```
ws30.200<- merge(ws30,ws200,all=TRUE)
ws30.500<- merge(ws30.200,ws500,all=TRUE)
```

Parameter `all=TRUE` will retain all the rows (relevés) and also all columns, but merge the latter if common species names are detected. In cases where a species is unique to one of the data frames then function `merge()` will fill missing values with `NAs` (an abbreviation for 'not available'). Although not really needed in the present case it is safe to replace `NAs` routinely by zeros. Finally, the old relevé names are restored:

```
ws30.500[is.na(ws30.500)]<- 0
rownames(ws30.500)<- rn.all
```

The new data frame `ws30.500` with 2178 relevés and 1262 species is now ready for analysis.

Because plots of different size were nested in the sampling plan it can be expected that the larger ones have higher species richness. To reveal such patterns adjustments of relevé vectors have to be avoided by using Euclidean distance as a resemblance measure and PCOA for ordination as shown in Figure 13.4(a–c). These are ordinations of the joint data frame `ws30.500` with 2178 relevés included. In Figure 13.4(a–c) points originating from plot size 30, 200 and 500 m^2, respectively, are highlighted. Plots of size 30 m^2 [Figure 13.4(a)] are clearly restricted to the lower part of the ordination whereas plots of size 500 m^2 [Figure 13.4(c)] spread across the entire range of the ordination. Plots of size 200 m^2 [Figure 13.4(b)] have intermediate spread. In conclusion, plot size affects the similarity of the relevés involved. Which plot size, then, yields the best fit with environmental factors? Is it 30 m^2, probably the most homogeneous ones, or is it 500 m^2 where more species are included?

Figure 13.4(d–f) attempts to find a graphical answer. For this the three different data frames are ordinated (PCOA) individually and trend surface of elevation is fitted as shown earlier (Figure 6.13), hoping that in an alpine environment this will reveal an interpretable pattern (Körner 1999). Clearly, Figure 13.4(d–f) shows the same overall pattern. But relevés from plot size 30 m^2 [Figure 13.4(d)] seem to perform worse at low elevations, namely between 600 m and 800 m.

A method to detect a possible difference in the fit of vegetation with the environment is NP-MANOVA (Section 7.5) by $\mathcal{R}$ function `adonis()`. From data frame `ws30` we get the following result:

```
o.adonis<- adonis(ws30^0.5~wssit$elev,method="euclid")
o.adonis

Call:
adonis(formula = ws30^0.5 ~ wssit$elev, method = "euclid")

            Df SumsOfSqs MeanSqs F.Model      R2 Pr(>F)
wssit$elev   1      2660  2659.9   45.16 0.05871  0.001 ***
Residuals  724     42645    58.9         0.94129
Total      725     45304                 1.00000
```

The same done with data frame ws200 yields:

```
Call:
adonis(formula = ws200^0.5 ~ wssit$elev, method = "euclid")

             Df SumsOfSqs MeanSqs F.Model      R2 Pr(>F)
wssit$elev    1      4984  4984.2  56.888 0.07285  0.001 ***
Residuals   724     63433    87.6         0.92715
Total       725     68418                 1.00000
```

And finally ws500 delivers:

```
Call:
adonis(formula = ws500^0.5 ~ wssit$elev, method = "euclid")

             Df SumsOfSqs MeanSqs F.Model      R2 Pr(>F)
wssit$elev    1      6525  6525.4  63.257 0.08035  0.001 ***
Residuals   724     74685   103.2         0.91965
Total       725     81211                 1.00000
```

There is a clear trend towards increasing fit with increased plot size, with an F-value starting at 45.16, followed by 56.88 and reaching 63.26 at plot size 500 m^2. The same pattern is seen in the proportion of variance explained by elevation, starting at 5.9%, going up to 7.3% and reaching 8.0% with plot size 500 m^2. It is noteworthy that error probability in this case is not helpful at all as all models are always highly significant. In conclusion, the largest plot size is the most adequate for ecological modelling, provided other site factors behave similarly.

The panels in Figure 13.4 all reside on function pco(). For Figure 13.4(a) this is:

```
de<- vegdist(ws30.500^0.5,method="euclid")
o.pco<- pco(de)
plot(o.pco$points[,1],o.pco$points[,2],pch=16,col="gray")
```

This plots all symbols as dots (pch=16) in grey. For Figure 13.4(a) we overlay the first 726 pertaining to the 30 m^2 plots with darker symbols (before closing the plot window):

```
x30<- o.pco$points[1:726,1] ; y30<- o.pco$points[1:726,2]
points(x30,y30,pch=16,col="black")
```

Ordinations of Figure 13.4(d–f) are made from data frames ws30, ws200 and ws500 respectively, the same way as Figure 13.4(a–c). Trend lines are added by function ordisurf(). For the first data frame this is:

```
de<- vegdist(ws30^0.5,method="euclid")
o.pco<- pco(de)
plot(o.pco$points[,1],o.pco$points[,2],pch=16,col="gray")
ordisurf(o.pco,wssit$elev,add=TRUE)
```

To add the trend lines the plot window with ordinations has to be kept open.

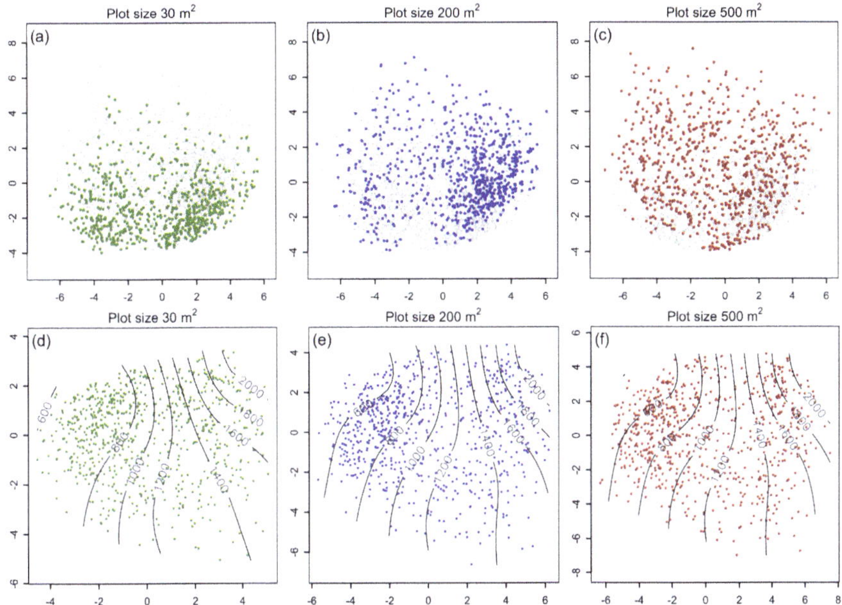

Figure 13.4: The effect of different plot size on similarity pattern. Panels (a–c): PCOA ordinations of joint data sets. Relevés with plot sizes of 30, 200 and 500 m^2, respectively, highlighted. Panels (d–f): Trend surface of elevation fitted to individual samples with different plot sizes.

13.3.3 Which factors reflect vegetation pattern?

This is the main question treated in Chapter 7. Because the present survey concerns a large, well-known area, an effort to interpret an ordinary vegetation map such as the one in Figure 13.2 appears promising. Even though the map considers discrete types only, the dispersion of some of the symbols accords with altitudinal zones: along the upper-left part in Figure 13.2 the Jura Mountains, ranging from south-west to north-east, can easily be identified. Dominating groups are numbers 1, 3 and 5. Parallel to this follows the Plateau (Mittelland), where types 2 and 3 prevail. In the Pre-Alps groups 4 and 5 are most common, whereas the high alpine zone is mainly above the timber line. Embedded in the Alps are central Alpine valleys, the driest locations in the investigation area with the highest abundance of groups 6 and 7. The sequence ends in the south where relevé group 8 corresponds to the Insubrian climate.

A quantitative overview of the performance of all independent variables in resolving the group structure of the vegetation yields variance ranking as introduced in Section 7.2.2. Table 13.4 results from function `srank()`:

```
o.srank<- srank(wssit,method="jancey",o.grel,y=1)
```

```
o.srank
```

All independent variables reside in data frame **wssit** and **o.grel** is group membership as computed in Section 13.3.1. In this specific analysis elevation and degree days (warmth) perform best.

Table 13.4: F-values of site factors based on eight forest vegetation types. All F-values are significant at the 1% error probability level.

Rank	No.	Variable	F-value
1	4	Elevation, DTM	273.695
2	3	Elevation, field	264.047
3	6	Degree days	194.162
4	9	Lowest monthly temperature	179.356
5	13	Soil depth (soil map)	84.306
6	15	Nutrients (soil map)	81.683
7	14	Water capacity (soil map)	79.837
8	2	y-coordinate (N–S gradient)	75.322
9	5	Slope (deg.)	47.134
10	18	pH (upper soil layer)	43.090
11	12	Moisture index July	38.928
12	8	Frost days in growing season	38.465
13	16	Water permeability (soil map)	33.254
14	7	Yearly precipitation	26.035
15	1	x-coordinate (E–W gradient)	18.100
16	11	Yearly water balance	15.322
17	10	Yearly radiation	11.518
18	17	Soil wetness (soil map)	8.540

Maps revealing far more detail than the one in Figure 13.2 are continuous, although for display they require separate layers for every vegetation type or every species considered. The multinomial logistic regression model (see Section 9.7) does this for all types (**o.grel**) simultaneously. For simplicity – and to keep the number of degrees of freedom as low as possible – I use four selected site factors as predictors (see Table 13.4 for the full list): (1) elevation taken from DTM (**elev**), (2) degree days (**degd.0**), (3) yearly precipitation (**prcp.yy**) and (4) pH of the soil (**pH.LFI**). A call of function **multinom()** computes the regression coefficients and function **predict()** yields occurrence probabilities for all types within all plots. To do all that it is probably safe to repeat classification of vegetation:

```
vdm<- as.dist((1 - cor(t(ws200^0.25)))/2)
o.hclr<- hclust(vdm,method="ward.D2")
o.grel<- as.factor(cutree(o.hclr,k=8))
```

Then follows a call of function **multinom()**. The fitted values are used for plotting the eight maps of vegetation types in Figure 13.5, each as a separate graph:

```
o.mu<- nnet::multinom(o.grel ~ elev+prcp.yy+ddeg.0+pH.LFI,data=wssit)
```

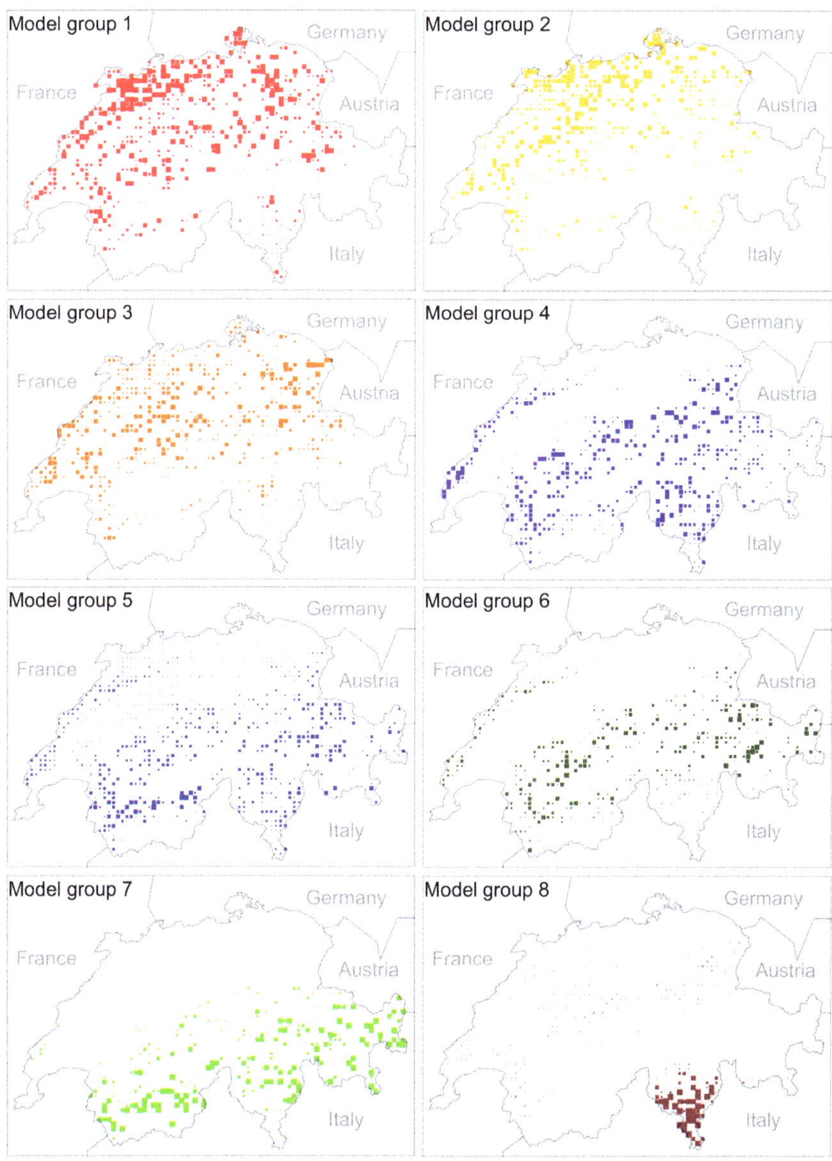

Figure 13.5: Vegetation probability map (eight groups from cluster analysis distinguished). Site factors 4, 6, 7 and 18 in Table 13.4 are used.

```
x<- wssit$x.coord ; y<- wssit$y.coord
fitval<- o.mu$fitted.values
for(i in 1:8) {
  plot(x,y,pch=16,cex=fitval[,i]^0.5,col="gray",asp=1)
```

```
  title(main=paste("Model group",i),cex.main=0.8,font.main=1)
}
```

In the maps shown in Figure 13.5, the surface of dots is proportional to the probability of occurrence. Symbols occur on forested plots only; that is, about 30% of the area of the country. In general the spatial distribution pattern accords with the topographic pattern dominated by the altitudinal gradients. Figure 13.5 reveals the continuous pattern of forest systems far more clearly than Figure 13.2. But how well do the surveyed and simulated vegetation maps agree? As explained in Section 9.7 we first derive the new discrete vegetation types from the fitted values of function `multinom()` using function `apply()` followed by a call to cost function `ccost()`:

```
newgr<- apply(o.mu$fitted.values,1,which.max)
o.ccost<- ccost(ws200,oldgr=o.grel,newgr=newgr,y=0.25)
o.ccost$ccost
```

```
[1] 51.39556
```

Whether an agreement of this extent is good or bad is a question of preference. However, the cost function is most useful for evaluating the underlying model. We can ask, for instance, how the agreement between field data and simulation changes when omitting pH in the model. For this, function `multinom()` is run again without the last term, `pH.LFI`. This alters the costs to $cf = 57.0682$, a clearly inferior agreement. Upon inspecting the confusion matrix in `o.ccost$conf` we can see that the entire group 3 vanishes when omitting `pH.LFI`. This suggests that the model with four site factors is the superior one. What is the highest possible fit that can be achieved by this method? It is the model that includes all 18 site variables in object `wssit` for which we write:

```
o.mu<- nnet::multinom(o.grel ~ .,data=wssit)
newgr<- apply(o.mu$fitted.values,1,which.max)
o.ccost<- ccost(ws200,oldgr=o.grel,newgr=newgr,y=0.25)
o.ccost$ccost
```

The dot after the tilde means ‘use all variables in object `wssit`’. The result is $cf = 44.07537$. All that, however, comes at the expense of a very complicated model.

Returning to the question posed in Section 13.3.2 we can now check the models for performance depending on plot size, that is 30, 200 and 500 m^2 when using objects `ws30`, `ws200` and `ws500`, respectively, for clustering followed by multinomial regression. The result is summarized in Table 13.5. This includes the now widely used AIC, displayed when typing `o.mu`. AIC is nothing else but residual deviance (Section 9.3) modified by a ‘penalty’ for higher degrees of freedom (Crawley 2005, p. 208).

Table 13.5: Performance of multinomial models of forest vegetation with variable plot size (initial classifications derived from 30, 200 and 500 m^2 plot samples respectively). Site factors 4, 6, 7 and 18 in Table 13.4 are used.

Data frame used	Plot size (m^2)	cf-value	AIC
ws30	30	46.31	1870.59
ws200	200	51.39	1679.966
ws500	500	42.60	1489.17

The largest plot size, 500 m^2, performs best in terms of multinomial regression. This not only suggests that plot sizes 30 and 200 m^2 (considered 'normal' in phytosociological literature as reported, for instance, by Mueller-Dombois and Ellenberg 1974) are below optimum for this kind of modelling, but using plots larger than 500 m^2 would probably further improve the result. Plots of 30 m^2, however, yield a better cost value than the same with 200 m^2, despite the model yielding the worst AIC coefficient. We can speculate, in this case, that the homogeneity of the smallest plots come as an advantage for the prediction of a crisp classification.

13.3.4 Is tree species distribution man-made?

A way to learn more about human impact is to have a closer look at species whose use in forestry is better documented than is the case for vegetation types. This question is related to the well-known 'potential natural vegetation' (PNV) issue introduced in a systematic manner by Tüxen (1956). His fairly complicated definition is further developed by Lindacher (1996). Lindacher also mentions extensions of the concept to adapt for effects like climate change and environmental pollution, addressed in the definitions of Kowarik (1987) for example. In many forests human influence dominates a given site and the vegetation may be *anthropogenic* rather than natural (Küchler 1988). Or, one may want to reconstruct the vegetation as it would be without *Homo sapiens* living on earth, as Neuhäusl (1984) explained.

In all references given above there is a general agreement that PNV could only be derived when land-use history of the past few thousand years is taken into account. As this is not known, all findings will remain hypotheses. Furthermore, tree species planted within their natural range do not really change the entire vegetation composition. Gobet *et al.* (2010) recently reported that fossil records suggest a much wider distribution of many tree species common in Europe than observed today. In view of all the uncertainties, Chiarucci *et al.* (2010) seriously question the underlying ideas of PNV and suggest abstaining from using it.

Hence, I restrict the evaluation to the discussion of four selected tree species: *Fagus sylvatica*, *Fraxinus excelsior*, *Larix decidua* and *Castanea sativa*. If the occurrence of any of these is man-made, this should become evident in either the *geographical*, the *compositional* or the *ecological* distribution pattern.

In Figure 13.6 *Larix decidua* is a species with a rather peculiar spatial distribution pattern. Under natural conditions it mainly grows in the central alpine belt from the west to the east of the country and the simulated distribution pattern (logistic regression with site factors elevation, precipitation, radiation and pH) confirms this. Isolated, scattered plots north of this belt mark the locations where plantations may have been made in the past. Clearly, this is still an interpretation rather than a proof. Equally striking is the pattern of *Castanea sativa*. This is restricted to southern Switzerland, but a few clumped plantations in the western part of the country emerge in the survey grid. Based on the simulation, these locations are not really favourable for *Castanea* and the population there may be of artificial origin. From *Fagus sylvatica* it is known that this species is hardly ever planted. And in fact the uppermost panels in Figure 13.6 show that the range of observations is roughly identical to the expectation based on site factors. Almost the same is observed for *Fraxinus excelsior* which is not as abundant as *Fagus*. If plantations exist, they probably occurred within the ecological range of the species and therefore remain undetected.

To generate maps of observed species distributions, the left-hand side of Figure 13.6, we plot all locations where the species is present using black dots:

```
x<- wssit$x.coord ; y<- wssit$y.coord
plot(x,y,pch=16,col="black",cex=sign(ws200[,77]),asp=1)
```

Parameter `cex` is symbol size and column number 77 is the scores vector of *Fagus sylvatica* in object `wssit`. Function `sign()` transforms scores to presence–absence. Addressing columns 346, 31 and 702 instead of 77 yields maps of *Fraxinus excelsior*, *Larix decidua* and *Castanea sativa*, respectively. The panels with species expectations on the right-hand side of Figure 13.6 result from GLMs as introduced in Section 9.3. The dependent variable is the species vector `sp` transformed to presence–absence:

```
sp<- sign(ws200[,77])
o.glm<- glm(sp~elev+prcp.yy+ddeg.0+pH.LFI,data=wssit,family=binomial)
plot(x,y,pch=16,cex=fitted(o.glm)^0.5,col="gray",asp=1)
```

The four dependent variables are the same as in Figure 13.5, that is (1) elevation taken from DTM (`elev`), (2) degree days (`degd.0`), (3) yearly precipitation (`prcp.yy`) and (4) pH of the soil (`pH.LFI`). The size (`cex`) of the grey dots (`pch=16`) is proportional to the fitted values of the model.

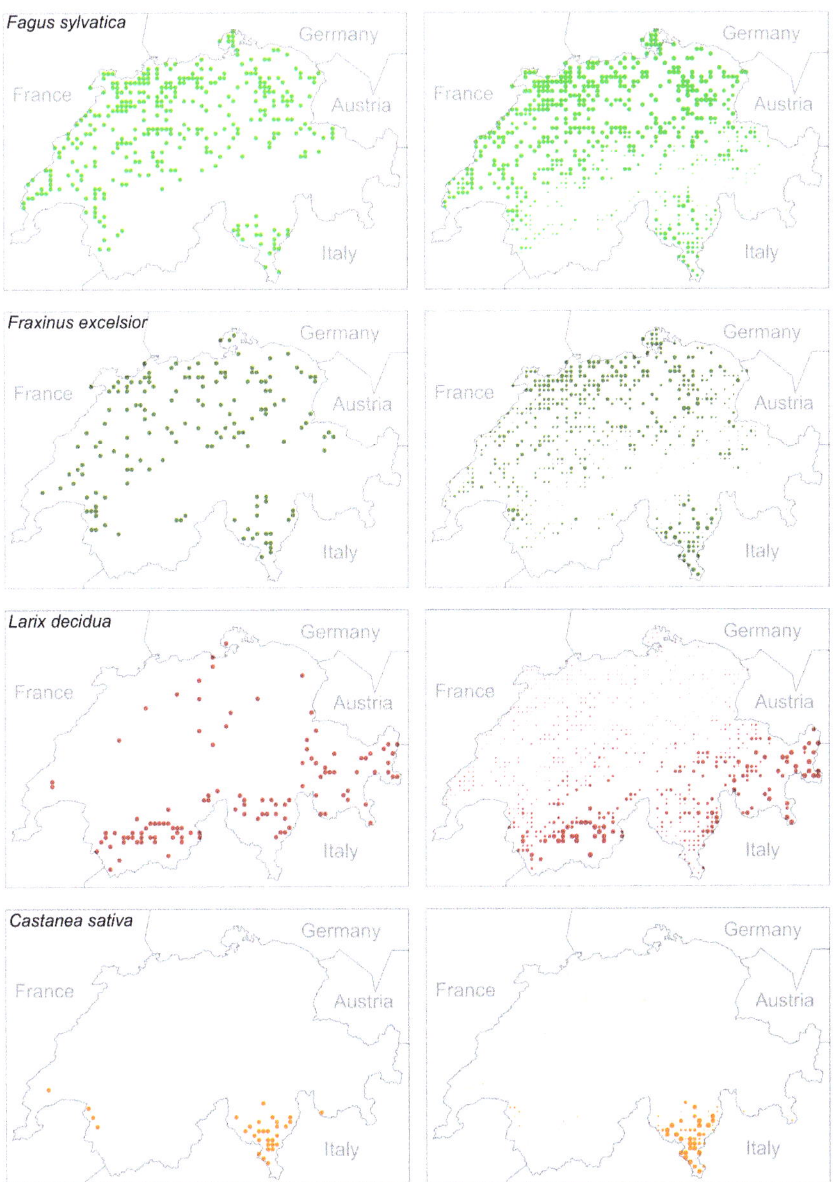

Figure 13.6: Observed (left-hand side) and potential (right-hand side) distribution of four tree species. The potential is the expectation of a logistic model (GLM) with site factors elevation, precipitation, radiation and pH.

Can tree plantations be inferred from the *compositional* pattern? A tool to investigate this is ordination. Under natural conditions plant species

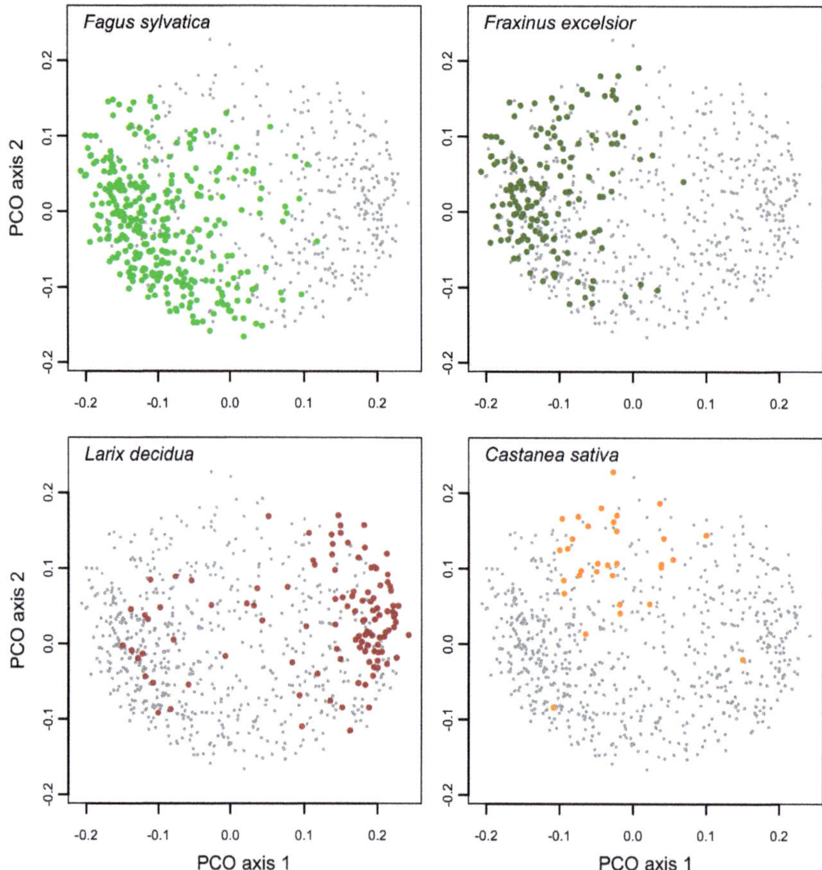

Figure 13.7: Ordination of forest stands. Four selected tree species are marked.

rarely exhibit a disjunct compositional distribution pattern. Where such a phenomenon occurs it is likely that plantations have taken place. In Figure 13.7 the occurrence of the same species as used in Figure 13.6 is shown within an ordination. In *Larix decidua* and *Castanea sativa* a compositional centre of occurrence exists, but with some scattered remote points: in these the two trees grow in common with an entirely different set of plant species, supporting the evidence from the geographical pattern that plantations exist. Unlike in the geographical map, a centre of occurrence of *Fraxinus excelsior* is now visible. In all remaining sites of the ordination it is absent. This supports the hypothesis that *Fraxinus excelsior* stands are either natural or are planted where they would occur under natural conditions. The same interpretation applies to *Fagus sylvatica*.

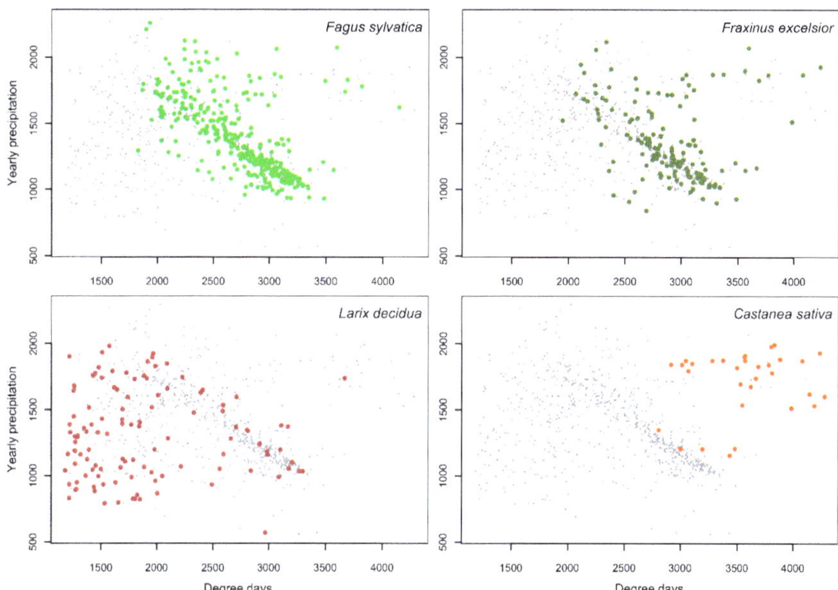

Figure 13.8: Ecograms of forest stands. Four selected tree species are marked.

Plotting species occurrence within an ordination is done as follows:

```
o.pco<- pco(as.dist((1-cor(t(ws200^0.25)))/2))
x<- o.pco$points[,1] ; y<- o.pco$points[,2]
plot(x,y,pch=16,col="gray",asp=1)
points(x,y,pch=16,cex=sign(ws200[,77]),col="black")
```

This is PCOA [function `pco()`] with all relevés plotted as grey dots, overlaid with black dots where *Fagus sylvatica* (species number 77) occurs.

Can tree plantations be inferred from the *ecological* pattern? Ecograms offer a similar kind of interpretation as maps and compositional ordinations (Figure 13.8). These point patterns show climatic conditions below the timber line. There is a pronounced main gradient from the lower-right corner (warm, dry) towards the upper-left (cold, wet). Along this gradient, warmth and water supply are strongly correlated. But there are two more areas following a different pattern. The first is the left-hand lower edge of the ecogram. These are the central alpine regions, with low temperature and dry conditions. The second is the upper-right corner, where the opposite holds and it is warm and wet. This is the climate of southern Switzerland, the Insubric area.

Looking at the same tree species as before, *Fagus sylvatica* spreads along the main gradient and also in the direction of the Insubrian conditions, avoiding the central alpine growth conditions (Figure 13.8). That is where

Larix decidua has its centre of distribution. If the previously stated hypothesis is correct then the few locations along the main altitudinal gradient are plantations (including one isolated stand in the Insubrian climate). Even more striking is the pattern of *Castanea sativa*, with its centre of distribution in the Insubrian part of the ecogram. Five stands along the main altitudinal gradient are really disjunct and definitely aberrant from the ecological point of view.

Plotting the ecograms in Figure 13.8 requires no further computations as the site factors are used directly as coordinates:

```
x<- wssit$ddeg.0 ; y<- wssit$prcp.yy
plot(x,y,pch=16,cex=1,col="grey")
points(x,y,pch=16,cex=sign(ws200[,77]),col="black")
```

First, all plots are displayed in grey, then all where *Fagus sylvatica* (number 77) occurs are added in black (or any other colour). For *Fraxinus excelsior*, *Larix decidua* and *Castanea sativa* the column numbers are 346, 31 and 702, respectively.

Table 13.6: Number of plots where selected tree species occur in the tree, shrub and herb layers.

Columns no.	Species	Tree layer	Shrub layer	Herb layer
25, 26, 27	*Abies alba*	203	180	304
260, 261, 262	*Acer platanoides*	15	9	51
266, 267, 268	*Acer pseudoplatanus*	160	183	395
66, 701, 68	*Alnus incana*	26	30	35
69, 63, 70	*Betula pendula*	54	30	33
71, 72, 73	*Carpinus betulus*	22	25	35
702, 703, 704	*Castanea sativa*	32	18	35
77, 78, 79	*Fagus sylvatica*	343	291	345
346, 347, 348	*Fraxinus excelsior*	152	139	332
31, 32, 33	*Larix decidua*	113	52	52
28, 29, 30	*Picea abies*	468	360	409
686, 687, 688	*Pinus cembra*	23	24	26
681, 682, 683	*Pinus mugo arborea*	7	6	5
34, 35, 36	*Pinus sylvestris*	69	14	15
219, 220, 221	*Prunus avium*	35	40	116
80, 81, 82	*Quercus petraea*	25	9	40
83, 84, 85	*Quercus robur*	37	12	80
46, 47, 48	*Salix caprea*	16	57	75
205, 206, 207	*Sorbus aria*	48	100	131
199, 200, 201	*Sorbus aucuparia*	42	127	320
39, 40, 41	*Taxus baccata*	4	10	8
288, 289, 290	*Tilia cordata*	8	7	15
283, 284, 285	*Tilia platyphyllos*	21	21	32
86, 87, 88	*Ulmus glabra*	24	47	56

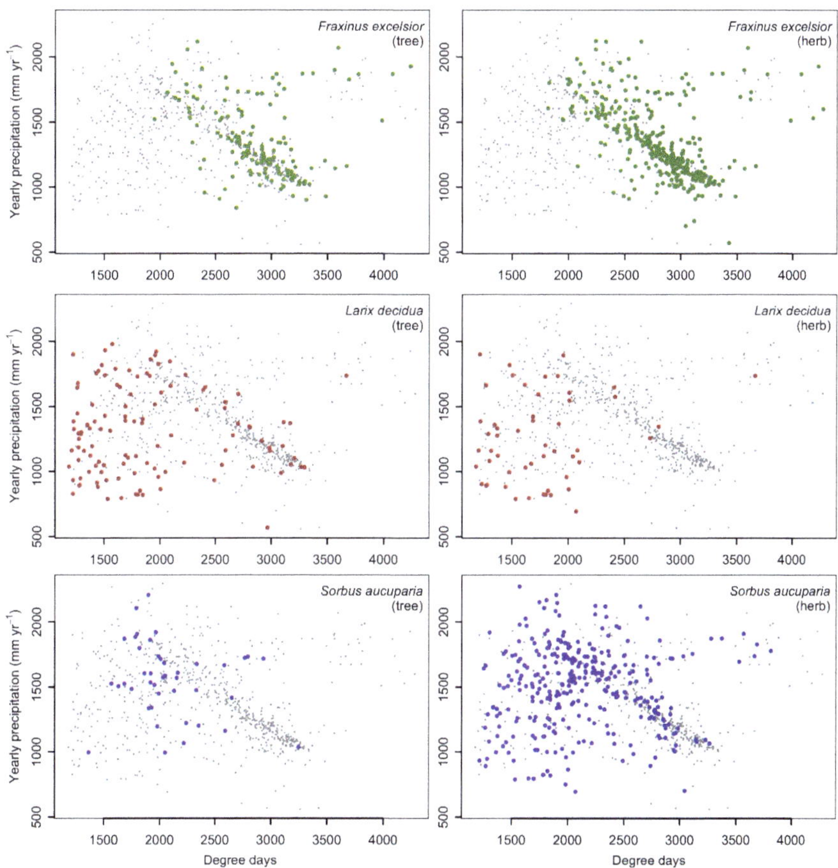

Figure 13.9: Distribution of three species in ecological space. Tree layer (left); herb layer (right).

13.3.5 Is the tree species pattern expected to change?

This question is based on the idea that the next generation of tree species is already present in the form of seedlings and saplings. As can be seen in Table 13.6 the tree species behave differently in this regard. Some species with large seeds, such as *Alnus incana*, *Pinus cembra* and *Fagus sylvatica*, show almost the same frequency for the tree and the herb layers. But most interesting are the species with a much higher frequency in the herb layer: these have the potential to extend their area in the face of changing climate much faster than species which must expand through propagation of seeds. Typical cases are *Abies alba*, *Quercus robur*, *Acer pseudoplatanus*, *Fraxinus excelsior* and *Sorbus aucuparia*.

Deriving species frequency as seen in Table 13.6 is easiest when done by column number. The trees of *Abies alba*, for example, are recorded in column number 25 of object ws200 and the frequency is obtained as follows:

```
sum(sign(ws200[,25]))
```

This is the vector sum of the signum-transformed species vector. The shrub layer of the same species is found in column 26, the herb layer in 27.

In Figure 13.9 three species are compared in this regard. *Fraxinus excelsior*, with almost twice as many plots with seedlings than plots with trees, has almost identical distribution patterns for both. The seeds of this species are very mobile and are wind-dispersed, but the seedlings have not so far established in areas of different ecological conditions. Mobility of seeds, however, suggests that extension of area may still be rapid in the case of climate change. *Larix decidua* is the opposite case: rejuvenation is almost exclusively restricted to sites where it occurs naturally; where it is planted, for example along the main altitudinal gradient, seedlings and saplings are very rare. Finally, *Sorbus aucuparia* is very abundant in the herb and shrub layers. The range of the seedlings exceeds the same of the trees considerably. Under changed conditions *Sorbus aucuparia* could rapidly grow up to the tree layer, thereby expanding its present area. Along the main altitudinal gradient the species is mainly lacking in the tree layer, probably due to competition by taller growing trees. For it to expand successfully, other species would have to reduce vitality.

Plotting the panels in Figure 13.9 is done the same way as for Figure 13.8. Column numbers of the species displayed are 346, 348, 31, 33, 199 and 201.

13.4 Conclusions

In this chapter patterns are identified and analysed in the biological, the geographical and the ecological space with emphasis on the interaction between vegetation and site. The temporal dimension is missing, but indirect reasoning is possible because tree species are recorded according to three different layers, as trees, shrubs and herbs respectively. A category of environmental factors, assessing the human influence, is also missing: management is hardly ever documented in forests all around the globe. But as shown in Section 13.3.4 some irregularities in the distribution patterns of tree species not explained by environmental factors give rise to speculations about tree plantations. All that is possible only because forest management in Switzerland is strongly regulated. It is for the same reason that corresponding patterns are found in vegetation types, individual species and site factors, relationships that ruthless timber harvest certainly would mask.

Although the analytical methods seem to work well, the success of the analysis in the first place is based on the quality of the data set, careful field work (see Wohlgemuth *et al.* 2008 for more details) and most importantly, the systematic sampling plan. Using a square grid may not be optimal to include rare vegetation types and rare species, but it circumvents the risk of bias as would be expected in preferential sampling. The sample size of $n = 726$ is still small considering the complexity of the investigation area. The role sample size plays could be investigated too: by forming subsets and comparing the corresponding patterns, an exercise demonstrated in Section 9.6.

The analysis of the data sets requires a variety of methods introduced in preceding chapters. Remarkable, again, is the subtle role of data transformation. In Section 13.3.2 the comparison of samples from different plot size is possible only when avoiding any vector transformation of relevés. Using correlation, as for instance in Figure 13.7, would adjust relevés to unit variance thereby hiding the differences that occur in vector length. This is one of the examples where absence of transformation is a means for answering specific questions.

When the creators of the data set decided to choose a variable plot size they were probably not aware of a topic addressed only recently: the 'dark' or 'hidden' diversity. This is discussed, for instance, in a review by Pärtel (2014). The issue is absent species, where absence can have various reasons. In the forest data set used here species occurring in the large plots, but not in the small ones, are a case in point. One can assume that these have the potential to occur in the small plots as well, but may either have been overlooked or be absent by chance. What is the role of these species? Do they confirm the same pattern as found in the full plots? Or could we expect a different, complementary distribution pattern?

Pärtel (2014) also mentions that ecological models are apt to suggest the presence of hidden species. Figure 13.6 gives an example. In many locations species are absent, but the probability to occur is still fairly high. Assuming the presence of these species also yields a surrogate for the dark species pool, although the reasons for their absence would still remain unknown.

14 Back to the roots?

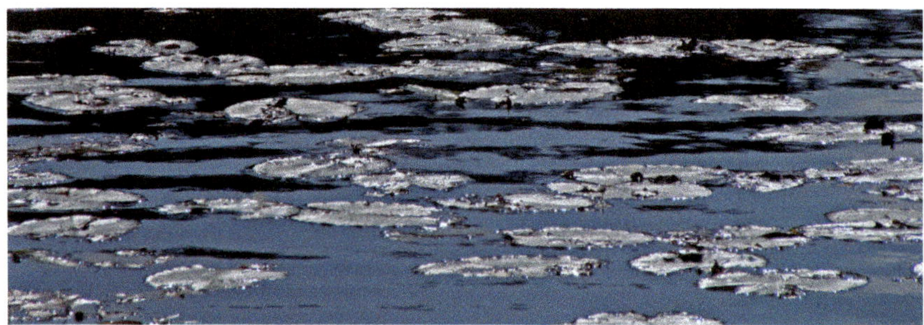

Plant ecology is known for pioneering work in mathematical analysis. Many of the methods have been developed to handle the properties of vegetation data and sometimes to fulfil the prejudices of practitioners. Newly emerging free software may redirect these towards ever-growing all-purpose statistical methods.

Throughout this book vegetation ecology is conceived of as a discipline analysing multiple vectors of species or plant traits jointly with environmental variables in space and time. As stated in similarity theory (Feoli and Orlóci 2011) it assumes regularities to occur in multivariate data sets in the form of recurring patterns caused by interactions (Clarke 1993). Mathematics has long been considered a tool of preference to handle vegetation data, which are large and hampered by statistical noise and nonlinear relationships. Standard statistical methods have not always yielded the expected results and this encouraged the development of approaches specific to plant ecology, one of the earliest known being the similarity coefficient of Jaccard introduced in the year 1901. Since then, not only have methods adapted to the needs of the discipline, but various specialized books on how to handle vegetation and site data have been released (Mueller-Dombois and Ellenberg 1974; Orlóci 1978; Gauch 1982; Pielou 1984; Digby and Kempton 1987; Jongman *et al.* 1995; Podani 1997; Borcard *et al.* 2011; Kent 2012; Legendre and Legendre 2012; Wildi 2013). In these we encounter common statistical methods, but also others hardly ever seen elsewhere. Botany, it seems, eventually gained the reputation of regularly inventing novel approaches.

Is there a 'botanical' approach? An opinion concerning this question is found, for example, in the paper of Groth *et al.* (1977), 'The Clustering of Galaxies', where they argue that cluster analysis is a 'botanical approach'. In recent years, however, software adapted to the specific requirements of vegetation data not only faced support, but also competition by the free software environment $\mathcal{R}$ (R Development Core Team, 2017), for instance. Statistical analysis underwent a globalization blurring the frontiers between disciplines and encouraging the use of a broad spectrum of methods. While revising this book I got the impression that vegetation ecology is shifting away from its ancient tradition and moving closer to standard methods used in other fields of science. There are various instances confirming this trend.

Sampling revisited. In the early times of vegetation science, about a century ago, emphasis in sampling was on the ingenuity of the investigator in recognizing patterns in the field directly, thereby unveiling an impressive spectrum of plant communities. Data collection was considered an act of documentation rather than a primary step towards analysis. As the guild of scientists grew it became evident that results obtained this way could not be reproduced and in the book of Mueller-Dombois and Ellenberg (1974) it is terminologically downgraded to 'subjective sampling' in contrast to the apparently more valuable 'objective sampling'. In Orlóci (1978) the now widely used term 'preferential sampling' is used implicitly, expressing that this practice no longer deserves the qualification of a 'scientific investigation.'

As shown in Chapter 13 systematic sampling facilitates the interpretation of various ecological investigations. The spatial patterns found there, although sometimes weak and blurred by statistical noise, are certainly credible whereas in preferential sampling there is an ongoing risk of finding patterns that turn out to be sampling artefacts.

Unlike in experimental design, efficiently sampling in space (and time) remains a difficult task (Orlóci and Pillar 1989). An alternative would be model-based sampling, that is, sub-sampling based on intermediate results (Albert *et al.* 2010; Pellissier *et al.* 2012). This in the literature as yet underrepresented idea is addressed in Chapter 1, but not demonstrated in any of the examples in this book.

Measuring species abundance. Plant ecology profited much from the establishment of a standard cover-abundance scale by Braun-Blanquet (1928, 1932). But this also caused confusion as it mixes symbols and ranks in a not really consistent manner (Mucina 1997; Podani 2006). What we observe these days is that most users of old data sets in a first step transform this to cover percentage and sometimes further to a rank scale or presence–absence (see Table 3.4). Refining the scale (van der Maarel 1979, for instance) does not alleviate the situation. Practitioners sometimes hesitate to use percentage directly because they are scared about limited precision imposed in estimations. Assessing a convention for rounding percentages could mitigate these concerns. Cover percentage offers the freedom to change the scores according to objective, to ranks, binary variables or other data types (Section 3.4). Table 7.8 confirms the widely known fact that transformations close to presence–absence yield the best results in ecological modelling. Nature too tells us that cover of many plants is rather ephemeral as it is prone to rapid seasonal change.

Investigating untransformed cover percentage scores, on the other hand, can be the main objective of analysis. This is the case in Markov chains (Section 10.6) where cover of all species within a plot adds up to 100%. Yet another example is Figure 10.6 where an ordination reflects temporal change in terms of cover, also summing up to 100% when including bare soil as a variable. In conclusion, using some raw measurements like cover, number of individuals or biomass in the course of sampling opens a chance to use a data set in the future in a different context for answering as yet unknown questions.

Assessing multivariate resemblance. The number of resemblance measures devised for vegetation data may easily exceed 100 (see Legendre and Legendre 2012, for instance). In most other fields of science the Kendall correlation coefficient prevails (Section 4.4). To my surprise I found that this coefficient (or its complement, correlation as distance, Equation 4.6) frequently performs better than, for instance, the Bray–Curtis distance in modelling, clustering and ordination. Faith *et al.* 1987 and Clarke (1993) argue that the Bray–Curtis coefficient is one of the most reliable performers. In Sections 5.8 and 6.8 we have noticed that this is true only because the Bray–Curtis coefficient intrinsically – and sometimes even unintendedly – transforms abundance scales close to presence–absence. Its use comes at the expense of lost flexibility by changing scores into the direction of presence–absence thereby correcting for improper adjustment of scale.

Equally important is a revival of Euclidean distance. This projects scores into geometric space, by default without any further adjustments. There are instances where this is indispensable. An example is Figure 10.6 where an ordination documents temporal change in terms of cover percentage (that is, surface cover).

Splitting the continuum. Although vegetation is now understood as a continuum, discrete classification has various practical applications (Mucina 1997). It is worth viewing this as a tool of communication on the one hand and an analytical means of data reduction on the other hand. The case study in Chapter 12 illustrates that these two objectives are difficult to reconcile. When using classification for ecological modelling, clustering methods based on the full similarity matrix of vegetation perform best, and among these, those dividing the vegetation continuum into groups of similar size. In Chapter 12 the winner is minimum-variance clustering. Many other methods may have their strengths in different situations. A contrasting example is space-for-time substitution (Section 10.7.2) where a minimum spanning tree is used to connect time series fragments. This method is related to single linkage cluster analysis which is usually considered unsuitable for vegetation classification.

Like any other analytical method clustering is also prone to bias in the data. A risk of bias occurs in attempts to combine data sets from different databases and using these for classification revisions (see Tíchy *et al.* 2014, for example). Whether the age of 'Big Data' has arrived in vegetation ecology is questionable (Wiser 2016). Mathematical methods in this case tend to identify patterns reflecting activity of scientists rather than vegetation process.

A drawback of ordinary classification is its crispness, to escape from which requires fuzzy clustering (Roberts 1986). This still lacks popularity as it is closely related to ordination. Crisp classification is useful for vegetation mapping. It has often been overlooked that fuzzy vegetation mapping would be an alternative to traditional methods allowing for far more detailed maps to be drawn, although requiring more space as well. Examples derived through ecological modelling are presented in Section 9.8 and in the case study of Chapter 13 (Figure 13.5).

Nonlinearity in ordinations. Although ordination is a key method in vegetation ecology, the lack of linearity in species correlations causes unpopular results, with the horseshoe or arch effect considered the worst side-effect (Section 6.5). The most frequently used example throughout this book, the Schlaenggli data set, illustrates the issue. An almost square sampling universe with a systematic sampling plan (Figure 2.7) exhibits arch-shaped patterns on the level of vegetation similarity irrespective of the ordination method used (Figure 6.15). Equally alarming are unexpected results when using cover percentage of species without further transformation. Misleadingly, these are sometimes attributed to principal component analysis. But as shown in Figure 6.15 the same happens when using principal coordinates analysis or nonmetric multidimensional scaling. Principal component analysis is a fundamental, perfectly understood method in science and the results agree with anticipations if the data are properly transformed. Figure 10.6 is an example where further transformations – including the use of distance measures applying these intrinsically – would ruin the results.

A method aimed at stretching arch-shaped point clouds graphically is detrended correspondence analysis (Section 6.5.4) and as shown in Figure 6.15 the outcome may even be similar to nonmetric multidimensional scaling. Flexible shortest path adjustment (Section 6.5.2) proved to be interesting in simple cases but its performance in complex situations is as yet unknown.

According to Clarke (1993) the key to high quality analysis in ecology is relying on ranks rather than metric scores. His preferred ordination method therefore is nonmetric multidimensional scaling (Section 6.5.3), even more so when combined with Bray–Curtis distance. Again, this should be used with care as the equilibrium states of the iterative process involved in nonmetric multidimensional scaling regularly resembles the initial configurations, and careful comparison with principal component analysis, for example, is recommended.

Vegetation and site. Relating vegetation and site on the level of classifications is usually hampered by skewed distributions of site measurements. Analysis of variance requires normally distributed predictors. Statistical inference therefore is hardly ever justified. But F-values can be used in comparative context as shown in Section 7.2.2 where they merely serve the ranking of variables – and they do this more efficiently than the frequently used IndVal coefficients (Wildi and Feldmeyer-Christe 2013).

Constrained ordination is an option with high illustrative potential and it is well suited for internal comparison of variables. Like any ordinary ordination it is sensitive to odd distributions and of course also to non-linear correlations. In this regard generalized linear models are a big step ahead. The methods replace presence–absence scores by a meaningful S-shaped or unimodal function, although this is convenience- rather than evidence-based. Generalized additive models are more flexible in this regard. Hitherto, however, they remain black boxes to be interpreted with much care.

Ongoing succession. Examples in Chapter 10 confirm the number of time steps being the bottleneck in temporal analysis. Where vegetation is exposed to seasonality, minimum time step length is usually one year and there is no way to accelerate the process. An exception is archaeological records, such as pollen data. In the example in Section 10.8 the number of time steps exceeds 100 allowing alternative temporal resolutions to be probed for. One objective of temporal analysis is to assess vegetation composition before and after an expected environmental impact. If a change occurs unexpectedly we may strive for the forces causing the process. It is therefore worthwhile to investigate process velocity. In Section 10.8 peaks in the velocity profile mark times of tree invasion. In the Vraconnaz study (Section 10.5) 20 consecutive states reveal a succession process proceeding in two surges, most likely caused by a changing species pool. Hence, a high quality temporal investigation should have sufficient temporal resolution as well as replicates in space.

In all examples presented in Chapter 10 investigations encompassing merely a few time steps do not justify predicting future change. Long-term investigations suggest that an equilibrium state probably only occurs when there is an unchanged species pool. In the age of global exchange this is unlikely to happen, suggesting that succession is a never-ending process. Vegetation monitoring therefore remains an everlasting task as long as society cares about biodiversity.

Bibliography

Achermann, G., Schütz, M. and Krüsi, B.O. 2000. Tall-herb communities in the Swiss National Park: long-term development of the vegetation. *Nationalpark-Forschung in der Schweiz* **89**: 67–88.

Adler, J. 2010. *R in a Nutshell.* O'Really Media, Inc., Sebastopol. CA 95472, U.S.A.

Albert, C.H., Yoccoz, N.G., Edwards Jr, T.C., Graham, C.H., Zimmermann, N. E. and Thuiller, W. 2010. Sampling in ecology and evolution – bridging the gap between theory and practice. *Ecography* **33**: 1028–1037.

Allen, T.F.H. and Starr, T.B. 1982. *Hierarchy: Perspectives for Ecological Complexity.* The University of Chicago Press.

Anand, M. 1997. The fundamental nature of vegetation dynamics: a chaotic synthesis. *Coenoses* **12**(2–3): 55–62.

Anderberg, M.R. 1973. *Cluster Analysis for Applications.* Academic Press, New York.

Anderson, M. J. 2001. A new method for non-parametric multivariate analysis of variance. *Australian Journal of Ecology* **26**: 32–46.

Austin, M.P. 1987. Models for the analysis of species response to environmental gradients. *Vegetatio* **69**: 35–45.

Austin, M.P. 2013a. Inconsistencies between theory and methodology: a recurrent problem in ordination studies. *Journal of Vegetation Science* **24**: 251–268.

Austin, M.P. 2013b. Vegetation and environment: discontinuities and continuities. In: van der Maarel, E. and Franklin, J. (eds.) *Vegetation Ecology.* 2nd Edition. Blackwell Publishing, Oxford. 71–106.

Batschelet, E. 1975. *Introduction to Mathematics for Life Sciences.* 2nd Edition. Springer-Verlag, Berlin.

Begon, M., Townsend C.R. and Harper, J.L. 2005. *Ecology: From Individuals to Ecosystems.* 4th Edition. Wiley-Blackwell Science Ltd. Cambridge, MA.

Belbin, L. and McDonald, C. 1993. Comparing three classification strategies for use in ecology. *Journal of Vegetation Science* **4**: 341–348.

Benzécri, J.P. 1969. Statistical analysis as a tool to make patterns emerge from data. In: S. Watanabe (ed.) *Methodologies of Pattern Recognition.* Academic Press, New York. 35–60.

Beran, J. 1994. *Statistics for Long-Memory Processes. Monographs on Statistics and Probability* 61. Chapman & Hall.

Binz, H.-R. and Wildi, O. 1988. Das Simulationsmodell MaB-Davos. *Schlussberichte zum Schweizerischen MaB-Programm* 33, Bern.

Borcard, D. and Legendre, P. 2012. Is the Mantel correlogram powerful enough to be useful in ecological analysis? A simulation study. *Ecology* **93**: 1473–1481.

Borcard, D., Gillet, F. and Legendre, P. 2011. *Numerical Ecology with R.* Springer New York.

Borhidi, A. 1995. Social behaviour types, the naturalness and relative ecological indicator values of the higher plants in the Hungarian flora. *Acta Botanica Hungarica* **39**: 97–181.

Box, E.O. 1981. *Macroclimate and Plant Forms. An Introduction to Predictive Modelling in Phytogeography.* Tasks for Vegetation Science, **1**. Dr. W. Junk Publishers, The Hague.

Box, E.O. 1995. Factors determining distributions of tree species and plant functional types. *Vegetatio* **121**: 101–116.

Box, E.O. 1996. Plant functional types and climate at the global scale. *Journal of Vegetation Science* **7**: 309–320.

Boyce, R.L. and Ellison, P. 2001. Choosing the best similarity index when performing fuzzy set ordination on binary data. *Journal of Vegetation Science* **5**: 439–440.

Bradfield, G.E. and Kenkel, N.C. 1987. Nonlinear ordination using flexible shortest path adjustment of ecological distances. *Ecology* **68**: 750–753.

Brassel, P. and Brändli U.B. (eds) 1999. *Schweizerisches Landesforstinventar.* Ergebnisse der Zweitaufnahme 1993–1995. Birmensdorf, Eidgenössische Forschungsanstalt für Wald, Schnee und Landschaft. Bern,

Bundesamt für Umwelt, Wald und Landschaft. Bern, Stuttgart, Wien. Haupt.

Braun-Blanquet, J. 1928. *Pflanzensoziologie.* Springer Verlag, Berlin.

Braun-Blanquet, J. 1932. *Plant Sociology: The Study of Plant Communities.* (Translated by G.D. Fuller and H.S. Conard.) McGraw-Hill, New York.

Braun-Blanquet, J. 1964. *Pflanzensoziologie.* 3. Aufl. Wien.

Bruelheide, H. 1995. Die Grünlandvegetation des Harzes und ihre Standortbedingungen: mit einem Beitrag zum Gliederungsprinzip auf der Basis von statistisch ermittelten Artengruppen. *Dissertationes botanicae* 244.

Bruelheide, H. 1997. Using formal logic to classify vegetation. *Folia Geobotabica & Phytotaxonomia* **32**: 41–46.

Bruelheide, H. and Flintrop, T. 1994. Arranging phytosociological tables by species-relevé groups. *Journal of Vegetation Science* **5**: 311–316.

Brzeziecki, B., Kienast, F. and Wildi, O. 1993. A simulated map of the potential natural forest vegetation of Switzerland. *Journal of Vegetation Science* **4**: 499–508.

Chapman, D.S. and Purse, B.V. 2011. Community versus single-species distribution models for British plants. *Journal of Biogeography* **38**: 1524–1535.

Chiarucci, A., Araújo, M.B., Decocq, G., Beierkuhnlein, C. and Fernández-Palacios, J.M. 2010. The concept of potential natural vegetation: an epitaph? *Journal of Vegetation Science* **21**: 1172–1187.

Clarke, K.R. 1993. Non-parametric multivariate analyses of changes in community structure. *Australian Journal of Ecology* **18**: 117–143.

Clements, F.E. 1916. *Plant Succession. An Analysis of the Development of Vegetation.* Washington, DC.

Connell, H.J. and Slatyer, R.O. 1977. Mechanisms of succession in natural communities and their role in community stability and organisation. *American Naturalist* **111**: 1119–1144.

Cornelissen, J.H.C., Lavorel, S., Garnier, E., Díaz, S., Buchmann, N., Gurvich, D.E., Reich, P.B., ter Steege, H., Morgan, D.H., van der Heijden, M.G.A., Pausas, J.G. and Poorter, H. 2003. A handbook of protocols for standardised and easy measurement of plant functional traits worldwide. *Australian Journal of Botany* **51**: 335–380.

Crawley, M.J. 2005. *Statistics. An Introduction Using R.* Wiley & Sons Ltd, Chichester.

De ath, G. 2002. Multivariate regression trees: A new technique for modeling species-environment relationships. *Ecology* **83**: 1105–1117.

De Cáceres, M., Chytrý, M., Agrillo E., Attorre, F., Botta-Dukát, Z., Capelo, J., Czúcz, B., Dengler, J., Ewald, J., ... , Wiser, S.K. 2015. A comparative framework for broad-scale plot-based vegetation classification. *Applied Vegetation Science* **18**: 543–560.

Dengler, J., Chytrý, M. and Ewald, J. 2008. Phytosociology. In: S.E. Jørgensen and B.D. Fath (eds) *General Ecology.* Vol. 4 of *Encyclopedia of Ecology.* Elsevier, Oxford. 2767–2779.

Dengler, J., Jansen, F., Glöckler, F., Peet, R.K., De Cáceres, M., Chytrý, M., Ewald, J., Oldeland, J., Lopez-Gonzales, G., Finckh, M., Mucina, L., Rodwell, J.S., Schaminée, J. and Spencer, N. 2011. The Global Index of Vegetation-Plot Databases (GIVD): a new resource for vegetation science. *Journal of Vegetation Science* **22**: 582–597.

Díaz, S. 1995. Elevated CO_2 responsiveness, interactions at the community level and plant functional types. *Journal of Biogeography* **22**: 289–295.

Diamond, J. 1999. *Guns, Germs and Steel: The Fates of Human Societies.* W.W. Norton & Company, New York.

Diekmann, M. 2003. Species indicator values as an important tool in applied plant ecology – a review. *Basic and Applied Ecology* **4**: 493–506.

Digby, P.G.N. and Kempton, R.A. 1987. *Multivariate Analysis of Ecological Communities.* Chapman & Hall, London.

Dray, S. and Legendre, P. 2008. Testing the species traits-environment relationships: the fourth-corner problem revisited. *Ecology* **89**: 3400–3412.

Dufrêne, M. and Legendre, P. 1997. Species assemblages and indicator species: the need for a flexible asymmetrical approach. *Ecological Monographs* **67**: 345–366.

Elith, J., Graham, C.H., Anderson, P.R., Dudík, M., Ferrier, S., Guisan, A., Hijmans, R.J., Huettmann, F., Leathwick, ... , Zimmermann, N.E. 2006. Novel methods improve prediction of species' distributions from occurrence data. *Ecography* **29**: 129–151.

Elith, J., Leathwick, J.R. and Hastie, T. 2008. A working guide to boosted regression trees.*Journal of Animal Ecology* **77**: 802–813.

Ellenberg, H. 1956. Aufgaben und Methoden in der Vegetationskunde. In: H. Walter. *Einführung in die Phytologie* IV/1, Stuttgart.

Ellenberg, H. 1974. *Zeigerwerte der Gefässpflanzen Mitteleuropas. Scripta Geobotanica.* Verlag Erich Goltze, Göttingen.

Ellenberg, H. and Klötzli, F. 1972. *Waldgesellschaften und Waldstandorte der Schweiz.* Mitteilungen der Eidgenössischen Anstalt für das forstliche Versuchswesen. **48**. Zürich.

Everitt, B.S., Landau, S., Leese, M. and Stahl, D. 2011. *Cluster Analysis.* 5th Edition. Wiley & Sons, Chichester.

Ewald, J. 2003. A critique for phytosociology. *Journal of Vegetation Science* **14**: 291–296.

Faith, D.P., Minchin, P.R. and Belbin, L. 1987. Compositional dissimilarity as a robust measure of ecological distance: a theoretical model and computer simulations. *Vegetatio* **69**: 57–68.

Feldmeyer-Christe, E., Ecker, K., Küchler, M., Graf, U. and Waser, L. 2007. Improving predictive mapping in Swiss mire ecosystems through re-calibration of indicator values. *Applied Vegetation Science* **10**: 183–192.

Feldmeyer-Christe, E., Küchler, M. and Wildi, O. 2011. Patterns of early succession on bare peat in a Swiss mire after a bog burst. *Journal of Vegetation Science* **22**: 943-954.

Feoli, E. and Orlóci, L. 1979. Analysis of concentration and detection of underlying factors in structured tables. *Vegetatio* **40**: 49–54.

Feoli, E. and Orlóci, L. 1991. The properties and interpretation of observations in vegetation study. In: E. Feoli and L. Orlóci (eds) *Computer Assisted Vegetation Analysis.* Kluwer, Dordrecht. 3–13.

Feoli, E. and Orlóci, L. 2011. Can similarity theory contribute to the development of a general theory of the plant community? *Community Ecology* **12**: 135–141.

Feoli, E. and Scoppola, A. 1980. Analisi informazionale degli schemi di dinamica della vegetazione. Un esempio sul popolamento vegetale delle dune di litorale di Venezia. *Giornale Botanica Italiano* **114**: 227–236.

Fischer, H.S. 1990. Simulating the distribution of plant communities in an alpine landscape. *Coenoses* **5**: 37-43.

Fisher, R.A. 1940. The precision of discriminant functions. *Annals of Eugenics* **10**: 422–429.

Fitzpatrick, M.C. and Hargrove, W.W. 2009. The projection of species distribution models and the problem of non-analog climate. *Biodiversity and Conservation* **18**: 2255–2261.

Floyd, R.W. 1962. Algorithm 97: shortest path. *Communications of the Association for Computing Machinery* **5**: 345.

Forrester, J.W. 1968. *Principles of Systems.* Wright-Allen Press, Cambridge, MA.

Fridley, J.D. 2003. Diversity effects on production in different light and fertility environments: an experiment with communities of annual plants. *Journal of Ecology* **91**: 396–406.

Gan, G., Ma, C. and Wu, J. 2007. *Data clustering. Theory, Algorithms and Applications. ASA-SIAM Series on Statistics and Applied Probability*, SIAM, Philadelphia.

Gauch, H.G. 1982. *Multivariate Analysis in Community Ecology.* Cambridge Studies in Ecology. Cambridge University Press, Cambridge.

Ghosh, S. and Wildi, O. 2007. Statistical analysis of landscape data: space-for-time, probability surfaces and discovering species. In: F. Kienast, O. Wildi and S. Ghosh (eds) *A Changing World: Challenges for Landscape Research. Springer Landscape Series* Vol. 8. Springer, Dordrecht. 209–221.

Gleason, H.A. 1926. The individualistic concept of the plant association. *Bulletin of the Torrey Botanical Club* **53**: 7–26.

Gleason, H.A. 1939. The individualistic concept of the plant association. *American Midland Naturalist* **21**: 92–110.

Gobet, E., Vercovi, E. and Tinner, W. 2010. Ein paläoökologischer Beitrag zum besseren Verständnis der natürlichen Vegetation der Schweiz. *Botanica Helvetica* **120**: 105–115.

Gower, J.C. 1966. Some distance properties of latent root and vector methods used in multivariate analysis. *Biometrika* **53**: 325–338.

Gower, J.C. and Ross, G.J.S. 1969. Minimum spanning tree and single linkage cluster analysis. *Applied Statistics* **18**: 54–64.

Graf, U., Wildi, O., Feldmeyer-Christe, E. and Küchler, M. 2010. A phytosociological classification of Swiss mire vegetation. *Botanica Helvetica* **120**: 1–13.

Green, R.H. 1979. *Sampling Design and Statistical Methods for Environmental Biologists.* Wiley-Interscience, New York.

Grime, J.P. 2001. *Plant Strategies, Vegetation Processes and Ecosystem Properties.* 2nd Edition. John Wiley & Sons, Chichester.

Groth, E.J., Peebles, P.J.E. and Soneira, R.M. 1977. The clustering of galaxies. *Scientific American* **237**: 76–98.

Grünig, A., Steiner, G.M., Ginzler, C., Graf, U. and Küchler, M. 2005. Approaches to Swiss mire monitoring. *Stapfia* **85**: 435–452.

Guisan, A. and Zimmermann, N.E. 2000. Predictive habitat distribution models in ecology. *Ecological Modeling* **135**: 147–186.

Guisan, A., Broennimann, O., Engler, R., Vust, M., Yoccoz, N.G., Lehmann, A. and Zimmermann, N.E. 2005. Using niche-based models to improve sampling of rare species. *Conservation Ecology* **20**: 502–511.

Hennekens, S.M. and Schaminée, J.H.J. 2001. TURBOVEG, a comprehensive data base management system for vegetation data. *Journal of Vegetation Science* **12**: 589–591.

Hill, M.O. 1973. Reciprocal averaging: an eigenvector method of ordination. *Journal of Ecology* **61**: 237–249.

Hill, M.O. 1979a. *DECORANA: A FORTRAN Program for Detrended Correspondence Analysis and Reciprocal Averaging.* Cornell University, Ithaca, NY.

Hill, M.O. 1979b. *TWINSPAN: A FORTRAN Program for Arranging Multivariate Data in an Ordered Two-way Table by Classification of the Individuals and Attributes.* Cornell University, Ithaca, NY.

Hill, M.O. and Gauch, H.G. 1980. Detrended correspondence analysis, an improved ordination technique. *Vegetatio* **42**: 47–58.

Hill, M.O., Mountford, J.O., Roy, D.B. and Bunce, R.G.H. 1999. Ellenberg's indicator values for British plants. *Institute of Terrestrial Ecology*, Huntingdon, UK.

Huisman, J., Olff, H. and Fresco, L.F.M. 1993. A hierarchical set of models for species response analysis. *Journal of Vegetation Science* **4**: 37–46.

International Statistical Institute 2009. Multilingual Glossary of Statistical Terms. http://isi.cbs.nl/glossary.htm (accessed 18 November 2012).

Jaccard, P. 1901. Étude comparative de la distribution florale dans une proportion des Alpes et du Jura. *Bull. Soc. Vaudoise Sci. Nat.* **37**: 547–579.

Jancey, R.C. 1979. Species ordering on a variance criterion. *Vegetatio* **39**: 59–63.

Jennings, M., Loucks, O., Peet, R., Faber-Langendoen, D., Glenn-Lewin, D., Grossmann, D., Damman, A., Barbour, M., Pfister, R., Walker, M., Talbot, S., Walker, J., Hartshorn, G., Waggoner, G., Abrams, M., Hill, A., Roberts, D., Tart, D. and Rejmanek, M. 2003. *Guidelines for describing associations and alliances of the US National Vegetation Classification Panel.* The Ecological Society of America Vegetation Classification Panel.

Jennings, M., Faber-Langendoen, D., Loucks, O., Peet, R. and Roberts, D. 2009. Standards for associations and alliances of the U.S. National Vegetation Classification. *Ecological Monographs* **79**: 173–199.

Jongman, R.H.G., ter Braak, C.J.F. and van Tongeren, O.F.R. 1995. *Data Analysis in Community and Landscape Ecology.* Cambridge University Press, Cambridge.

Kaufman, L. and Rousseeuw, P.J. 1990. *Finding Groups in Data: An Introduction to Cluster Analysis.* Wiley, New York.

Keddy, P. A. 1992. Assembly and response rules: two goals for predictive community ecology. *Journal of Vegetation Science* **3**: 157–164.

Keller, W., Wohlgemuth, T., Kuhn, N., Schütz, M. and Wildi, O. 1998. *Waldgesellschaften der Schweiz auf floristischer Grundlage. Mitteilungen der Eidgenössischen Forschungsanstalt für Wald, Schnee und Landschaft (WSL)* **73**, Vol. 2.

Kent, M. 2012. *Vegetation Description and Analysis.* 2nd Edition. Wiley-Blackwell, Chichester.

Kienast, F., Wildi, O. and Ghosh, S. (eds) 2007. *A Changing World: Challenges for Landscape Research.* Springer Landscape Series, Vol. 8. Springer, Dordrecht.

Kleyer, M., Dray, S., de Bello, F., Lepš, J., Pakeman, R.J., Strauss, B. Thuiller, W. and Lavorel, S. 2012. Assessing species and community functional responses to environmental gradients: which multivariate methods? *Journal of Vegetation Science* **23**: 805–821.

Körner, C. 1999. *Alpine Plant Life. Functional Plant Ecology of High Mountain Ecosystems.* Springer. Berlin.

Kowarik, I. 1987. Kritische Anmerkungen zum theoretischen Konzept der potentiellen natürlichen Vegetation mit Anregungen zu einer zeitgemässen Modifikation. *Tüxenia* **7**: 53–67.

Küchler, A.W. 1988. Mapping dynamic vegetation. In: A.W. Küchler and I.S. Zonneveld (eds) *Vegetation Mapping. Handbook of Vegetation Science* **10**: 13–23.

Küchler, M. 2009. *VEGEDAZ*. WSL Swiss Federal Institute for Forest, Snow and Landscape Research, Birmensdorf, Switzerland.

Krüsi, B.O., Schütz, M., Bigler, C., Grämiger, H. and Achermann, G. 1998. Huftiere und Vegetation im Schweizerischen Nationalpark von 1917 bis 1997. Teil 1: Einfluss auf die botanische Vielfalt der subalpinen Weiden; Teil 2: Einfluss auf das Wald-Freilandverhältnis. In: R. Cornelius and R. Hofmann (eds) *Extensive Haltung robuster Haustierrassen, Wildtiermanagement, Multi-Spezies-Projekte – Neue Wege in Naturschutz und Landschaftspflege?* Zoo- und Wildtierforschung, Berlin. 62–74.

Lance, G.N. and Williams, W.T. 1966. Computer programs for classification. *Proceedings of the ANCCAC Conference*, Canberra, Paper 12/3.

Landolt, E. 1977. *Oekologische Zeigerwerte zur Schweizer Flora. Veröff Geobot Inst ETH, Stiftung Rübel* **64**.

Landolt, E., Bäumler, B., Erhardt, A., Hegg, O., Klötzli, F., Lämmler, W., Nobis, M., Rudmann-Maurer, K., Schweingruber, F. H., Theurillat, J.-P., Urmi, E., Vust, M. and Wohlgemuth, T. 2010. *Flora indicativa. Ecological Indicator Values and Biological Attributes of the Flora of Switzerland and the Alps.* Haupt, Bern.

Lawesson, J.E., Fosaa, A.M. and Olsen, E. 2003. Calibration of Ellenberg indicator values for the Faroe Islands. *Applied Vegetation Science* **6**: 53–62.

Legendre, P. and Anderson, M.J. 1999. Distance-based redundancy analysis: testing multi-species responses in multi-factorial experiments. *Ecological Monographs* **69**: 1–24.

Legendre, P. and Fortin, M.-J. 1989. Spatial analysis and ecological modeling. *Vegetatio* **80**: 107–138.

Legendre, P. and Gallagher, E.D. 2001. Ecologically meaningful transformations of ordination of species data. *Oecologia* **129**: 271–280.

Legendre, P. and Legendre, L. 1998. *Numerical Ecology.* 2rd Edition. Elsevier, Amsterdam.

Legendre, P. and Legendre, L. 2012. *Numerical Ecology.* 3rd Edition. Elsevier, Amsterdam.

Legendre, P., Galzin, R. and Harmelin-Vivien, M. 1997. Relating behaviour to habitat: solutions to the fourth-corner problem. *Ecology* **78**: 547–562.

Lengyel, A. and Podani, J. 2015. Assessing the relative importance of methodological decisions in classifications of vegetation data. *Journal of Vegetation Science* **26**: 804–815.

Lengyel, A., Chytrý, M. and Tichý, L. 2011. Heterogeneity–constrained random resampling of phytosociological databases. *Journal of Vegetation Science* **22**: 175–183.

Lepš, J. and Šmilauer, P. 2003. *Multivariate Analysis of Ecological Data using CANOCO.* Cambridge University Press, Cambridge.

Lindacher, R. 1996. Verifikation der potentiellen natürlichen Vegetation mittels Vegetationssimulation am Beispiel der TK 6434 'Hersbruck'. *Hoppea, Denkschrift der Bayerischen Botanischen Gesellschaft* **57**: 5–143.

Lippe, E., de Smitt, J.T. and Glenn-Lewin, D.C. 1985. Markov models and succession: a test from a heathland in the Netherlands. *Journal of Ecology* **73**: 775–791.

Lischke, H. 2005. Modeling tree species migration in the Alps during the Holocene: what creates complexity? *Ecological Complexity* **2**: 159–174.

Lischke, H., Lotter, A.F. and Fischlin, A. 2002. Untangling a Holocene pollen record with forest model simulations and independent climate data. *Ecological Modeling* **150**: 1–21.

Loehle, C. 1983. Evaluation of theories and calculation tools in ecology. *Ecological modeling* **19**: 239–247.

Lotka, H.J. 1925. *Elements of Physical Biology.* Williams and Wilkins.

Lotter, A.F. 1999. Late-glacial and Holocene vegetation history and dynamics as shown by pollen macrofossil analyses in annually laminated sediments from Soppensee, central Switzerland. *Vegetation History and Archaeobotany* **8**: 165–184.

Lötter, M.C., Mucina, L. and Witkowski, E.T.F. 2013. The classification conundrum: species fidelity as leading criterion in search of a rigorous method to classify a complex forest data set. *Community Ecology* **14**: 121–132.

MacArthur, R. and Levins, R. 1967. The limiting similarity, convergence and divergence of coexisting species. *American Naturalist* **101**: 377–385.

MacLulich, D.A. 1936. Fluctuations in numbers of varying hares. *Science* **38**: 162.

Mantel, N. 1967. The detection of disease clustering and a generalized regression approach. *Cancer Research* **27**: 209–220.

Marsili-Libelli, S. 1989. Fuzzy clustering of ecological data. *Coenosis* **2**: 95–106.

Maynard Smith, J. 1974. *Models in Ecology.* Cambridge University Press, London.

McGill, B.J., Enquist, B.J., Weiher, E. and Westoby, M. 2006. Rebuilding community ecology from functional traits. *TRENDS in Ecology and Evolution* **21**: 178–185.

Meadows, D.H., Meadows, D.L. and Randers, J. 1972. *The Limits to Growth.* Universe Books.

Mellert, K.H., Fensterer, V., Küchenhoff, H., Reger, B., Kölling, C., Klemmt, H. J. and Ewald, J. 2011. Hypothesis-driven species distribution models for tree species in the Bavarian Alps. *Journal of Vegetation Science* **22**: 635–646.

Minchin, P. R. 1987. An evaluation of the relative robustness of techniques for ecological ordination. *Vegetatio* **69**: 89–107.

Mucina, L. 1997. Classification of vegetation: past, present and future. *Journal of Vegetation Science* **8**: 751–760.

Mucina, L., Bültmann, H., Dierßen, K., Theurillat, J.-P., Raus, T., Čarni, A., Sumberová, K., Willner, W., Dengler, J., ... , Tichý, L. 2016. Vegetation of Europe: hierarchical floristic classification system of vascular plant, bryophyte, lichen, and algal communities. *Applied Vegetation Science* **19**: 3–264.

Mueller-Dombois, D. and Ellenberg, H. 1974. *Aims and Methods of Vegetation Ecology.* John Wiley & Sons, New York.

Neuhäusl, R. 1984. Umweltgemässe natürliche Vegetation, ihre Kartierung und Nutzung für den Umweltschutz. *Preslia* **56**: 205–212.

Nishisato, S. 1980. *Analysis of Categorial Data: Dual Scaling and its Applications.* Mathematical Expositions No. 24. University of Toronto Press, Toronto.

Noy-Meir, I. and Whittaker, R.H. 1977. Continuous multivariate methods in community analysis: some problems and developments. *Vegetatio* **33**: 79–98.

Odum, E.P. 1971. *Fundamentals of Ecology.* 3rd Edition. W. B. Saunders, Philadelphia.

Oldeland, J., Dorigo, W., Lieckfeld, L., Lucieer, A. and Jürgens, N. 2010. Combining vegetation indices, constrained ordination and fuzzy classification for mapping semi-natural vegetation units from hyperspectral imagery. *Remote Sensing of Environment* **144**: 1155–1166.

Orlóci, L. 1967. An agglomerative method for classification of plant communities. *Journal of Ecology* **55**: 193–206.

Orlóci, L. 1973. Ranking characters by a dispersion criterion. *Nature* **244**: 371–373.

Orlóci, L. 1978. *Multivariate Analysis in Vegetation Research.* 2nd Edition, Junk Publishers, The Hague.

Orlóci, L. 1991a. *CONAPACK: A Program for Canonical Analysis of Classification Tables. Ecological Computations Series:* Vol. 4. SPB Academic Publishing, The Hague.

Orlóci, L. 1991b. On character-based plant community analysis: choice, arrangement, comparison. *Coenoses* **6**: 103–107.

Orlóci, L. 1993. The complexities and scenarios of ecosystem analysis. In: G.P. Patil and C.R. Rao (eds) *Multivariate Environmental Statistics.* Elsevier Scientific, New York. 423–432.

Orlóci, L. 2001. Prospects and expectations: reflections on a science in change. *Community Ecology* **2**: 187–196.

Orlóci, L. and Kenkel, N. 1985. *Introduction to Data Analysis.* International Co-operative Publishing House, Burtonsville, MD.

Orlóci, L. and Orlóci, M. 1985. Comparison of communities without the use of species: model and example. *Annali di Botanica (Roma)* **43**: 275–285.

Orlóci, L. and Pillar, V. de Patta 1989. On sample size optimality in ecosystems survey. *Biometrie-Praximetrie* **29**: 173–184.

Orlóci, L., Anand, M. and He, X. 1993. Markov chain: a realistic model for temporal coenosere? *Biometrie- Praximetrie* **33**: 7–26.

Parker, V.T. and Pickett, S.T.A. 1998. Historical contingency and multiple scales of dynamics within plant communities. In: D.L. Peterson and V.T. Parker (eds) *Ecological Scale: Theory and Applications.* Columbia University Press. 171–191.

Pärtel, M. 2014. Community ecology of absent species: hidden and dark diversity. *Journal of Vegetation Science* **25**: 1154–1159.

Peet, R. and Roberts, D.W. 2013. *Classification of natural and seminatural vegetation.* In: van der Maarel, E. and Franklin, J. (eds), Vegetation ecology. 2nd Edition, pp. 28–70. Wiley-Blackwell, Chichester, UK.

Pellissier, V., Bergès, L., Nedeltcheva, T., Schmitt, M.-C., Avon, C., Cluzeau, C. and Duponey J.-L. 2012. Understorey plant species show long-range spatial patterns in forest patches according to distance-to-edge. *Journal of Vegetation Science* **24**: 9–24.

Peres-Neto, P.R. and Jackson, D.A. 2001. How well do multivariate data sets match? The advantages of a Procrustean superimposition approach over the Mantel test. *Oecologia* **129**: 169–178.

Peres-Neto, P.R., Jackson, D.A. and Somers, K.M. 2005. How many principal components? Stopping rules for determining the number of non-trivial axes revisited. *Computational Statistics & Data Analysis* **49**: 974–997.

Pickett, T.A. 1989. Space-for-time substitution as an alternative to long-term studies. In: E. Likens (ed.) *Long-term Studies in Ecology: Approaches and Alternatives.* Springer, New York. 110–135.

Pielou, E.C. 1984. *The Interpretation of Ecological Data.* John Wiley & Sons, Ltd, New York.

Pignatti, S. 2005. Valori di bioindicazione delle piante vascolari della flora dItalia. *Braun-Blanquetia* **39**: 1–97.

Pillar, V. de Patta 1999. On the identification of optimal plant functional types. *Journal of Vegetation Science* **10**: 631–640.

Pillar, V. de Patta and Sosinski, E.E. 2003. An improved method for searching plant functional types by numerical analysis. *Journal of Vegetation Science* **14**: 323–332.

Pillar, V. de Patta, Duarte, L.S., Sosinski, E.E. and Joner, F. 2009. Discriminating trait-convergence and trait-divergence assembly patterns in ecological community gradients. *Journal of Vegetation Science* **20**: 334–348.

Podani, J. 1997. SYN-TAX 5.1: New version for PC and Macintosh computers. *Coenoses* 12: 149–152.

Podani, J. 2006. Braun-Blanquet's legacy and data analysis in vegetation science. *Journal of Vegetation Science* **22**: 113–117.

Podani, J. and Feoli, E. 1991. A general strategy for the simultaneous classification of variables and objects in ecological data tables. *Journal of Vegetation Science* **2**: 435–444.

Podani, J. and Miklós, I. 2002. Resemblance coefficients and the horseshoe effect in principal coordinates analysis. *Ecology* **83**: 3331–3343.

Poore, M.E.D. 1955. The use of phytosociological methods in ecological investigations. I–III. *Journal of Ecology* **43**: 226–244, 245–269, 606–651.

Poore, M.E.D. 1962. The method of successive approximation in descriptive ecology. *Advances in Ecological Research* **1**: 35–68.

R Development Core Team 2012: A Language Environment for Statistical Computing. R version 2.15.2. R Foundation for Statistical Computing 2012, Vienna, Austria.

R Development Core Team 2017: A Language Environment for Statistical Computing. R version 3.3.3. R Foundation for Statistical Computing 2017, Vienna, Austria.

Rao, C.R. 1964. The use and interpretation of principal component analysis in applied research. *Sankhyaá, Series A* **26**: 329–358.

Raunkiaer, C. 1937. *The Life Forms of Plants.* (Translated from the original, published in Danish, 1907). Oxford University Press, Oxford.

Rényi, A. 1961. On measures of entropy and information. In: J. Neyman (ed.) *Proceedings of the 4th Berkeley Symposium on Mathematical Statistics and Probability.* University of California Press, Berkeley. 547–561.

Ridgeway, G. 1999. The state of boosting. *Computing Science and Statistics* **31**: 172–181.

Ripley, B.D. 1996. *Pattern Recognition and Neural Networks.* Cambridge University Press.

Roberts, D.W. 1986. Ordination on the basis of fuzzy set theory. *Vegetatio* **66**: 123–131.

Roberts, D.W. 2015. Vegetation classification by two new iterative reallocation optimization algorithms. *Plant Ecology* **216**: 741–758.

Rodwell, J.S. 2006. *NVC Users' Handbook*, 68 pages, ISBN 978 1 86107 574 1

Rousseeuw, P.J. 1987. Silhouettes: A graphical aid to the interpretation and validation of cluster analysis. *Journal of Computational and Applied Mathematics* **20**: 53–65.

Roweis, T.S. and Saul, L.K. 2000. Nonlinear dimensionality reduction by locally linear embedding. *Science* **290**: 2323–2326.

Sampford, M.R. 1962. *An Introduction to Sampling Theory with Applications to Agriculture.* Oliver and Boyd, Edinburgh.

Shepard, R.N. 1962. The analysis of proximities: multidimensional scaling with an unknown distance function. *Psychometrika* **27**: 125–139.

Sneath, P.H.A. and Sokal, R.R. 1973. *Numerical Taxonomy: The Principles and Practice of Numerical Classification.* W.H. Freeman, San Francisco.

Sørensen, T. 1948. A method of establishing groups of equal amplitude in plant sociology based on similarity of species content. *Royal Danish Academy of Sciences and Letters* **5**: 1–34.

Swan, J.M.A. 1970. An examination of some ordination problems by use of simulated vegetational data. *Ecology* **51**: 89–102.

Tenenbaum, J.B., de Silva, V. and Langford, J.C. 2000. A global geometric framework for nonlinear dimensionality reduction. *Science* **290**: 2323–2326.

ter Braak, C.J.F. 1986. Canonical correspondence analysis: a new eigenvector technique for multivariate direct gradient analysis. *Ecology* **67**: 1167–1179.

Tichý, L. 2002. JUICE, software for vegetation classification. *Journal of Vegetation Science* **13**: 451–453.

Tichý, L., Chytrý, M. and Botta-Dukát, Z. 2014. Semi-supervised classification of vegetation: preserving the good old units and searching for new ones. *Journal of Vegetation Science* **25**: 1504–1512.

Tobias, J.A., Cornwallis, C.K., Derryberry, E.P., Claramunt, S., Brumfield, R.T. and Seddom, N. 2014. Species coexistence and the dynamics of phentotypic evolution in adaptive radiation. *Nature* **506**: 359–363.

Tüxen, R. 1956. Die heutige potentielle Vegetation von Oberfranken. *Angewandte Pflanzensoziologie (Stolzenau)* **13**: 5–42.

Usher, M.B. 1981. Modeling ecological succession, with particular reference to Markovian model. *Vegetatio* **46**: 11–18.

van der Maarel, E. 1979. Transformation of cover-abundance values in phytosociology and its effects on community similarity. *Vegetatio* **39**: 97–114.

van der Maarel, E. 2005 *Vegetation Ecology.* Blackwell Publishing, Malden, Oxford, Victoria.

van der Maarel, E., Janssen, J.G.M. and Louppen, J.M.W. 1978. TABORD, a program for structuring phytosociological tables. *Vegetatio* **38**: 143–156.

van der Maarel, E. and Franklin, J. (eds) 2013. *Vegetation Ecology.* 2nd Edition. Blackwell Publishing, Oxford.

van der Valk, A.G. 1992. Establishment, colonization and persistence. In: Glenn-Lewin, D.C., Peet, R.K. and Veblen, T.T. (eds) *Plant succession: theory and prediction.* Chapman & Hall, London, UK.

van Groenewoud, H. 1965. *Ordination and classification of Swiss and Canadian coniferous forests by various biometric and other methods.* Dissertation ETHZ 3700. Berichthaus Zürich.

Venables, W.N. and Ripley, B.D. 2010. *Modern Applied Statistics with S.* 4th Edition. Springer.

Volterra, V. 1926. Fluctuations in the abundance of a species considered mathematically. *Nature* **118**: 558–560.

von Wehrden, H., Hanspach, J., Bruelheide, H. and Wesche, K. 2009. Pluralism and diversity: trends in the use and application of ordination methods 1990–2007. *Journal of Vegetation Science* **20**: 695–705.

Walther, G.-R., Post, E., Convey, P., Menzel, A., Parmesan, C., Beebee, T.J.C., Fromentin, J.-M., Hoegh-Guldberg, O. and Bairlein, F. 2002. Ecological responses to recent climate change. *Nature* **416**, 389–395.

Wagner, H.H. 2004. Direct multi-scale ordination with canonical correspondence analysis. *Ecology* **85**: 342–351.

Ward, J.H. 1963. Hierarchical grouping to optimize an objective function. *Journal of the American Statistical Association* **58**(301): 236–244.

Wartenberg, D., Ferson, S. and Rohlf, F.J. 1987. Putting things in order: a critique of detrended correspondence analysis. *American Naturalist* **129**: 434–448.

Watkinson, A. R. 1987. Plant population dynamics. In: Crawley, M.J. (ed.), *Plant Ecology*: 137–184. Blackwell, Oxford Scientific Publications.

Watt, A.S. 1947. Pattern and process in the plant community. *Journal of Ecology* **35**: 1–22.

Weber, H.E., Movarec, J. and Theurillat, J.-P. 2000. International code of phytosociological nomenclature. *Journal of Vegetation Science* **11**: 739–768.

Wildi, O. 1976. Untersuchung von Vegetationsgrenzen mit Hilfe dynamischer Modelle. *Berichte der Deutschen Botanischen Gesellschaft* **89**: 365–370.

Wildi, O. 1977. *Beschreibung exzentrischer Hochmoore mit Hilfe quantitativer Methoden.* Veröffentlichungen des Geobotanischen Institutes der ETH, Stiftung Rübel **60**.

Wildi, O. 1984. Species selection by interactive ranking. *Vegetatio* **56**: 161–166.

Wildi, O. 1989. A new numerical solution to traditional phytosociological tabular classification. *Vegetatio* **81**: 95–106.

Wildi, O. 1990. A multiple scale sampling design for long term monitoring. Proceedings of the International Conference and Workshop, Vol. 2. Bethesda, MD. 975–982.

Wildi, O. 2001. Statistical design and analysis in long term vegetation monitoring. In: C.A. Burga and A. Kratochwil (eds) *Biomonitoring: General and Applied Aspects on Regional and Global Scales.* Kluwer, Dordrecht. Tasks for Vegetation Science. Vol. **35**: 17–39.

Wildi, O. 2002. Modeling succession from pasture to forest in time and space. *Community Ecology* **3**(2): 181–189.

Wildi, O. 2013. *Data analysis in vegetation ecology.* 2nd Edition. Wiley-Blackwell, Chichester.

Wildi, O. 2016. Why mean indicator values are not biased. *Journal of Vegetation Science* **27**: 40–49.

Wildi, O. and Feldmeyer-Christe, E. 2013. Indicator values (IndVal) mimic ranking by F-ratio in real-world vegetation data. *Community Ecology* **14**: 139–143.

Wildi, O. and Orlóci, L. 1991. Flexible gradient analysis: a note on ideas and an application. In: E. Feoli and L. Orlóci (eds) *Computer Assisted Vegetation Analysis.* Kluwer, Dordrecht. 265–271.

Wildi, O. and Orlóci, L. 1996. *Numerical Exploration of Community Patterns.* 2nd Edition, SPB Academic Publishing, The Hague.

Wildi, O. and Orlóci, L. 2007. Essay on the study of the vegetation process. In: F. Kienast, O. Wildi and S. Ghosh (eds) *A Changing World: Challenges for Landscape Research.* Springer Landscape Series, Dordrecht. 195–207.

Wildi, O. and Schütz, M. 2000. Reconstruction of a long-term recovery process from pasture to forest. *Community Ecology* **1**: 25–32.

Wildi, O. and Schütz, M. 2007. Scale sensitivity of synthetic long-term vegetation time series derived through overlay of short-term field records. *Journal of Vegetation Science* **18**: 471–478.

Wildi, O., Feldmeyer-Christe, E., Ghosh, S. and Zimmermann, N.E. 2004. Comments on vegetation monitoring approaches. *Community Ecology* **5**: 1–5.

Wiser, S. K. 2016. Achievements and challenges in the integration, reuse and synthesis of vegetation plot data. *Journal of Vegetation Science* **27**: 868–879.

Wissel, C. 1989. *Theoretische Ökologie.* Springer-Verlag, Berlin.

Wohlgemuth, T., Moser, B., Brändli, U.-B., Kull, P. and Schütz, M. 2008. Diversity of forest plant species at the community and landscape scales in Switzerland. *Plant Biosystems* **142**: 604–613.

Zelený, D. and Schaffers, A.P. 2012. Too good to be true: pitfalls of using Ellenberg indicator values in vegetation analyses. *Journal of Vegetation Science* **23**: 419–431.

Zimmermann, N.E. and Kienast, F. 1999. Predictive mapping of alpine grassland in Switzerland: species versus community approach. *Journal of Vegetation Science* **10**: 469–482.

Appendix A: Functions in package dave

Table 1: Main functions in the R package dave. Requires packages dabdsv, vegan and tree to be loaded in addition to standard downloads of R (R Development Core Team 2016).

Functions	Purpose	Value, S3 class
aocc()	Analysis of concentration	"aocc"
ccost()	Cost function to compare two classifications	"ccost"
centroid()	Computes group centroids of relevé tables	"centroid"
davesil()	Silhouette plots from object "hclust"	"davesil"
dircor()	Direction dependent autocorrelation	"dircor"
fitmarkov()	Fitting a first-order Markov series	"fitmarkov"
fspa()	Flexible shortest path adjustment	"fspa"
Mtabs()	Ordering vegetation tables in Mulva style	"Mtabs"
mxplot()	Draw mean similarities between groups	"mxplot"
orank()	Ranking by orthogonal functions	"orank"
outlier()	Statistic of outlier relevés	"outlier"
overly()	Overlay of vegetation time series	"overly"
pcobiplot()	PCOA followed by correlating species	"pcobiplot"
pcaser()	Connecting time series in PCA plot	"pcaser"
pcovar()	Computing and comparing six PCOA plots	"pcovar"
SNPtm()	Dynamic time model of a subalpine pasture	"SNPtm"
SNPsm()	Dynamic spatial model of a subalpine pasture	"SNPsm"
speedprof()	Velocity profile of a vegetation time series	"speedprof"
srank()	Ranking species by indicator- or F-values	"srank"
vvelocity()	Phase space plot of a vegetation time series	"vvelocity"

Appendix B: Data sets used

Table 2: Data sets included in the R package dave. These data frames are accessed through the names listed in the first column.

Name	Rows; columns	Comments	Reference
EKs	2533; 11	Data set of Swiss forests, site	Ellenberg and Klötzli (1972)
EKv	2533; 1259	Data set of Swiss forests, vegetation	Ellenberg and Klötzli (1972)
ltim	19; 1	Heathland succession data, time scale	Lippe *et al.* (1985)
lveg	19; 9	Heathland succession data, vegetation	Lippe *et al.* (1985)
mveg	25; 94	Ellenberg's (1956) meadow data (vegetation only)	Mueller-Dombois and Ellenberg (1974)
nsit	11; 8	Artificial data of European beech forests, site	Wildi and Orlóci (1996)
nveg	11; 21	Artificial data of European beech forests, vegetation	Wildi and Orlóci (1996)
psit	145; 1	Pollen profile from Soppensee, Switzerland, time scale	Lotter (1999)
pveg	145; 14	Pollen profile from Soppensee, Switzerland, tree species	Lotter (1999)
ssind	119; 9	Indicator values for sveg	Landolt *et al.* (2010)
sspft	119; 23	Species traits for sveg	Landolt *et al.* (2010)
ssit	63; 20	Wetland gradient, site	Wildi (1977)
sveg	63; 119	Wetland gradient, vegetation	Wildi (1977)
sn7sit	97; 2	7 selected time series, Swiss National Park, site	Wildi and Schütz (2000)
sn7veg	97; 6	7 selected time series, Swiss National Park, vegetation	Wildi and Schütz (2000)
sn59sit	751; 2	59 time series, Swiss National Park, site	Wildi and Schütz (2000)
sn59veg	751; 6	59 time series, Swiss National Park, vegetation	Wildi and Schütz (2000)
tsit	16; 2	Time series Tr6, Swiss National Park, site	Wildi and Schütz (2000)
tveg	16; 6	Time series Tr6, Swiss National Park, vegetation	Wildi and Schütz (2000)
vrveg	231; 154	Peat bog Vraconnaz, vegetation	Feldmeyer-Christe *et al.* (2011)
vrsit	231; 26	Peat bog Vraconnaz, site	Feldmeyer-Christe *et al.* (2011)
wetsit	1500; 69	Subsample of wetland plots, site	Graf *et al.* (2010)
wetveg	1500; 1164	Subsample of wetland plots, vegetation	Graf *et al.* (2010)
ws30	726; 1262	Swiss forest grid, 30 m^2 plots, vegetation	Wohlgemuth *et al.* (2008)
ws200	726; 1262	Swiss forest grid, 200 m^2 plots, vegetation	Wohlgemuth *et al.* (2008)
ws500	726; 1262	Swiss forest grid, 500 m^2 plots, vegetation	Wohlgemuth *et al.* (2008)
wssit	726; 20	Swiss forest grid, site (all plots)	Wohlgemuth *et al.* (2008)

Index